Phytoremediation of Metal-Contaminated Soils

NATO Science Series

A Series presenting the results of scientific meetings supported under the NATO Science Programme.

The Series is published by IOS Press, Amsterdam, and Springer in conjunction with the NATO Public Diplomacy Division

Sub-Series

I. Life and Behavioural Sciences	IOS Press
II. Mathematics, Physics and Chemistry	Springer
III. Computer and Systems Science	IOS Press
IV. Earth and Environmental Sciences	Springer

The NATO Science Series continues the series of books published formerly as the NATO ASI Series.

The NATO Science Programme offers support for collaboration in civil science between scientists of countries of the Euro-Atlantic Partnership Council. The types of scientific meeting generally supported are "Advanced Study Institutes" and "Advanced Research Workshops", and the NATO Science Series collects together the results of these meetings. The meetings are co-organized bij scientists from NATO countries and scientists from NATO's Partner countries – countries of the CIS and Central and Eastern Europe.

Advanced Study Institutes are high-level tutorial courses offering in-depth study of latest advances in a field.
Advanced Research Workshops are expert meetings aimed at critical assessment of a field, and identification of directions for future action.

As a consequence of the restructuring of the NATO Science Programme in 1999, the NATO Science Series was re-organised to the four sub-series noted above. Please consult the following web sites for information on previous volumes published in the Series.

http://www.nato.int/science
http://www.springer.com
http://www.iospress.nl

Series IV: Earth and Environmental Sciences – Vol. 68

Phytoremediation of Metal-Contaminated Soils

edited by

Jean-Louis Morel
Institut National Polytechnique de Lorraine,
Nancy, France

Guillaume Echevarria
Institut National Polytechnique de Lorraine,
Nancy, France

and

Nadezhda Goncharova
International Sakharov Environmental University,
Minsk, Belarus

Springer

Published in cooperation with NATO Public Diplomacy Division

Proceedings of the NATO Advanced Study Institute on
Phytoremediation of Metal-Contaminated Soils
Trest, Czech Republic
18-30 August, 2002

A C.I.P. Catalogue record for this book is available from the Library of Congress.

ISBN-10 1-4020-4687-1 (PB)
ISBN-13 978-1-4020-4687-2 (PB)
ISBN-10 1-4020-4686-3 (HB)
ISBN-13 978-1-4020-4686-5 (HB)
ISBN-10 1-4020-4688-X (e-book)
ISBN-13 978-1-4020-4688-9 (e-book)

Published by Springer,
P.O. Box 17, 3300 AA Dordrecht, The Netherlands.

www.springer.com

Printed on acid-free paper

Printed in the Netherlands.

TABLE OF CONTENTS

LIST OF AUTHORS

Balík J. - Czech University of Agriculture in Prague, 165 21 Prague – Suchdol, Czech Republic

Barbu C.H. - "Lucian Blaga" University Bd. Victoriei 10, Sibiu 2400, Romania

Belluck D. - FHWA/US Department of Transportation, 400 7th Street, S.W. Washington, D.C.20590, USA

Benjamin S. - National Wildlife Biologist,US Department of Agriculture Farm Service Agency, 1400 Independence Ave. SW, Room 4714 – S Mail Stop 0513, Washington D.C. 20250, USA

Benlloch M. - Agricultural University of Plovdiv, 12 Mendeleev Str., 4000-Plovdiv, Bulgaria

Bleeker P. - Department of Ecology and Physiology of Plants, Vrije Universiteit, De Boelelaan 1087, 1081 HV Amsterdam, The Netherlands

Boronin A.M. - Institute of Biochemistry and Physiology of Microorganisms, Russian Academy of Sciences (RAS), Pushchino, Moscow oblast, 142290, Russia

Boruvka L. - Department of Soil Science and Geology Czech University of Agriculture in Prague, Czech Republic

David S. - HDR Engineering, Inc., 6190 Golden Hills Drive, Minneapolis, Minnesota 55416, USA

Dickinson N. - Liverpool John Moores University, School of Biological and Earth Sciences, JMU, Byrom Street, Liverpool, United Kingdom

Fernández A. - Agricultural University of Plovdiv, 12 Mendeleev Str., 4000-Plovdiv, Bulgaria

Francova K. - Department of Biochemistry and Microbiology, Faculty of Food and Biochemical Technology, ICT Prague, Technická 5, 166 28 Prague 6, Czech Republic

Geebelen W. - Limburgs Universitair Centrum, Environmental Biology, Universitaire Campus, Building D, B-3590 Diepenbeek, Belgium.

Goncharova N. - Laboratory of Environmental Monitoring, International Sakharov Environmental University, 23 Dolgobrodskaya Str., 220009 Minsk, Belarus

Keller C. - Swiss Federal Institute of Technology (EPFL) ENAC – ISTE – Laboratory of Soil Science Bât. GR, Ecublens, 1015 Lausanne, Switzerland

King R. - School of Biological Sciences, Jones Building, Liverpool University, L69 3GS, United Kingdom

Kochetkov V.V. - Institute of Biochemistry and Physiology of Microorganisms, Russian Academy of Sciences (RAS), Pushchino, Moscow oblast, 142290 Russia

Lepp N. - Liverpool John Moores University, Byrom Street, Liverpool, L3 3AF, United Kingdom.

Lopareva A. - Laboratory of Environmental Monitoring, International Sakharov Environmental University, 23 Dolgobrodskaya Str., 220009 Minsk, Belarus

Macek T. - Department of Natural Products, Institute of Organic Chemistry and Biochemistry, Academy of Sciences of the Czech Republic, Flemingovo n. 2, 166 10 Prague 6, Czech Republic Department of Biochemistry and Microbiology, Faculty of Food and Biochemical Technology, ICT Prague, Technická 5, 166 28 Prague 6, Czech Republic

Mackova M. - Department of Natural Products, Institute of Organic Chemistry and Biochemistry, Academy of Sciences of the Czech Republic, Flemingovo n. 2, 166 10 Prague 6, Czech Republic Department of Biochemistry and Microbiology, Faculty of Food and Biochemical Technology, ICT Prague, Technická 5, 166 28 Prague 6, Czech Republic

Mench M. - INRA Agronomy, Centre Bordeaux-Aquitaine, 71 avenue E. Bourlaux, BP81, F-33883 Villenave d'Ornon, France.

Oprean L. - "Lucian Blaga" University Bd. Victoriei 10, Sibiu 2400, Romania

Reeves R.D. - Phytochemical Consultant, 54 Jickell Street, Palmerston North, New Zealand

Royle A. - Biological and Earth Sciences, John Moores University, Byrom Street, Liverpool, L3 3AF, United Kingdom

Ruttens A. - Limburgs Universitair Centrum, Environmental Biology, Universitaire Campus, Building D, B-3590 Diepenbeek, Belgium.

Sancho Emir-Uco E.D. - Agricultural University of Plovdiv, 12 Mendeleev Str., 4000-Plovdiv, Bulgaria

Sand S. - Regional Environment Protection Agency, Sibiu, Romania

Shilev S. - Agricultural University of Plovdiv, 12 Mendeleev Str., 4000 - Plovdiv, Bulgaria

Sizova O.I. - Institute of Biochemistry and Physiology of Microorganisms, Russian Academy of Sciences (RAS), Pushchino, Moscow oblast, 142290, Russia

Sura M. - Department of Natural Products, Institute of Organic Chemistry and Biochemistry, Academy of Sciences of the Czech Republic, Flemingovo n. 2, 166 10 Prague 6, Czech Republic

Száková J. - Czech University of Agriculture in Prague, 165 21 Prague – Suchdol, Czech Republic

Tlustos P. - Czech University of Agriculture in Prague, 165 21 Prague – Suchdol, Czech RepublicPavlíková D Czech University of Agriculture in Prague, 165 21 Prague – Suchdol, Czech Republic

Vacha R. - Research Institute for Soil and Water Conservation in Prague, Czech Republic

Vandenhove H. - SCK-CEN, Radiation Protection Research Department Boeretang 200, 2400 Mol, Belgium

Vangronsveld J. - Limburgs Universitair Centrum, Environmental Biology, Universitaire Campus, Building D, B-3590 Diepenbeek, Belgium.

PREFACE

Phytoremediation, the use of plants to remediate environmental media, is being pursued as a new approach for the cleanup of contaminated soils and waters, including groundwater. Plant-assisted bioremediation, sometimes referred to as a type of phytoremediation, involves the interaction of plant roots and the microorganisms associated with these root systems to remediate soils containing elevated concentrations of organic compounds. These techniques could provide cost-effective methods of remediating soils and groundwater contaminated with metals, radionuclides, and various types of organics, with fewer secondary wastes and less environmental impact than would be generated using traditional remediation methods.

All plants extract necessary nutrients, including metals, from their soil and water environments. Some plants, called hyperaccumulators, have the ability to store large amounts of metals, even some metals that do not appear to be required for plant functioning. In addition, plants can take up various organic chemicals from environmental media and degrade or otherwise process them for use in their physiological processes.

Phytoremediation technologies are in the early stages of development, with laboratory research and limited field trials being conducted to determine processes and refine methods. Additional research, including genetic engineering, is being conducted to improve the natural capabilities of plants to perform remediation functions and to investigate other plants with potential phytoremediation applications.

Large areas in Western and Eastern countries are polluted with heavy metals and radionuclides in natural, rural, urban or industrial areas.

When regulations mandate the cleanup of such sites, owners require that remediation technologies adopted be able to meet or exceed regulatory standards (very often, these standards consist of threshold concentrations of total metals in mg kg^{-1} of soil). In some cases, e.g., in buildable areas in the centre of cities, lucrative re-uses of the land after clean-up make it possible to mobilize significant financial resources, permitting one to achieve rapid and intensive remediation. However, in areas where there is no extreme pollution or no immediate demand for the land, it is often far more difficult to come up with a satisfactory budget. In these latter cases, low-cost, medium-term remediation technologies are the only options that can be considered. Phytoremediation, i.e., the use of plants to remediate contaminated soils, is particularly well suited for these situations and, in this context, has attracted considerable attention in the last few years.

Four main aspects of phytoremediation have been developed:

1. Phytostabilisation, which consists of the installation of a plant cover on the contaminated soil to prevent from water and wind erosion of contaminated soil particles (dissemination of pollution).

2. Phytodegradation of persistent organic pollutants that are recalcitrant to natural degradation. The rhizosphere of plants is a microcosm in which microbial activity is enhanced and this leads to a more active degradation of such recalcitrant molecules.

3. Phytovolatilisation of elements (e.g., Se, Hg) that are transformed into volatile methyl compounds by microorganisms in the rhizosphere of plants or inside plant cells, and are subsequently eliminated from the soils.

4. Phytoextraction of trace metals. It is based on the properties of plants to take up metals from the soils and to transfer them into their shoots (aerial biomass). Two different approaches have been explored to date: the first one consists of growing successively several crops of hyperaccumulator plants that have the natural ability of transferring metals at very high concentrations in their shoots (i.e., more than 10,000 mg Zn kg^{-1} DW and 1,000 mg Cd kg^{-1} DW for the species *Thlaspi caerulescens*. The second one uses highbiomass accumulating plants (willow, cash crops) to remove significant amounts of trace metals from soils.

Research conducted during the last decade has demonstrated that phytoremediation is a promising, cost-effective technique that makes it possible to remediate polluted soils with a wide range of metal concentrations and for which the demand for re-use allows some time for cleanup (7 to 12 years). Phytoremediation also has the added advantage that it maintains or even soil fertility, unlike intensive physico-chemical treatments that often leave the soil unertile and biologically inactive.

At this point, phytoremediation improves technologies are rapidly emerging as a practical, cost-effective option at a large number of sites.

As such, they have attracted considerable attention from researchers and consulting engineers in NATO countries, and to some extent in Partner countries as well. Regulators have not taken these technologies much into consideration yet, but research results obtained in various countries suggest that these technologies should be adopted increasingly in the many contaminated sites for which cleanup budget is scarce and time is not a significant constraint. Given this situation, it seems very timely for a NATO Advanced Study Institute (ASI) to concentrate on these technologies, to facilitate the (otherwise haphazard and severely hindered) transfer of knowledge about these technologies between Western and Eastern countries, and also to point out the questions that remain currently unresolved and that should become further fields of research. Most of the contaminated sites are polluted with both metals and persistent organic pollutants (POPs). Therefore, besides to metals, special attention will be devoted to the effect of phytoremediation technologies on the elimination of POPs.

This ASI was proposed for the students to benefit from the most recent research on, and assessments of, the technical and economical feasibility of phytoextraction of trace elements from contaminated sites.

Lecturers were all leading scientists in this field: some of them were among the first ones to have suggested the use of hyperaccumulator plants for extraction of metals from soils. Members of an EU funded research project on the feasibility of phytoextraction of metals (PHYTOREM project) from contaminated soils were also among the lecturers.

The ASI was organised in two parts: (1) an overview of the fundamental aspects of the mechanisms of acquisition, transfer and accumulation of metals in plants and especially for hyperaccumulator plants and (2) the latest results from on site phytoextraction pilot studies that allow to state the future needs for research in this field and the already existing technologies. Special efforts to make the lectures very didactic, and ample time for questions and debates, were ensure the success of this ASI. Participants were chosen among scientists but also from non-scientists (consulting engineers, officials of regulatory agencies, human health specialists and toxicologists) of NATO and partner countries that deal with contaminated sites (government officials, environmental agencies). A field trip to heavily contaminated sites in Bohemia was help visualise some of the problems that phytoremediation has to face and also the practical solutions that it may provide.

Manuscripts based on the lectures presented at the ASI were revised to take into account ASI participants' comments and suggestions, and went through a round of peer review and editing.

Discussion summaries and practical recommendations, emanating from the working group and round table discussions, are provided in separate chapters at the end of the book.

Our hope is that the various chapters in this book will provide to individuals who could not attend the ASI in Trest a chance to reflect on some of the issues described during the lectures, as well as a feel for some of the discussions that took place during the workshop. This, hopefully, will encourage readers to discuss these issues further with the authors of the various chapters.

Publication of this book would not have been possible without the extremely helpful assistance and the encouragements of Dr Philippe Baveye from Cornell University, USA and Darya Bairasheuskaya doctoral student at the International Sakharov Environmental University provided, commendable assistance with the editing and formatting of the chapters. Sincere gratitude is also expressed toward Mrs. Annelies Kersbergen, NL, Springer, Mrs Deniz Beten, director of NATO's Scientific Affairs Division, and to her administrative assistant, Mrs. Lynn Campbell- Nolan and Mrs Susan Williamson, for their infallible and stimulating support.

Finally we thank all the participants in theirs ASI for their valuable contributions and for the many stimulating discussions which ensued.

The editors:
Jean-Louis Morel
Guillaume Echevarria
Nadezhda Goncharova

CHAPTER 1

WHY REMEDIATE?

BELLUCK, D.A.,* BENJAMIN, S.L.,**^ and S. DAVID***
**Minnesota Department of Transportation, 395 John Ireland Boulevard, St. Paul, MN 51555, USA*
***Risk Writers, Ltd., 3108 46th Ave. South, Minneapolis, MN 55406, USA*
****HDR Engineering, Inc., 6190 Golden Hills Drive, Minneapolis, Minnesota 55416, USA*
^Address correspondence to Risk Writers, Ltd., 3108 46th Avenue South, Minneapolis, MN 55406 USA, riskwriters@earthlink.net

Abstract

Recognition of massive environmental contamination, including soil and groundwater, resulted in passage of numerous United States environmental statutes during the 1970's and 1980's. Based on federal statutes, most enacted their own statutes, rules and regulations requiring the remediation of soils, when those soils would directly or indirectly threaten human health and, in many cases, the environment. The primary mechanism to determine whether site soils are unacceptably contaminated is quantitative risk assessment and its offshoot, risk-based numerical criteria (i.e. concentrations of a substance that, when exceeded could result in unacceptable human health or resource contamination requiring additional investigation or mandatory cleanup of the contaminated soil). There are numerous criteria used in answering the question, "Why Remediate." The simplest is that a government agency has so ordered.

This paper will discuss the methods used to determine the necessity to remediate. It will also discuss new and some less rigorous methods to arrive at an economic, yet defensible decision.

1. INTRODUCTION

Discovery of massive contamination of groundwater and soil at sites scattered across the United States resulted from passage and implementation of the Resource Conservation and Recovery Act (RCRA) of 1976 and the Comprehensive Environmental Response, Compensation, and Liability Act of 1980 (CERCLA, commonly known as Superfund). The true nature and

J.-L. Morel et al. (eds.), Phytoremediation of Metal-Contaminated Soils, 1–23.

extent of contamination was underestimated at the time these bills were enacted into law.

The number of suspected or known contaminated soils sites in a given area is typically dependent on systematic soil sampling, site investigations, reporting, data compilation and registries (and disputes and incidents). For many countries, only rudimentary data is available.

Even in highly industrialized countries, estimates are variable and data may not be reliable or reflect actual soils contamination levels. As a result, the worldwide extent of soil contamination is not precisely known (you can give an order of magnitude). However, data compiled for and presented in this paper shows that it is a worldwide problem of unknown but major dimensions.

In the U.S.A., initial cost estimates for Superfund were $1.6 billion (average cost of $3.6 million per site for cleanup of 400 contaminated sites). By 1990, the USEPA estimated total cleanup costs of $27 billion (average cost of $26 million per site). Later, USEPA estimated that the Superfund site list could grow to more than 2,000 sites costing $100-$500 billion. More recent estimates indicate cleanup costs based on stringent standards could exceed $1 trillion, when including Department of Defense, Department of Energy, state governments, and privately owned sites (National Research Council 1999).

Billions of Euros and U.S. dollars are spent each year in Europe and the United States to clean up contaminated soils (Bardos, et al., 2001).

The last two decades have seen a rise in the number of government agencies requiring and overseeing soil remediation and a concomitant rise in the number of consulting professionals performing soil remediation activities. While many of these professionals may know how to remediate, they may not fully understand why they remediate. This paper is intended to fill that gap and provide persons interested or working in the field of remediation. These include traditional remediation technologies, as well as alternative methods, e.g. phytoremediation. Also this paper presents some background to understand the history behind the first remediation laws and the technical basis behind the contaminant concentration cleanup requirements or goals established for a given contaminated soils site.

2. ENVIRONMENTAL CONTAMINATION IN UNITED STATES LEADS TO SITE INVESTIGATION AND REMEDIATION LAWS

In 1979, recognition of impacts to human health and the environment due to exposure to contaminated soils led the United States federal government to propose a hazardous waste cleanup fund (USEPA June 13, 1979). In this news release, USEPA noted:

"A rash of recent incidents resulting from improper disposal of hazardous wastes has made it tragically clear that faulty hazardous waste management practices, both past and current, present a grave threat to public health and to the environment." Furthermore, "Hazardous wastes, if not properly disposed of, are capable of causing death, acute poisoning, cancer, miscarriages, and birth defects. There may be several thousand potentially dangerous dump sites around the country."

About a year later (USEPA September 11, 1980), USEPA stated that passage of Superfund legislation is "one of the most badly needed pieces of environmental legislation of the decade." When passed by the Congress, this bill would allow for the clean up of thousands of abandoned hazardous waste dump sites across the country and immediate clean-up of the nearly 14,000 spills of oil and hazardous substances into waterways and on land which occur each year. Citing community contamination cases, such as Love Canal (a New York State community discovered buried leaking drums resulting in President Jimmy Carter declaring a state of emergency (USEPA, 2002), polluted water supplies, serving 300,000 Iowa residents, and the "Valley of the Drums" in Kentucky, where up to 20,000 to 30,000 discarded and leaking barrels of chemical wastes were counted, the USEPA Administrator stated: "The situation concerning hazardous waste disposal sites is grim."

At this time, USEPA estimated that the number of hazardous waste sites in the United States ranged from 32,000 to 50,000, with 1,200 to 2,000 of these sites posing significant health or environmental problems (USEPA September 11, 1980). Of the 250 hazardous waste disposal sites evaluated by 1980, the USEPA, found 32 sites where 452 drinking water wells had to be closed because of chemical contamination and an additional 130 sites, where water supplies and groundwater had been contaminated but wells have not been closed. Of these sites, USEPA reported that 27 sites were associated with "actual damages to human health (kidneys, cancer, mutations, aborted pregnancies, etc.), 41 sites where soil contamination made the land unfit for livestock or human uses, and at least 36 sites where income loss could be expected as a result of loss of livestock, fish kills, crop damage and similar losses." Based on these incomplete findings (USEPA September 11, 1980), USEPA concluded that: "This legacy of many years of uncontrolled hazardous waste disposal may well be the most serious environmental problem facing the nation today." An optimistic USEPA noted that Superfund is "the only major environmental legislation for years which does not establish a new regulatory regime or impose new structures and rules on the nation at large."

In 1980, the Congress of the United States passed Superfund (USEPA 2002) (Comprehensive Environmental Response, Compensation, and Liability Act (CERCLA) 42 U.S.C. s/s 9601 et seq. 1980). Looking back on the three years it took Congress to pass the Superfund legislation, Senator Robert T. Stafford (Stafford, 1981) noted that eighty percent of Americans wanted some type of legislation and that the "Surgeon General of the United States considers toxic chemicals to pose the major threat to health in the United

States for the decade of the 1980s." Stafford also provided a glimpse of chemical contamination in the United States to illustrate this perceived health threat including:

- PCBs leaking from an out-of-service transformer entered the food chain and spread through 19 States and two foreign countries resulting in contamination of hundreds of thousands of hogs, chickens, turkeys, and a large quantity of other foodstuffs requiring their destruction;
- An estimated one-third to one-half of the drinking water and irrigation wells in the San Joaquin Valley were contaminated by the pesticide dibromochloropropane (DBCP) that was known to cause sterility in males and suspected also of causing cancer; and
- Portions of Lakes Ontario and Erie were closed to commercial fishing because of chemical contamination and stocked Coho-salmon sport fishing banned.

Another law related to soils contamination, the Resource Conservation and Recovery Act (RCRA) was passed in 1976. It addressed treatment, storage, and disposal of hazardous waste at operating facilities. The 1984 RCRA amendments required RCRA facilities to conduct corrective action at solid waste management units and areas of concern. Under RCRA, these facilities must investigate and clean up environmental contamination (National Research Council 1999).

It is clear from the historical literature, provided by USEPA, that citizens and legislators recognized that environmental contamination, including contamination of soils, was detrimental to human health and the environment. The numerous "horror stories" cited by USEPA and the call to action by United States citizens resulted in the passage of the federal Superfund law. The next section will discuss how CERCLA and RCRA treat soil contamination.

3. RCRA AND CERCLA PROVISIONS FOR SOIL CLEANUP

Although RCRA and CERCLA have different regulatory requirements and mechanisms, both statutes have similar processes for selecting soil cleanup remedies. For example, these two statutes have similar threshold criteria (i.e. overall protection of human health and the environment and compliance with ARARs, site investigation requirements, remedy selection techniques, and soil cleanup goals.

Prior to 1996, soil cleanup goals were calculated and selected on a case-by-case basis using risk assessment techniques. In 1996, USEPA developed soil screening guidance and Soil Screening Levels (SSLs) (National Research Council 1999). According to USEPA (1996) "SSLs are risk-based concentrations derived from standardized equations combining exposure information assumptions with EPA toxicity data." One type of SSL would establish a threshold for the protection of human health resulting from the exposure to soil contaminants via ingestion, inhalation or dermal routes while another type of SSL protects humans against ingestion of groundwater polluted by soil contaminants. In general, SSLs are calculated to an 1 x 10-6 or E-6 cancer risk (one excess cancer death per one million people) or a non-carcinogen Hazard Index of 1 (National Research Council 1999).

Laws like RCRA and CERCLA provide for the cleanup of contaminated site soils under federal government control or oversight. At the state level, environmental protection and associated agencies either implement and administer federal soil contaminant investigation and cleanup statutes or use similar state statutes to achieve the same outcome. Both the federal and state soil investigation and cleanup statutes rely on technical professionals in government bureaucracies to fill in the blanks between "go forth and remediate" and the technical methods and administrative methods to implement this mission.

4. REMEDIATION PRIMER

Soil contamination occurs when either a solid or liquid substance mixes with soils and becomes physically or chemically attached to soil particles or trapped in the spaces in between (USEPA 1999), resulting in an actual or calculated threat to human health or the environment. For example, water added to a soil would not result in a contaminated mixture, nor would fertilizer when added at nutrient rates. Conversely, if the fertilizer were over applied, contamination could result. Thus, in sufficient quantities, chemicals become soil contaminants, when they can harm humans, plants, animals, leach contaminants to ground water at unacceptable levels and unacceptably degrade the soil resource.

Once it is determined that a soil is unacceptably contaminated, cleaning it up can be done by: 1) soil excavation, treatment and/or disposal; 2) in situ soil treatment; 3) soil containment to prevent soil contaminant movement (USEPA 1999), or 4) limiting the exposure to the contaminated soil. The remainder of this paper will provide a primer on important aspects of remediation and the extent of contamination of soils worldwide.

4.1. Why is it necessary to remediate contaminated soils?

The simple answer it that once a contaminated soil is identified, government laws, rules and regulations often require an investigation to determine whether the pollutants and contaminants constitute unacceptable levels of contamination. Contaminated land remediation is performed to address environmental risks, site user risks, as well as financial and legal liabilities (Scottish Environmental Protection Agency 2002). Soil contamination is managed to render the soil acceptable for a given future purpose (and to avoid the contamination of the surrounding ecosystems). This is true for sites where fresh chemical releases have occurred or where chemicals have been "weathering" at a site for many decades (Loehr, 1996).

4.2. What motives drive the remediation process?

Many countries have legislation requiring remediation of land identified as "contaminated". Typical "drivers" of this legislation include human and environmental health protection, the ability for redevelopment of contaminated sites (including economic support for land regeneration), the repair of previous remediation work or redevelopment projects where past actions have been inadequate, to limit potential liabilities, and to increase land asset values (Bardos, et al., 2001). Numerous other reasons for remediating contaminated soils are found in Table 1. Currently, concern for human health risks form the primary basis for remediation decisions in the United States and the European Community (Bardos, et al., 2001). The extent of human exposure, morbidity and mortality from unmitigated Superfund waste sites is unknown (Florida Center for Public Management 1997), although some morbidity/mortality data may be available for particular sites.

4.3. How are contaminated soils found?

Existing contaminated soils sites are scrutinized because brownfields are increasingly being investigated for reuse. According to the USEPA, 'brownfield site' means real property, the expansion, redevelopment, or reuse of which may be complicated by the presence or potential presence of a hazardous substance, pollutant, or contaminant (USEPA, 2002). In 1998, the United Kingdom announced that 60% of all new homes and commercial property would be built on Brownfield sites. Many contaminated soils are found in the normal course of government soil sampling activities, site specific investigation of suspected contaminated sites, or private party "due diligence" investigations associated with normal business or land transaction activities.

Table 1. Reasons for soils remediation.

*Government orders site remediation based on exceedence of one or more risk-based numerical limits.
*Government orders site remediation based on risk assessment findings of unacceptable risks to human health or the environment.
*Obtain financing for site purchases.
*Court Ordered.
*Public Relations.
*Regulatory Compliance.
*Violation of government statutes.
*Unacceptable risks or hazards to humans or the environment.
*Reduce future toxic tort liability.
*Restore contaminated urban land and buildings to productive use.
*Soil contaminants moving off-site in contaminated media (e.g. dust, groundwater).
*Stop contaminant migration.
*Limit environmental liability in terms of current and future legislation.
*Protect and ensure the long-term sustainability of strategically important soil resources.
*Land use changes resulting in increasing levels of human exposure (e.g. industrial to residential).
*Groundwater, drinking water, air, and food contamination threat.
*Occupational health protection.
*Nearness of contaminated soils to sensitive resource or receptors.
*Avoid future criminal liability.
*Site closure or transfer requirement.
*Change in land ownership.
*Remove as much as possible ongoing legal environmental liability from ownership of a contaminated site.
*Unknown future use of site.
*Fire or explosion hazard.
*Corrosive and attack building materials or services.
*Increase land value.
*Restore ecological functionality.
*Restore landscape value.
*Improve soil fertility.
*Recycle wastes.
*Legal drivers that require local authorities to inspect an area from time to time for the purpose of identifying contaminated land.
*Soil contaminant levels exceed risk management criteria.
*Voluntary
*Political Benefit

Table 2. Examples of soil polluting activities from localized sources as percent of total (European Environmental Agency 2001).

Country	Industrial Activities	Municipal Waste Disposal	Industrial Waste Disposal	Accidents	Other
Austria	47	43	10		
Belgium		39	44		17
Denmark	70	30			
Finland	22	9	2	1	66
Germany	65	35			
Liechtenstein	20			80	
Netherlands	30	10		15	45
Sweden	55	20	20		5

Soil contamination results from many activities (Table 2). Due diligence reviews (i.e. assess potential environmental liabilities attached to a piece of land) are performed by buyers and sellers "To gather information on actual and/or potential environmental impairment of a property in order to make informed decisions regarding purchase, sale, redevelopment, or regulatory compliance (Lane and Schmitter, 1998).

4.4. How are contaminated soils determined to need remediation?

The typical drivers for the need to remediate are questions like:

1. Is it safe?
2. Is the health of my family threatened?
3. Will the economic value of my property be diminished?
4. Will I be held liable for the property if the owner defaults on their loan?

Currently liability issues are driving clean up at most sites where contamination is recognized.

The rigor undertaken to answer these questions will vary pending the political and regulatory landscape of the specific site. There are three basic methods used to determine if a soil requires remediation.

The first two are risk assessment based methods and the third uses a simple local yardstick to determine if a soil has elevated levels of one or more contaminants. The first method, human health and ecological risk assessment, is complex, time consuming, and can be quite expensive. Following the standard USEPA risk assessment paradigm, risk assessors evaluate the hazards of the chemicals in a contaminated soil, determine the environmental fate and transport of these chemicals on-and off-site, assess potential exposures and finally determine the potential risks or hazards posed by these chemicals to human or ecological receptors. Risk or hazard levels are compared to acceptable levels for a given political entity (e.g. nation, state, province, etc.) to determine if these chemicals, alone or combination, exceed a given acceptable risk or hazard level. If there is an exceedence, soil cleanup or remediation may be ordered.

The second method is to compare observed soil contaminant concentrations against lists of acceptable soil contaminant concentrations that have been previously prepared by government agencies using risk assessment methods, algorithms, and variable values appropriate to or required by their localities. If these soil concentrations, alone or in combination, exceed acceptable levels, soil cleanup or remediation may be ordered. The third method is to compare soil contaminant levels to local soil background levels. When soil contaminant levels exceed soil background levels, soil cleanup or remediation may be ordered.

4.5. How much remediation is enough?

Also known as the "how clean is clean" question, the level of remediation required at a given site can vary significantly. This is because the final cleanup values are not always based directly on a calculated value. They may involve a risk management judgment. Unless stipulated by law, government risk managers (i.e. persons who decide how much remediation is needed to ensure regulatory compliance and acceptable risks or hazards at a contaminated soils site) are not required to only use a risk level concentration generated by a risk assessment or a government agency pre-calculated risk level concentration in making their risk management decisions. Government risk managers use the process of risk management (whereby decisions are made to accept a known or assessed risk (Bardos, et al., 2001) to determine final cleanup levels and how much residual chemical contamination will remain on a site after remediation activities are complete.

While risk assessments produce numerical risk levels or risk-based cleanup concentrations, they do not include numerous other factors that must go into a final risk management decision, "How clean is clean?". Important to the decision are the political factors, social factors (blight removal, community concerns, site amenity value, infrastructure provision, Bardos, et al., 2001], economics [impacts on local businesses and inward investment, local employment and site occupancy, Bardos, et al., 2001], efficient and cost effective cleanup, stakeholder acceptability, etc.). In order to reach an informed remediation decision, risk managers must integrate scientific and nonscientific findings to generate final soils remediation values that credibly support both the soil remediation numerical values used for a site and the acceptability of residual chemicals and their risks remaining at a "cleaned-up" contaminated soils site.

A good question is how to determine the acceptable, residual risk? And it is a periodic question, because the residual risk can change with elapsed time, and therefore must be periodically assessed. How to forecast the costs related to this point.

The ideal decision-making process of a risk manager integrates site-specific information, general information about legislation, laws, regulations, policies, and rules, regulatory guidelines, scientific data, engineering feasibility, and other information to make a transparent, consistent and reproducible decision that can withstand peer and legal scrutiny. In other words, the final cleanup level for each contaminant need not be solely based on a risk-based numerical value, but may be the result of a decision-making process that attempts to achieve an optimal or best solution based on scientific and non-scientific variables.

4.6. What equations and variable values are used to calculate soil cleanup values?

The number and types of risk assessment based equations used to calculate soil cleanup values vary between government entities depending on their laws, rules and regulations as well as whether they are producing generic (one number fits all soils for a given exposure scenario) or site specific cleanup values. In their simplest form, generic cleanup values use a single equation into which are fit default variable values. For example, USEPA provides the following equation and default values for calculating generic ingestion noncarcinogen and carcinogen surface soil SSLs (USEPA, 1996):

A. Ingestion of Noncarcinogenic Contaminants in Residential Soil

$$\text{Screening Level (mg/kg)} = \frac{\text{THQ x BW x AT x 365 d/y}}{\text{1/RfDo x 10-6 kg/mg x EF x ED x IR}}$$

Parameter/Definition (units)	Default Value
THQ/target hazard quotient (unitless)	1
BW/body weight (kg)	15
AT/averaging time (yr)	6 (for noncarcinogens, averaging time equals to exposure duration)
RfDo/oral reference dose (mg/kg-d?)	chemical-specific
EF/exposure frequency (d/yr)	350
ED/exposure duration (yr)	6
IR/soil ingestion rate (mg/d)	200

N.B. Childhood only exposure scenario

B. Equation for Ingestion of Carcinogenic Contaminants in Residential Soil

$$\text{Screening Level (mg/kg)} = \frac{\text{TR x AT x 365 d/yr}}{\text{SFo x 10-6 kg/mg x EF x IF soil/adj}}$$

Parameter/Definition (units)	Default
TR/target cancer risk (unitless)	10-6
AT/averaging time (yr)	70
SFo /oral slope factor (mg/kg-d)-1	chemical-specific
EF/exposure frequency (d/yr)	350
IFsoil/adj /age-adjusted soil ingestion factor (mg-yr/kg-d)	114

N.B. Time weighted average soil ingestion rate.

These relatively simple equations and their default values provide one mechanism for calculating generic cleanup values for a given exposure scenario. The generic cleanup numbers generated by these simple equations are usually very conservative, relying on Reasonable Maximum Exposures. More complex default values include those provided in Table 3.

By definition, site-specific soil cleanup values require potentially costly site-specific inputs. These inputs can be used in equations that are more complex than those presented above to derive site specific values that are more representative of site location, activities, physical, chemical and biological attributes of that site. Presenting these equations is beyond the scope of this brief paper, the reader is referred to documents such as USEPA (2001) for a more detailed review of this subject.

Once a soil cleanup value is calculated and determined to be appropriate for contaminated site use, the next step is to determine the techniques to use for reducing site contamination.

4.7. What techniques are used to remediate contaminated soils?

In the past, soil remediation was primarily done by physical removal of soils from contaminated sites for landfilling, incineration or chemical stripping of contaminants from soil resulting soils sufficiently clean to leave on site. In many cases, significant additional damage would result. Evaluation of alternative methods, that were less invasive, providing the similar clean-up results with less damage, were not only allowed, but required. With the advent of bioremediation, (e.g. use of living organisms to clean up oil spills or remove other pollutants from soil, water, or wastewater [USEPA, 1997]) and its subset phytoremediation (e.g. low-cost remediation option for

Table 3. Examples of default exposure factors that can be used to calculate soil cleanup values.

Scenario	Residential	Non-Residential (Commercial/Industrial)		Construction	
Receptor	On-site Resident	Outdoor Worker	Indoor Worker	Construction Worker	Off-site Resident
Exposure Frequency (d/yr)	350	225	250	site-specific	site-specific
Exposure Duration (yr)	30 [6 (child) for non-cancer effects]	25	25	site-specific	site-specific
Event Frequency (events/d)	1	1	NA	1	NA
Soil Ingestion Rate (mg/d)	200 (child) 100 (adult)	100	50	330	NA
Ground Water Ingestion Rate (L/d)	2	2	2	NA	NA
Inhalation Rate (m^3/d)	20	20	20	20	20
Surface Area Exposed (cm^2)	5,700 (adult) 2,800 (child)	3,300	NA	3,300	NA
Adherence Factor (mg/cm^2)	0.2 (child) 0.07 (adult)	0.2	NA	0.3	NA
Body Weight (kg)	15 (child) 70 (adult)	70	70	70	70
Lifetime (yr)	70	70	70	70	70

sites with widely dispersed contamination at low concentrations [USEPA, 1997]), site cleanup managers have a new option that allows for site cleanup without necessarily disrupting soil profiles and function.

Given the high costs of traditional remediation techniques, increasing consideration is being given to phytoremediation because of the potential of remediating a site at a relatively low cost compared to conventional remediation methods (SRI Consulting, 2002). Phytoremediation is significantly cheaper than conventional remediation technologies for shallow contamination, and in the case of phytoextraction, target cleanup concentrations are close to the concentrations at the contaminated site (SRI Consulting, 2002). A report by the McIlvaine Company (1998) noted that world bioremediation and phytoremediation of contaminated sites will grow from $870 million to $1.1 billion in 2002. This market share represents only about 5% of the $25 billion spent on site remediation. In 1997 12.5 million cubic yards of soil were treated by bioremediation and phytoremediation methods.

The largest application of bioremediation and phytoremediation is for landfill and hazardous waste dumpsites. Military applications rank second in the United States in contrast to Europe and Africa where petroleum contaminated sites are the second largest application (McIlvaine Company, 1998). In the United States, growth predictions for the phytoremediation market range from between $200 and $370 million by 2005. Because of the difficulty in patenting phytoremediation methods, it is not clear that the phytoremediation market will remain distinct from the general remediation market (SRI Consulting, 2002).

5. EXTENT OF SOIL CONTAMINATION IN URBAN AND RURAL SOILS

In 1966, the National Research Council (page 21) “Ironically, laws passed between 1952 and 1977 to control air and water pollution (in the United States) caused many industries and municipalities to turn to land disposal for wastes, an action that has contributed to some of the most difficult problems of ground water and soil contamination now faced.” More than forty years of federal government regulatory action to control toxic and hazardous substances has revealed the lack of available knowledge regarding the extent and severity of hazardous materials impacts on human and environmental health (National Research Council, 1966, page 8). It is still not precisely known how many potentially toxic substances are in use in the United States, how many business enterprises are involved in hazardous waste management, the total volume of chemical wastes produced in this country each year, and the total number of sites used to manage hazardous waste (National Research Council, 1966, page 8). Prior to RCRA, up to 90% of hazardous waste was disposed of improperly (National Research Council, 1966, page 10).

While scientists have an understanding of the general nature of intended or accidental release of chemicals to the environment, little is usually known about the exact levels of soil contamination in a given locality because routine surveys are not conducted to determine soil degradation (United Nations Environment Program, 2002). In the European Union, ... "no quantitative information is yet available about the scale and severity of contamination from activities such as waste disposal, industry and accidents..."that could pose risks to soil. "Activities in the EU that result in soil pollution vary widely from country to country" (European Environmental Agency, 2001). Although 300,000 sites across the EU have been identified as definitely or potentially contaminated, it is estimated there are a total of 1.5 million contaminated areas (European Environmental Agency No Date).

Table 4 presents a brief summary of known or potentially contaminated soil sites, by nation. The manner in which data is reported makes it difficult to compile a single list of attributes with numerical values. As a result, the data is presented in much the way it was provided by the cited report authors. It is clear from this table and supporting documentation that soils contamination is a worldwide problem whose dimensions are now only starting to be understood. In some locations, soils contamination can be staggering in its nature and extent. It is reasonable to assume that as countries increase the environmental surveillance of soils, the nature and extent of soils contamination will grow. For many countries, the authors of this paper were unable to find any information on the nature and extent of soils contamination.

6. SOURCES OF SOIL CONTAMINATION IN URBAN AND RURAL AREAS

Soils contamination can result from numerous anthropogenic activities. For example, chemicals can contaminate soils through spills, leaks, tank and pipeline ruptures (Loehr, 1996). Air pollution, agricultural and forestry practices, applying wastes to land, and various routine industrial activities can cause soil contamination. Additionally, improper application of chemicals, or the application of chemicals with as-yet unknown effects, or the desire to reclassify a site to alternate uses may also lead to the designation of a contaminated site. Potential negative impacts from these activities include altered rates of nutrient cycling, altered rates of organic matter decomposition, reduced microbial biodiversity and biomass, loss of organic matter and resultant damage to soil physical structure, mobilization of potentially toxic elements, loss of ecosystem functions in the soil, and damaging effects on the water and air environments (Scottish Environmental Protection Agency, 2001).

Table 4. Survey of known or potentially contaminated soil sites worldwide.*

Country	Number of Potentially Contaminated Sites		Number of Known Contaminated Sites		Other Information On Contaminated Sites
	Identified (screening completed)	Estimated Total	Identified (risk assessment completed)	Estimated Total	
Albania	NI(a)	NI	78(a)	NI(a)	
Armenia	NI	NI	NI	NI	No inventory of contaminated sites (u, nn)
Azerbaijan	NI	NI	NI	NI	No estimates of contaminated sites, 86,415 kilometers squared of contaminated land(nn)
Australia	NI	NI	NI	NI	10 000 to over 60,000(h,i) South Australia –588 registered sites (cc)
Austria	28,000(a)	~80,000(a)	135(a)	~1,500(a)	Identified approximately 29,000 sites out of an estimated 80,000 sites (e)
Belgium (Flemish Region)	5,528(a)	~9,000(a)	7,870 (a)	NI(a)	
Belarus	NI	NI	NI	NI	23% of country contaminated with Cesium-137 and lesser percentages from other radioactive contaminants related to Chernobyl accident (nn)
Bosnia-Herzegovina	NI	NI	NI	NI	3,000 square kilometers are contaminated, 6 registered sites (nn)
Bulgaria	NI	NI	NI	NI	43,000 ha with 7,700 heavily polluted (t) 44,913 ha of agricultural land polluted by industrial activities (nn)
Canada	NI	NI	NI	NI	25,000-30,000 sites (y) 876 confirmed sites in Manitoba (aa) 2,731 in British Columbia (mm)
Denmark	37,000(a)	~40,000(a)	3,673(a)	~14,000(a)	In 1997, just over 4,000 sites (g,h)
Estonia	~755(a)	NI(a)	NI(a)	NI(a)	842 contaminated sites (w) 1,565 military sites (nn)
European Countries/ European Community	NI	NI	NI	NI	Greater than 1.5 million contaminated sites (a) 300,000 to 1.5 million contaminated sites (ii) ~750,000 suspected highly contaminated sites in Europe (jj)

Country	Number of Potentially Contaminated Sites		Number of Known Contaminated Sites		Other Information On Contaminated Sites
	Identified (screening completed)	Estimated Total	Identified (risk assessment completed)	Estimated Total	
Finland	10,396(a)	25,000(a)	1,200(a)	NI(a)	
France	NI(a)	700,000-800,000(a)	896(a)	NI(a)	
Georgia	NI	NI	NI	NI	Greater than 27 ha (nn)
Germany	202,880(a)	~240,000 (a)	NI(a)	NI(a)	1993 estimate =139,000 (240,000 not unrealistic). Estimate ~10% may need remedial actions. (d) 260,000 (q) >190,000 registered potentially contaminated sites (jj)
Hungary	NI(a)	NI(a)	600(a)	10,000(a)	173 registered sites (nn)
Iceland	NI(a)	300-4000 (a)	2(a)	NI(a)	
Ireland	NI(a)	2,000 (a)	NI(a)	NI(a)	
Italy	8,873(a)	NI(a)	1,251(a)	NI(a)	
Japan	NI			NI	870 cases of soil contamination (q) 320,000 suspicious site (r) at least 15,000 sites and up to hundreds of thousands when small-scale contamination is included (ll)
Kazakhstan	NI	NI	NI	NI	No inventory of contaminated sites (s)
Latvia	NI	NI	NI	NI	111 sites in pilot study, 53 military sites (nn)
Lithuania	~1,700(a)	NI(a)	NI(a)	NI(a)	2,743 military sites (nn)
Luxemburg	616(a)	NI(a)	175(a)	NI(a)	
F.Y.R. Macedonia	NI	NI	NI	NI	No sites registered (nn)
Moldova	NI	NI	NI	NI	No estimates (nn)
Netherlands	NI(a)	110,000-120,000(a)	NI(a)	NI(a)	110,000 (q)
Norway	2,121(a)	NI(a)	NI(a)	NI(a)	
Poland	NI	NI	NI	NI	850,000 ha contaminated (nn)

Country	Number of Potentially Contaminated Sites		Number of Known Contaminated Sites		Other Information On Contaminated Sites
	Identified (screening completed)	Estimated Total	Identified (risk assessment completed)	Estimated Total	
Romania	NI	NI	NI	NI	960,000 ha of farmland and 50,000 ha of forest need immediate contamination cleanup (v) 1,634 sites with estimated land area of 164,000 ha (nn)
Russia	NI	NI	NI	NI	No estimate available.
Spain	4,902(a)	NI(a)	370(a)	NI(a)	Initial identification of 370 contaminated sites (e)
Slovak Republic	NI	NI	NI	NI	1.4% of land resources area is contaminated and 0.4% heavily contaminated. (p)
Slovenia	NI	NI	NI	NI	No estimates (nn).
Sweden	7,000(a)	NI(a)	12,000(a)	22,000(a)	22,000 contaminated sites (Z)
Switzerland	35,000(a)	50,000(a)	~3,500(a)	NI(a)	In 1996 almost 20,000 landfills, just under 14,000 polluted industrial sites and 1,500 accident sites. Estimate 50,000 total polluted sites (f) 1,400 in Geneva region (ff)
UK	NI(a)	~100,000 (a,j)	NI(a)	~10,000(a)	5,000-20,000 (n) 5,000 to10,000 hectares of land might be contaminated in Scotland (n) 75,000-100,000 (bb) 5,000-20,000 sites in England and Wales (dd)
Ukraine	NI	NI	NI	NI	No estimates (nn)

Country | Number of Potentially Contaminated Sites: Identified (screening completed), Estimated Total | Number of Known Contaminated Sites: Identified (risk assessment completed), Estimated Total | Other Information On Contaminated Sites

Country	Number of Potentially Contaminated Sites		Number of Known Contaminated Sites		Other Information On Contaminated Sites
	Identified (screening completed)	Estimated Total	Identified (risk assessment completed)	Estimated Total	
United States	NI	NI	NI	NI	"…number of contaminated sites in the United States range in the hundreds of thousands…" (b) 300,000 to 400,000 contaminated sites (c) California has the third highest number of contaminated sites identified for remediation in the USA (k). Department of Defense has about 2,700. (l) Approximately 1,300 Superfund sites, 1,500-3,500 Resource Conservation and Recovery Act (RCRA)-related cleanup projects, and about 320,000 tank sites could require remediation. (m) 500,000 (q) 450,000-600,000 brownfield sites (x) Over 380,000 properties (ee) 1,200 federal identified sites, 31,000 federal sites needing future study, 12,000 non-federal sites, 24,000-32,000 by 2020 will need state or private remediation (gg) 100,000-500,000 Brownfield sites (hh) As many as 600,000 brownfield sites (kk)

*It is often difficult to differentiate in source documents between sites with only soil contamination and soil/water contamination.
NI = No information.

Sources:

(a) European Environment Agency 2001;
(b) National Research Council 1966;
(c) National Research Council 1997;
(d) Technical University of Hamburg-Harburg 1998;
(e) Danish Environmental Protection Agency 2002;
(f) Swiss Agency for the Environment, Forests and Landscape, SAEFL No Date;
(g) *Jensen 1998;*
(h) Parliament of Australia 1997;
(i) Commonwealth of Australia No Date;
(j) Department of the Environment, Transport and the Regions 2000;
(k) Trade Partners UK 2002;
(l) Ruby et al., 2001;
(m) Beyster et al., 1994;
(n) The Information Centre of the Scottish Parliament 2000;
(o) Halm 2002;
(p) Slovak Environmental Agency 2002;
(q) Kawamura 1999;
(r) Van Hall Institut December 5, 2001;
(s) United Nations Economic and Social Council 2001a;
(t) United Nations Economic and Social Council 2001b;
(u) United Nations Economic and Social Council 2001c;
(v) United Nations Economic and Social Council 2001d;
(w) United Nations Economic and Social Council 2001e;
(x) International Economic Development Council No Date;
(y) Mills July 16, 2002;
(z) Karolinska Institute of Environmental Medicine 2002;
(aa) Manitoba Environment 1997.
(bb) London Hazards Centre 1994.
(cc) Environment Protection Authority 1998.
(dd) Environment Agency 2002.
(ee) Syms and Simmons 1999.
(ff) City of Geneva & Canton of Geneva 2002.
(gg) Hansen 1993.
(hh) U.S. Department of Housing and Urban Development March 24, 1995.
(ii) Commission of European Communities 2002.
(jj) International Brownfields Exchange 2002.
(kk) Cope March 7, 2001.
(ll) Industry Canada 1999.
(mm) Environment Canada 1996.
(nn) Andersen 2000.

7. CONCLUSIONS

Recognition of massive environmental contamination resulted in the United States passing numerous environmental protection statutes that were then reflected in similar state statutes. Some of these statutes required contaminated soils to be investigated and remediated to levels where they would not be a human or environmental health concern. Risk assessment became the tool of choice to determine unacceptable risks, regulatory compliance, and to prioritize which sites required further investigation and the timeliness of cleanup actions.

Unfortunately, a lack of systematic soils sampling, site investigations, reporting, data compilation and registries has resulted in a poor understanding of the nature and extent of soils contamination. As is clear, significant soils contamination is a worldwide problem whose scope is only now beginning to be understood. The non-uniform manner in which contaminated soils are reported makes it difficult to grasp the true nature of soils contamination in a given area, region, country or continent. While the extent of worldwide soils contamination is not known, this is a worldwide problem whose dimensions will most likely grow as more and better data are compiled.

The reasons for "Why Remediate?" are numerous, but generally focus on human and environmental health protection, regulatory compliance, liability limitation, or due diligence. This paper has compiled numerous examples of reasons to remediate contaminated soils. Once a threshold requiring remediation is exceeded, a risk management decision is needed on the most effective way to implement this decision. Remediation is often a resource intensive activity. As a result, cost and time for remediation become important factors in determining which remediation alternative will be selected for a given site. Bio-remediation and phytoremediation are emerging as important and cost effective alternative methods to remediate certain types of sites and its use is expected to increase in the coming years.

REFERENCES

1. Andersen, J. N. 2000. Management of Contaminated Sites and Land in Central and Eastern Europe. Danish Environmental Protection Agency, Ministry of Environment. 2000.03.09. Energy. http://www.mst.dk/200004pubs/87-7909-888-6/default_eng.htm
2. Bardos, P., Lewis, A., Nortcliff, S., Mariotti, C., Marot, F., and T. Sullivan. 2001.
3. Review of Decision Support Tools and Their Use in Europe; Report of CLARINET Working Group 2. Final Report. September 2001. http://www.r3environmental.co.uk/WG2%20Report%20web%20version.PDF
4. Beyster, M.A., Ishikawa, Y., Shellenberg, T., and H. Schwendener. December 1994.
5. Environmental Trends and Implications: Land Use Issues. SRI Consulting. http://scup.sric.sri.com/Enframe/Report.html?report=ENVIR000&show=Abstract.html
6. City of Geneva & Canton of Geneva. 2002. Polluted Sites. Editors, town, etc.
7. Commission of European Communities. 2002. Communication From the Commission to the Council, The European Parliament, The Economic and Social Committee and The Committee of the Regions. Towards a Thematic Strategy for Soil Protection. C5-0328/02. 4/16/2002.
8. Commonwealth of Australia. No Date. Environmental Management of Commonwealth Land Site, Contamination and Pollution Prevention.

9. http://www.anao.gov.au/WebSite.nsf/Publications/4A256AE90015F69B4A25690A0025B181
10. Cope, G. March 7, 2001. A Smarter Partnership: Removing Barriers to Brownfields Cleanups. Testimony before the Subcomittee on Environment and Hazardous Materials, United States House of Representatives.
11. Danish Environmental Protection Agency. 2002. Groundwater protection in selected countries. Environmental Project no. 667. Danish Ministry of the Environment. http://www.mst.dk/homepage/default.asp?Sub=http://www.mst.dk/udgiv/publications/2002/87-7972-025-0/html/indhold_eng.htm
12. Department of the Environment, Transport and the Regions. 2000. CIRMCompendium Of Projects Database. Oct. 11, 2000.
13. Department of the Environment, Transport and the Regions. 2000. http://www.databases.odpm.gov.uk/construction/project_detail.asp?PactReference=CC1895
14. Environment Agency. 2002. Land Quality, Contaminated Land. United Kingdom. Editors, town, etc
15. Environment Canada. 1996. Fraser Basin Contaminated Sites, Progress Report 1995-1996. Pacific &Yukon Region. Environmental Assessment and Waste Prevention Section.
16. Environment Protection Authority. 1998. State Of The Environment Report For South Australia. 1998. Summary Report. In Cooperation With The Department For Environment, Heritage And Aboriginal Affairs, Natural Resources Council. Department For Environment, Heritage And Aboriginal Affairs.
17. European Environmental Agency. No Date. Soils. http://themes.eea.eu.int/Specific_media/soil/
18. European Environment Agency. 2001. Soil degradation (Chapter 3.6). Environment In The European Union At The Turn Of The Century. Environmental Assessment Report No 2. January 1, 2001. http://reports.eea.eu.int/92-9157-202-0/en/
19. European Environmental Agency. 2001. YIR01SO01 Soil polluting activities from localised sources. Indicator Fact Sheet Signals 2001 – Soil. http://themes.eea.eu.int/Specific_media/soil/indicators/sources/YIR01SO01.pdf
20. Florida Center for Public Management . 1997. Restoration of Contaminated Sites,
21. Restoration of Sites, Environmental Issue Context Environmental Issue Context, National Air and Radiation Indicators Project. http://www.pepps.fsu.edu/NARIP/
22. Halm, D. 2002. Groundwater Risk Assessment – Part 1, Introduction into Groundwater Risk Assessment at Contaminated Sites (GRACOS). 1 st IMAGETRAIN Advanced Study Course, Katowice, Poland, June 24- 28, 2002, June 24, Afternoon Session.
23. Hansen, D.R. 1993. Environmental Regulations and Just Compensation: The National Priorities List as a Taking. NYU Environmental Law Journal.
24. Industry Canada. 1999. Japan Environmental Remediation Market Assessment.
25. International Brownfields Exchange. 2002. Experiences in Risk Based Decision Making, Lessons and International Best Practices. International Brownfields Exchange 2000-2002. IBE 2000/02.
26. International Economic Development Council. No Date. Economic Development Reference Guide.
27. *Jensen, S. 1998.* Contaminated Sites in Denmark are being catalogued. Danish Environment, Internet Edition, December 1998.
28. http://www.mst.dk/magazine/issue7/catalogue/text.htm
29. Karolinska Institute of Environmental Medicine. 2002. Metals and Health, Contaminated Soil.
30. Kawamura, M. 1999. Land Transactions and Soil Contamination-The Growing Emphasis on Performance and Quality Issues of Land. NLI Research, NLI Institute, No. 133.
31. Lane, E.O. and R.D. Schmitter. 1998. Identification Of Environmental Issues And Overview Of Remediation Approaches. Redeveloping Brownfields, March 18-19, 1998. http://www.hsrc.org/hsrc/html/tosc/sswtosc/IEI/tsld003.htm
32. Loehr, R.C. 1996. The Environmental Impact Of Soil Contamination: Bioavailability, Risk Assessment, and Policy Implications. Reason Public Policy Institute. Policy Study No. 211. August 1996. http://www.rppi.org/environment/ps211.html
33. London Hazards Centre. 1994. London Hazards Centre Factsheet, Contaminated Land.

34. Manitoba Environment. 1997. State of the Environment Report for Manitoba, 1997.
35. McIlvaine Company. 1998. $1 Billion Bio and Phytoremediation Market by 2002. http://www.mcilvainecompany.com/news%20releases/NR487.html
36. Mills, B. July 16, 2002. Thousands of Contaminated Sites Missing From Government's New Listing. Bob Mills, M.P.
37. National Research Council. 1966. Hazardous Materials in the Hydrologic Environment: The Role of Research by the U.S. Geological Survey. National Academy Press. http://books.nap.edu/books/NI000131/html/21.html#page_middle
38. National Research Council. 1997. Innovations in ground water and soil cleanup: from concept to commercialization. National Academy Press, Washington, D.C.
39. National Research Council. 1999. Groundwater & Soil Cleanup, Improving Management of Persistent Contaminants. National Academy of Sciences, National Academy Press, Washington, D.C.
40. Parliament of Australia. 1997. Environmental Management of Commonwealth Land, Chapter 1: The Review of Environmental Management of Commonwealth Land. House of Representatives.
41. Ruby, M.V., Lowney, Y.W., Salatas, J., and Susan Brauning. 2001. Metal Drivers of Risk-Based Soil Remediation at DoD Sites. The Annual International Conference on Contaminated Soils, Sediments and Water. http://www.umasssoils.com/posters2001/heavymetposter.htm
42. Scottish Environmental Protection Agency. 2001. State of the Environment, Soil Quality Report. http://www.sepa.org.uk/publications/stateoftheenvironment/soil/soil_report.pdf
43. Scottish Environmental Protection Agency. 2002. Why Remediate Contaminated Land. www.sepa.org.uk/contaminated-land/introduction/1.3.htm
44. Slovak Environmental Agency. 2002. Increased demand for high quality information Zuzana Lieskovská, Juraj Bebej/Slovak Environmental Agency (SEA). Report on the State of Environment in the Slovak Republic. http://www.sazp.sk/index_en.html
45. SRI Consulting. 2002. PEP Review 97-14, Phytoremediation. May 2002. http://pep.sric.sri.com/Public/Reports/Phase_97/RW97-14/RW97-14.html
46. Stafford, R.T. 1981. Why Superfund Was Needed. EPA Journal. June 1981. http://www.epa.gov/history/topics/cercla/04.htm
47. Swiss Agency for the Environment, Forests and Landscape, SAEFL. No Date. Contaminated Sites, Statistics. Accessed January 4, 2002. http://www.umweltschweiz.ch/buwal/eng/fachgebiete/fg_altlasten/service/statistik/
48. Syms, P.M. and R.A. Simmons. 1999. Contaminated Land Registers: An Analysis of the UK and USA Approaches to Public Management of Contaminated Sites. Abstract from Land Contamination & Reclamation 7(2):121-132.
49. Technical University of Hamburg-Harburg. 1998. Research Centre SFB 188 (1998 - 2000), Treatment of Contaminated Soils, Summary. April 1998. http://www.tuharburg.de/aws/sfb/sfb98-1.html
50. The Information Centre of the Scottish Parliament. 2000.
51. Contaminated land. Research Note 27/00. 18 April 2000.
52. Trade Partners UK. 2002. Environment Market in the USA. http://www.tradepartners.gov.uk/environment/usa/profile/overview.shtml
53. United Nations Economic and Social Council. 2001a. Environmental Performance Review of Kazakhstan.
54. United Nations Economic and Social Council. 2001b. Environmental Performance Review of Bulgaria.
55. United Nations Economic and Social Council. 2001c. Environmental Performance Review of Armenia.
56. United Nations Economic and Social Council. 2001d. Environmental Performance Review of Romania.
57. United Nations Economic and Social Council. 2001e. Environmental Performance Review of Estonia.
58. United Nations Environment Program. 2002. Global Environment Outlook 3. http://www.grida.no/geo/geo3/english/pdf.htm
59. U.S. Department of Housing and Urban Development. March 24, 1995. Achieving Environmental Justice—A Developing Strategy.

60. USEPA. June 13, 1979. Administration Proposes Hazardous Waste Cleanup Fund. http://www.epa.gov/history/topics/cercla/01.htm
61. USEPA. September 11, 1980. Costle Presses for Immediate Passage of Superfund. http://www.epa.gov/history/topics/cercla/03.htm
62. USEPA. 1996. Soil Screening Guidance: Fact Sheet. Publication 9355.4-14FSA. EPA/540/ F-95/041. July 1996. http://www.epa.gov/superfund/resources/soil/fact_sht.pdf
63. USEPA. 1997. Terms of Environment. Revised December 1997.
64. USEPA. 1999. Soil Contamination. Superfund for Students and Teachers. December 7, 1999. http://www.epa.gov/superfund/students/wastsite/index.htm
65. USEPA. 2001. Supplemental Guidance For Developing Soil Screening Levels For Superfund Sites, Peer Review Draft. Solid Waste and Emergency Response. OSWER 9355.4-24. March 2001.
66. USEPA. 2002. Brownfields Glossary of Terms. The Brownfields Site definition is found in Public Law 107-118 (H.R. 2869) - "Small Business Liability Relief and Brownfields Revitalization Act" signed into law January 11, 2002. http://www.epa.gov/brownfields/ glossary.htm
67. USEPA. November 5, 2002. USEPA Timeline. http://www.epa.gov/earthday/timeline.htm
68. Van Hall Institut. December 5, 2001. Soil Remediation Urgency and Deadline Based on Risk Assessment (The Netherlands).

CHAPTER 2

HYPERACCUMULATION OF TRACE ELEMENTS BY PLANTS

R. D. REEVES
Phytochemical Consultant
54 Jickell Street, Palmerston North, New Zealand

Abstract

The plant species of both natural and human-influenced metalliferous soils are being increasingly studied, both as ecosystems in there own right, and with a view to identifying accumulator plants that may be valuable for various forms of phytoextraction. To date, more than 440 hyper-accumulator species have been described, three-quarters of these being Ni accumulators from the extensive occurrences of Ni-rich ultramafic soils found in many parts of world. The history of the discovery of hyper-accumulation in these species is outlined, and a broad assessment is given of the possibility of using them for metal extraction in various ways, including remediation of contaminated soils, extraction of metal ions from contaminated waters and phytomining from large naturally metalliferous areas.

1. INTRODUCTION

Over the last 150 years data have been accumulating about the inorganic composition and nutritional needs of many thousands of plant species. For a long time, most of this work was centred on crops of importance to animals and mankind, and was limited by available analytical techniques to the study of major elements, such as calcium, potassium, magnesium, nitrogen and phosphorus. During the twentieth century, with the improvement of the specificity and sensitivity of analytical techniques (and particularly with the development of atomic spectroscopic methods such as flame emission, atomic absorption, and plasma emission), more reliable data on minor and trace elements have become available. It has become possible to monitor the concentrations not only of the above elements, but of others that have been shown to be essential for plant growth, such as B, Cl, Cu, Fe, Mn, Mo and Zn. Some other elements, such as Ni, Se, Co and V, have either been shown to be

J.-L. Morel et al. (eds.), Phytoremediation of Metal-Contaminated Soils, 25–52.

essential for some species, or at least to have a beneficial effect on some plants, although they may not be universally essential.

Metalliferous soils, with abnormally high concentrations of some of the elements that are normally present as minor constituents (200-2000 mg/kg, e.g. Mn) or trace constituents (0.01-200 mg/kg, e.g. Zn, Cu, Ni, Cr, Pb, As, Co, Se, Cd), vary widely in their effects on plants. These effects depend on plant species, the particular assemblage of enriched elements, and the physical and chemical characteristics of the soil. There are some extreme examples of soils being toxic to almost all species of higher plants, but in other cases a characteristic flora of metal-tolerant species may develop. The metal-rich environment generally has only a small effect on the elemental composition of most metal-tolerant species: it is common to observe a slightly higher than normal concentration of the elements with which the soil is enriched.

Some plant species have evolved to occupy only the ecological niche presented by the metalliferous environment. Where these species are shown to be endemic to areas of well-defined soil composition, they have been used as geobotanical indicators in mineral exploration. However, some metal-tolerant plant species are also characterised by the ability to accumulate very high concentrations of certain elements, far in excess of normal physiological requirements (if any), and far in excess of the levels found in the majority of other species tolerant of metalliferous soils. These 'accumulator' plants, which are an important resource for researchers investigating various strategies of phytoremediation, are the main subject of this chapter.

2. HISTORICAL RECORDS OF METAL-ACCUMULATING PLANTS

It was known as early as 1855 that the 'calamine violet', *Viola calaminaria*, found on zinc-rich soils near Aachen, Germany, contained an unusually high zinc concentration [1]. Analytical data on this plant and other species of the so-called calamine soils in this area (on either side of the modern-day border between Germany and Belgium), including '*Thlaspi alpestre* var. *calaminare*', were provided by F. Risse and reported by J. Sachs [2]. The Zn levels reported then in the dry matter of *V. calaminaria* (typically about 800-1000 mg/kg or 0.08-0.10%) are not now regarded as particularly remarkable, although higher values have since been reported. However, those of the *Thlaspi*, quoted in 1865 as about 1.2% in dry leaves and 0.13% in roots, and reinforced by data of many workers since that time, are an extraordinary response by a plant not showing apparent ill effects from such abnormal uptake of what is normally a 'trace' element. The remarkable behaviour of many species of the genus *Thlaspi* is discussed in more detail below.

Observations of unusual accumulation of other metals emerged only during the twentieth century. There are records of strikingly high plant levels of lead dating from the 1920s, selenium from the 1930s, nickel from the 1940s, cobalt and copper from the 1960s, and cadmium, arsenic and manganese from the 1970s. These are discussed in turn below.

3. HYPERACCUMULATION

The term *hyperaccumulation*, describing a highly abnormal level of metal accumulation, was first applied by Jaffré *et al.* [3] in the title of their paper on nickel concentrations in the New Caledonian tree *Sebertia acuminata*. A similar concept had, however, been used previously by Jaffré and Schmid [4], who referred to certain high-nickel plants from the ultramafic soils of New Caledonia as *hypernickelophores*, i.e. "extreme nickel-bearers".

The specific use of 'hyperaccumulation' to denote a defined concentration (greater than 1000 mg Ni/kg) was introduced by Brooks *et al.* [5] in discussing Ni concentrations in species of *Homalium* and *Hybanthus* from various parts of the world. This work illustrated the possibility of detecting the occurrence of ultramafic rocks and soils from the nickel content of leaf fragments from herbarium specimens of plants of these two widely distributed genera. The Ni concentration threshold of >1000 mg/kg in dry leaf tissue was selected because this was 100-1000 times higher than that normally found in plants on soils not of ultramafic origin, and 10-100 times higher than that found for most other plants on Ni-rich ultramafic soils. This criterion is not entirely arbirary, as it has been noted that in many ultramafic floras, Ni concentrations of 100-1000 mg/kg are quite rare, and accumulation to >1000 mg/kg seems to represent a distinct form of plant response, implying some characteristic and unusual metabolic behavior, i.e. it is usually easy to characterise the species as 'hyperaccumulators' or as 'normal' plants on this basis.

To define Ni hyperaccumulation more precisely, Reeves [6] suggested the following definition: *a hyperaccumulator of Ni is a plant in which a Ni concentration of at least 1000 mg/kg has been recorded in the dry matter of any above-ground tissue in at least one specimen growing in its natural habitat.* This indicates that the classification should not be based on analyses of whole plants or subterranean plant parts, largely because of the difficulty of ensuring that the samples are free of soil contamination, but also because plants that immobilize metals in the root system and fail to translocate them further are of less interest for many purposes than those that actively accumulate metals into all their tissues.

Furthermore, (1) some specimens of a species may be found with >1000 mg/kg whereas others occur with <1000 mg/kg, (2) plant tissue other than leaves (e.g. latex) can be found with very high metal concentrations, or (3) a species may take up high levels of a metal under artificial conditions, such as through massive metal-salt amendments to an experimental soil or a nutrient solution. Plants qualify as hyperaccumulators under cases (1) and (2), but not under (3). The "forced" uptake of added metals from nutrient solution often leads eventually to plant mortality or other interruptions to full plant development. Such experiments often have little relevance to the behaviour of naturally occurring metallophyte populations, even though they may be of interest for some phytoremediation strategies. One further situation should be considered, however: a plant species that is never found as a hyperaccumulator in nature may hyperaccumulate when given the opportunity on a real metalliferous (but not artificially amended) soil. Such a species might be described as a 'facultative hyperaccumulator', particularly if it can be shown to complete its life cycle successfully on such a soil.

Some literature reports of high metal concentrations in plants prove on close examination to have been expressed on an ash weight basis. Such data have usually been obtained by direct analysis of plant ash, e.g., by arc emission spectroscopy or X-ray fluorescence. This basis gives values that are generally 10- to 20-fold higher than those on a dry-weight basis, and have led to exaggerated claims of metal accumulation by plants that are in fact behaving normally.

The definitions of hyperaccumulation have been extended to other elements. The 1000 mg/kg criterion was used in succession by Malaisse *et al.* [7] for Cu accumulation, by Brooks *et al.* [8] for Co and by Reeves and Brooks [9] for Pb. Jaffré [10] used the term *hypermanganèsophore* for plants with >10,000 mg/kg Mn, and Baker and Brooks [11] chose the same level to define Zn hyperaccumulation. Although some may consider this to be unduly restrictive, and a lower threshold of (say) 3000 mg/kg might be used, it is certainly true that 'normal' plant behaviour towards these elements involves higher levels (about 20-400 mg/kg), than are generally found for Ni, Cu and Co.

Existing tabulations of data on Se accumulation in plants, such as that of Reeves and Baker [12] have paid attention only to species in which concentrations >1000 mg/kg have been recorded. However, as most Se levels in plants are <1 mg/kg, even 100 mg/kg could be considered remarkable. Becasue Cd normally also occurs in plants at levels well below 1 mg/kg, there should again be interest in any species that accumulates this element to levels above 100 mg/kg. In recent times, there has also been renewed interest in As accumulation by plants, and concentrations above 1000 mg/kg in the plant dry matter must be deemed noteworthy.

From time to time, other bases for defining hyperaccumulation are suggested. These include criteria based on concentration ratios [shoot]/[root], with hyperaccumulation being defined (for example) as an occurrence of this ratio exceeding 1.0. However, a plant with 30 mg/kg Zn in shoot and 15 mg/kg in the root is not showing exceptional behaviour and such a combination of concentrations is not of great physiological interest, nor does it suggest any use for phytoremediation. In addition, the analysis of root material collected from the field is burdened with contamination difficulties. Other suggestions have included the use of the ratio [plant]/[soil] exceeding 1.0, where [soil] implies the total concentration of the element in the rooting zone of the plant. Again, this may not focus on the most significant occurrences of unusual metal uptake. For example, a plant with 2000 mg/kg Ni on a serpentine soil with 2500 mg/kg is still interesting and unusual, and behaving abnormally; on the other hand, a plant with 15 mg/kg Cu on a soil with 10 mg/kg is not extraordinary. Furthermore, in the search for hyperaccumulator plants, it is often the case that appropriate soil samples have not been collected with the plant material, making the use of a specific plant/soil concentration ratio impossible. Thus, for all their limitations, the concentration-based criteria outlined above seem to bring the focus onto plants that are most worthy of attention for unusual aspects of their physiology and biochemistry, and for their potential use in phytoremediation.

4. METALLIFEROUS SOILS

There are many different types of metalliferous soils, both naturally occurring and anthropogenic. The most important types of naturally metalliferous soil, which may support some hyperaccumulator plant species, are (1) serpentine soils, derived from Fe- and Mg-rich ultramafic rocks, which occur in many parts of the world, and which are significantly enriched in Ni, Cr, and Co; (2) soils enriched in Pb and Zn, the so-called "calamine" soils, which usually contain elevated Cd also, and sometimes carry high concentrations of As and/or Cu; (3) Cu- and Co-containing soils, such as those of the Shaban Copper Arc in the Democratic Republic of Congo, derived from argillites and dolomites containing sulfides of these metals, together with important secondary minerals (oxides, basic carbonates, silicates); (4) soils derived from various Se-rich rock types, in which Se concentrations are often >10 mg/kg and can sometimes exceed 50 mg/kg.

Serpentine soils (type (1) above) occur over extensive areas in many countries of the world, particularly those that have seen intense tectonic activity, and are an important feature of the geology of places such as Japan, the Philippines, parts of Indonesia, California, Turkey, Greece and the Balkan lands, parts of the European Alps, central Brazil, Cuba, Puerto Rico and parts of Central America. Notable Se-rich soils (type (4) above) have arisen particularly from the Cretaceous shales of the midwestern United States. These two types of soils often support characteristic metal-tolerant floras, which may include some hyperaccumulators of Ni and Se, respectively.

The other two types of soils noted above are more limited in their extent as natural surface exposures. However, metalliferous soils resulting from human activity, sometimes with characteristic floras, can be found in more restricted areas in many parts of the world. These include soils developed on wastes from mining, storage, transport and smelting of ores of such metals as Ni, Co, Cr, Cu, Pb, Zn, Cd and As. In addition, there are other human uses of metals or metal compounds that have led to local soil contamination that is now in need of some form of remediation. Such activities include agricultural and domestic uses (e.g. Cu from sprays, As, Cu and Cr from timber preservatives, Cd from the use of high-Cd phosphatic fertilisers) and industrial processes (e.g. the many uses of Pb in paints, batteries and organolead compounds).

5. ASPECTS OF PLANT ANALYSIS

Most of the plant analytical data in recent years has been generated by using atomic spectroscopic methods, especially atomic absorption (AAS) and plasma emission (ICP-ES) spectroscopy. Where extreme accumulation of metals is involved there is sufficient sensitivity in flame AAS that it is not generally necessary to resort to electrothermal methods of atomisation. However, for the determination of As (and sometimes Se) by AAS it is usual to employ the technique in which the compounds are reduced by sodium borohydride solution in acidic media, to give the volatile hydrides AsH_3 or SeH_2 for flushing into a heated tube in which they are decomposed to give the element atomic vapour. (Reduction of Hg compounds to elemental Hg can be carried out with the same reagent.) Where multielement information is desired, ICP-ES is more commonly used. With standard equipment it is possible to obtain data for 20-30 or more elements in the same sample, with a sample volume of 3-4 mL, in a process that takes only 1-2 minutes per sample. Sensitivity for most elements is in the 'ppb' (ng/mL) range, and is more than adequate for studies of most elements in contaminated soils and accumulator plants.

Contamination of plant material from soil or from aerial fallout can be a problem, especially when the soil contains particles of nearly pure minerals of the metal being investigated, such as carbonates, silicates, or sulfates of Pb, Zn or Cu. In the sampling of plant material adjacent to smelters, metal vapours and very fine dusts may deposit on plant material and produce contamination even more concentrated than that provided by dusts from the soil. Because any contamination of plant material from such sources is likely to pass easily into solution under the conditions used for plant analysis, special care is needed in the washing processes. Brief agitation of the plant material for a few minutes in water with detergent can be used, but it is impossible to be sure in all cases that particulate deposits have been removed. Where this appears to be a problem, it is possible to suspend leaf material in a non-aqueous inert solvent (e.g. hexane) in an ultrasonic bath to aid the removal of fine material from the leaf surfaces.

In the most extreme cases of genuine plant hyperaccumulation, a false indication caused by soil contamination is unlikely, as the soil concentration of the element concerned may actually be lower than the plant concentration.

The following section deals with particular elements that are accumulated to an abnormal degree by some plant species, and a later section is devoted to plants of the genus *Thlaspi* because of the recent upsurge of interest in unusual aspects of their behaviour.

6. HYPERACCUMULATION OF PARTICULAR ELEMENTS

6.1. Nickel

The first record of extraordinary Ni concentrations in a plant species was that of Minguzzi and Vergnano [13] who found concentrations of 1000-10,000 mg/kg (0.1-1.0%) in *Alyssum bertolonii* from serpentine soils of central Italy. Reports during 1961-1973 noted several other species showing similar behaviour: two more *Alyssum* species, *A. murale* from Armenia [14] and *A. serpyllifolium* subsp. *lusitanicum* from Portugal [15], *Pearsonia metallifera* from Zimbabwe [16], and *Hybanthus floribundus* from Western Australia [17, 18].

Since 1974, work by T. Jaffré and others in New Caledonia, and by R.R. Brooks, R.D. Reeves and co-workers, studying the serpentine vegetation of many other parts of the world, has widened our knowledge of the range of occurrence of Ni hyperaccumulation, both geographically and in terms of distribution in the plant kingdom. Among the most noteworthy findings are the following.

(a) A comprehensive survey of nearly all the 170 species of the genus *Alyssum* [19, 20] established that there were about 48 Ni hyperaccumulators, all within one section (Odontarrhena) of the genus, distributed from Portugal across Mediterranean Europe to Turkey and Armenia. Most are serpentine-endemic species, and many are of very restricted distribution. Later work has increased the number of Ni-accumulating *Alyssum* taxa to about 52, including a recently discovered new species [21].

(b) An extensive study of the New Caledonian serpentine flora [10] showed that about 45 species (in 8 different families) from the widespread ultramafic soils of this island were Ni hyperaccumulators. These included the remarkable tree *Sebertia acuminata* with its latex containing more than 20% Ni in the dry matter [3].

(c) Detailed investigation of the plants of temperate-zone serpentine soils has led to the discovery of Ni hyperaccumulation in a total of about 35 further species in the Brassicaceae. They occur in the genera *Cardamine* in Italy [22]; *Peltaria (Leptoplax)* in Greece [23]; *Thlaspi* in Europe [22, 24-27], the western USA [28], Turkey and Japan [29]; *Bornmuellera* in Greece and Turkey [30]; *Cochlearia (Pseudosempervivum)* and *Aethionema* in Turkey [29, 31]; and *Streptanthus* in California [32].

(d) Studies of the plants of tropical serpentine soils of Brazil, Queensland, Southeast Asia (especially various islands of Indonesia and the Philippines) and Sri Lanka have led to the identification of about another 35 species as Ni hyperaccumulators. More detailed accounts of this work can be found elsewhere [6, 33-37].

(e) It has been discovered that the ultramafic soils of Cuba are host to the largest number of Ni hyperaccumulators found in any one country. A survey of much of the Caribbean serpentine flora has revealed 130 such species in Cuba (mainly from the genera *Buxus, Phyllanthus, Leucocroton, Euphorbia, Pentacalia, Senecio, Psychotria, Ouratea* and *Tetralix*), as well as one species of *Phyllanthus* from the Dominican Republic [38, 39] and one species of *Psychotria* from Puerto Rico [37].

(f) In the Asteraceae, Ni hyperaccumulators have been found not only in Cuba [39], but have also been reported from Italy [24], Newfoundland [40], South Africa [41, 42] and Turkey [31, 43].

About 360 species worldwide are now known to act as Ni hyperaccumulators. The plant families most strongly represented are the Brassicaceae (with 89 taxa), Euphorbiaceae (87), Asteraceae (44), Flacourtiaceae (19), Buxaceae (17), and Rubiaceae (15). About 90 other species are from more than 30 different families, distributed widely throughout the plant kingdom, emphasizing the way in which this property must have evolved independently many times. (It is possible that this trait has been gained and lost several times even within a genus, and current DNA studies are partly designed to shed some light on this point.) Earlier summaries [11, 33, 38, 39] and the original literature can be consulted for details of species in genera containing many Ni hyperaccumulators, and for authorities for the species names.

The large number of hyperaccumulators of Ni (compared with those of other elements) may be partly the result of the attention to analytical work on ultramafic floras, but is almost certainly also a reflection of the areal extent and large number of tracts of nickeliferous soils. Continuous ultramafic areas of tens or even hundreds of square kilometres can be found in countries such as New Caledonia, Cuba and Turkey. Where such areas have been continuously available for plant colonization for many millions of years (not having suffered from glaciation, complete erosion, or burial beneath volcanic or sedimentary material), this has presented a major opportunity for the evolution of a characteristic flora with numerous endemic species, including some that have developed Ni accumulation. The influence of age of exposure of the soils appears to be particularly influential in the case of the Cuban serpentines [38, 39].

Table 1 gives examples of Ni hyperaccumulators and their maximum recorded levels of Ni accumulation; it is designed to illustrate both the geographic range of occurrence of these species, and their range within the plant kingdom. The overwhelming predominance of dicotyledonous species should be noted. There is no doubt that further Ni hyperaccumulators remain to be discovered. Some of these will be species not yet discovered or properly described, and others will be known species that have never been analyzed.

6.2. Zinc, Lead, and Cadmium

The discovery of zinc accumulation in certain *Viola* and *Thlaspi* species in 1855-65 has been mentioned already, and the development of further interest in *Thlaspi* species is described in section 7. A number of other species have been reported with more than 10,000 mg/kg Zn, notably *Arabidopsis halleri* (formerly *Cardaminopsis*) from western and central Europe. Some of these species are detailed in Table 2.

Lead is present in most soils and rocks at concentrations below 50 mg/kg, and generally shows relatively little mobility in the soils and into vegetation which on normal soils is typically found with <10 mg/kg Pb. Even where concentrations of 1-10 mg/kg are measured in above-ground plant parts, some of this almost certainly comes from various forms of environmental and laboratory contamination. When Pb does enter the plant roots in larger concentrations from Pb-enriched soils, significant translocation to the upper parts of the plant is uncommon.

Table 1. Some hyperaccumulators of nickel.

FAMILY Species	Location	Maximum Ni conc.(mg/kg)	References
ACANTHACEAE			
Ruellia geminiflora	Brazil (Goiás)	3330	44
ASTERACEAE			
Berkheya coddii	South Africa	11,600	41
Pentacalia (10 species)	Cuba	16,600	38
Senecio (9 species)	Cuba, S. Africa, Canada	11,000	38, 40, 42
BRASSICACEAE			
Alyssum (52 taxa)	S & SE Europe, Turkey, Armenia, Iraq, Syria	1280-29,400	13-15, 19-22, 25, 30, 45-47
Bornmuellera (6 taxa)	Greece, Albania, Turkey	11,400-31,200	30
Cochlearia aucheri	Turkey	17,600	29
Peltaria emarginata	Greece	34,400	23
Streptanthus polygaloides	USA (CA)	14,800	32
Thlaspi (23 taxa)	Europe, Greece, Turkey, Japan, USA (CA, OR, WA)	2000-31,000	22, 24-29
BUXACEAE			
Buxus (17 taxa)	Cuba	1320-25,420	37, 48
CLUSIACEAE			
Garcinia (4 species)	Cuba	7440	38
CUNONIACEAE			
Geissois (7 species)	New Caledonia	1000-34,000	10, 49
DICHAPETALACEAE			
Dichapetalum gelonioides ssp. *tuberculatum*	Philippines, Sabah	26,600	35, 50
EUPHORBIACEAE			
Euphorbia (5 taxa)	Cuba, Brazil (Goiás)	1800-9340	37
Leucocroton (28 taxa)	Cuba	2260-27,240	37, 48
Phyllanthus (42 taxa)	New Caledonia, Sabah, Philippines (Palawan, Mindanao), Cuba, Dominican Republic, Indonesia (Japan)	1090-60,170	10, 35, 37, 51, 52
FABACEAE			
Pearsonia metallifera	Zimbabwe	15,350	16, 53
FLACOURTIACEAE			
Homalium (7 species)	New Caledonia	1160-14,500	4, 5, 10
Xylosma (11 species)	New Caledonia	1000-3750	10, 54

FAMILY Species	Location	Maximum Ni conc.(mg/kg)	References
MELIACEAE			
Walsura monophylla	Philippines (Palawan)	7090	35, 50
MYRTACEAE			
Mosiera (4 taxa)	Cuba	1210-5030	38
OCHNACEAE			
Brackenridgea palustris ssp. *foxworthyi*	Philippines (Palawan)	7600	35, 50
Ouratea (2 species)	Cuba	1700-8190	38
RUBIACEAE			
Ariadne shafer (2 subspp.)	Cuba	13,070-22, 360	38
Phyllomelia coronata	Cuba	25,540	38
Psychotria (9 taxa)	Cuba, Puerto Rico, Sabah, New Caledonia, Sulawesi	1820-38,530	4, 37, 38, 55
SAPOTACEAE			
Planchonella oxyedra	Indonesia (Obi)	19,600	56
Sebertia acuminata	New Caledonia	11,700 a	3
STACKHOUSIACEAE			
Stackhousia tryonii	Australia (Queensland)	21,500	34
TILIACEAE			
Tetralix (5 species)	Cuba	2670-13,610	38
Trichospermum kjellbergii	Indonesia (Obi)	3770	56
TURNERACEAE			
Turnera subnuda	Brazil (Goiás)	6130	44, 57
VELLOZIACEAE			
Vellozia sp.	Brazil (Goiás)	3080	57
VIOLACEAE			
Hybanthus (6 taxa)	New Caledonia, Sri Lanka, Western Australia	1860-17,600	4, 5, 17, 18, 36
Rinorea bengalensis	Southeast Asia	17,500	58

a Latex contains up to 26% Ni (dry wt. basis)

Some examples of very high lead concentrations in plants in their natural habitats (invariably taken from areas of Pb/Zn mineralization, or mine or smelter wastes) are shown in Table 2. Because of difficulties with removing contaminating dust and soil, and because in some cases (especially in the vicinity of smelters) there is a possibility of very fine particulate material entering the plant leaves directly through the stomata, it is not certain that all these reported high values represent genuine uptake through the root system of the plant, and translocation to the leaves. It was well established by Baker *et al.* [59] that *Thlaspi caerulescens* can accumulate Pb to high concentrations (>1000 mg/kg on a whole plant basis) from nutrient solutions with low concentrations of added Pb, but most of this becomes fixed in the root system, with limited transport to the leaves (e.g. 28,900 mg/kg in the root; 262 mg/kg in the leaves). It has also proved possible to induce a wider variety of plants to accumulate Pb from contaminated soils with the addition of complexing agents such as EDTA [60]. The role of such soil amendments is one of the central (and most controversial) current issues in phytoremediation.

6.3. Cobalt and Copper

Most of the cases of hyperaccumulation of these elements have been reported from the metalliferous soils of the Democratic Republic of Congo (formerly Zaïre), where the two metals occur together at elevated levels in the soils, but in widely varying proportions. A few other records of plants with >1000 mg/kg Cu from copper-mineralized areas [67, 78, 79] should be reinvestigated in detail.

Normal concentrations of cobalt and copper in plants are in the ranges 0.03-2 and 5-25 mg/kg, respectively. The tupelo or black gum of the southeastern United States (*Nyssa sylvatica*) is remarkable in being able to accumulate as much as 845 mg/kg Co from normal soils [80-82]. However, even on cobalt-enriched soils, such as those derived from ultramafic rocks, plant Co rarely exceeds 20 mg/kg. Plant Cu concentrations are controlled within a remarkably narrow range, and Cu concentrations above 100 mg/kg are rarely found in carefully washed plant leaves, even in the presence of high soil Cu.

Duvigneaud [83] found moderate concentrations of Co in *Crotalaria cobalticola* on cobalt-rich soils in Zaïre, and Brooks *et al.* [8] recorded concentrations up to 3000 mg/kg in this species. F. Malaisse, R. R. Brooks, and their co-workers carried out extensive studies of the vegetation of many sites of mining and smelting activity throughout Zaïre through plant and soil sample collections and analysis, and identified 30 hyper-accumulators of cobalt and 32 of copper, with 12 species being common to the two lists. It is important to note that many of these species do not invariably show

Table 2. Some hyperaccumulators of zinc, cadmium and lead.

FAMILY Species	Locality	Max. conc. (mg/kg) a			References
		Zn	Cd	Pb	
BRASSICACEAE					
Arabidopsis halleri b	Germany	13,620			64
Thlaspi brachypetalum	France	15,300			26
T. caerulescens	W. & C. Europe	43,710	2130	2740	2, 26, 64-70
T. ochroleucum	Thasos, Greece	4130		1210	71
T. cepaeifolium ssp. *rotundifolium*	S. Austria, N. Italy	18,500		8200	9, 72
T. praecox	Bulgaria	15,500			26
T. stenopterum	Spain	11,500			26
T. tatrense	Slovakia	20,100			26
CARYOPHYLLACEAE					
Minuartia verna	Yugoslavia; UK	11,400		20,000	70, 73, 74
Polycarpaea synandra	Queensland, Australia	6960		1044	75
DICHAPETALACEAE					
Dichapetalum gelonioides c	Sumatra; Mindanao; Sabah	30,000			35
PLUMBAGINACEAE					
Armeria maritima var. *halleri*	Germany			1600	66
POACEAE					
Agrostis tenuis	UK			13,490	74, 76
Arrhenatherum elatius	France			1500	77
Festuca ovina	UK			11,750	74
POLYGONACEAE					
Rumex acetosa	UK	11,000		5450	73
VIOLACEAE					
Viola calaminaria	Belgium, Germany	10,000			2, 66

a Maximum values for various elements, not all found in the same specimen
b Formerly *Cardaminopsis*
c Subspecies *tuberculatum*, *sumatranum*, and *pilosum*

high metal concentrations (unlike many serpentine-endemic Ni hyper-accumulators which always contain high Ni), nor are high concentrations of both Cu and Co commonly found in the same specimen. *Pandiaka metallorum* (Amaranthaceae) for example, has been recorded from these metalliferous soils with Co ranging from 31-2131 mg/kg and Cu from 72-6270 mg/kg [84, 85]. Many other similar wide variations can be found (even within a restricted metalliferous area), the reasons for which are not well understood, and may be only partly due to local variations in soil metal concentrations. The Co and Cu hyperaccumulators have been found in more than a dozen plant families; some examples are given in Table 3, and a more complete listing can be found in Reeves and Baker [12]. It should also be noted that most of the Cu and Co accumulators are not restricted to metalliferous soils. Great care therefore needs to be taken in selecting seed of any of these species for studies connected with metal-accumulation experiments or with studies of their potential for phytoremediation.

Ni hyperaccumulators show a wide variation in their concomitant Coaccumulating ability from their natural serpentine soils, ranging from <1 mg Co/kg in some cases to occasional values in the 10-100 mg/kg range and >1000 mg/kg in just one recorded instance. One specimen of the Cuban *Phyllanthus williamioides* was found with 430 mg/kg Ni, 4200 mg/kg Mn, and 1140 mg/kg Co, but others were recorded with >1% Ni and much lower Co. Co concentrations above 1000 mg/kg have been induced in leaves of Ni hyperaccumulators such as *Alyssum* and *Thlaspi* species in pot trials with Co-amended soils or in Co-enriched nutrient solutions [59, 86, 87].

6.4. Manganese

Normal levels of manganese in plant dry matter fall within the rather wide range of 20-500 mg/kg, and occasionally exceed 1000 mg/kg in plants on normal soils (e.g. the value of 5270 mg/kg in *Vaccinium myrtillus* reported by Denaeyer-De Smet [95]). Symptoms of deficiency can often be detected when the leaf concentration is <10 mg/kg, while toxic levels fall in the range 1000-12,000 mg/kg, depending on species. Manganese is an essential element, activating some of the enzymes involved in the tricarboxylic acid cycle, and a central role for manganese cluster complexes in the oxidation of water to oxygen has been demonstrated.

Relatively little work has been done on the manganese concentrations of plants on soils formed over areas of Mn deposits (most importantly the oxide minerals pyrolusite and psilomelane). Some species have been found with 1000-5000 mg/kg Mn on soils with manganese mineralization (bearing >1% Mn) and on soils of lower Mn concentration.

Table 3. Some hyperaccumulators of copper and cobalt (all species from Democratic Republic of Congo [Zaïre] unless otherwise noted).

FAMILY Species	Maximum concentration (mg/kg)		References
	Co	Cu	
AMARANTHACEAE			
Pandiaka metallorum	2131	6270	84, 85, 88, 89
ASTERACEAE			
Anisopappus davyi	2889	3504	84, 85, 89, 90
COMMELINACEAE			
Cyanotis longifolia	4197		84, 89
CYPERACEAE			
Ascolepis metallorum	1118	1211	85, 88, 90
Bulbostylis pseudoperennis a	2127	7783	8, 84, 85
EUPHORBIACEAE			
Phyllanthus williamioides b	1140		39
FABACAEAE			
Crotalaria cobalticola	3010		8, 85
Vigna dolomitica		3000	91
LAMIACEAE			
Aeollanthus subacaulis var. *linearis*a, c	5176	13,700	7, 8, 85, 91, 92
Haumaniastrum robertii	10,230	2070	85, 88, 89, 93, 94
POACEAE			
*Eragrostis racemosa*a		2800	91
PTERIDACEAE			
Actiniopteris sp.		3535	85, 90
SCROPHULARIACEAE			
Buchnera henriquesii	2435	3520	8, 85
Sopubia neptunii	2476		85
TILIACEAE			
Triumfetta dekindtiana		1283	85
T. welwitschii var. *descampii*	2380		85
VELLOZIACEAE			
Xerophyta retinervis var. *equisetoides*	1520		85

a Nomenclature and/or identification changed since original publication (F. Malaisse, personal communication).

b From El Toldo, Moa, Cuba.

c Spelling of generic name revised as shown; Cu up to 13,700 mg/kg in whole plant or corms; 2,150-3,500 mg/kg in leaves and stems.

Ultramafic soils may have 1000-5000 mg/kg Mn, which is not regarded as strongly abnormal, but it is from these soils that most records of high plant Mn have come. Jaffré [10] found that 98 out of 445 species (22%) on ultramafic soils of New Caledonia had mean Mn concentrations above 1000 mg/kg; 6 species had means exceeding 10,000 mg/kg, and 9 had at least one specimen above this level. Baker and Brooks [11] chose this level to define Mn hyperaccumulators, and listed 8 such species. Table 4 lists those in which the maximum value found has exceeded 10,000 mg/kg (1%). Comparable values have not been found in studies of the ultramafic vegetation of Cuba or Brazil. The present author, in unpublished work in 1991-2, found Mn at levels of 2300-2600 mg/kg in *Austromyrtus bidwillii* (Myrtaceae) on the serpentinites of the Marlborough region in Queensland, Australia, but Bidwill [96] has reported values of up to 19,200 mg/kg in leaves of this species, in samples collected both from serpentine soils and from krasnozem soils over tertiary basalts.

Table 4. Some hyperaccumulators of manganese (species from New Caledonia unless noted otherwise).

FAMILY Species	Maximum (mean) concentration (mg/kg)	References
APOCYNACEAE		
Alyxia rubricaulis	14000 (10270)	10, 97, 99
CELASTRACEAE		
Maytenus bureaviana	33750 (19230)	10, 97
M. pancheriana	16370	10
M. sebertiana	22500	10
CLUSIACEAE		
Garcinia amplexicaulis	10500 (2230)	10
MYRTACEAE		
Austromyrtus bidwillii a	19200 (16500)	96
Eugenia clusioides	10880 (6810)	10
Eugenia sp. b	13700	100
PROTEACEAE		
Beaupreopsis paniculata	12000 (6900)	10
Macadamia angustifolia	11590 (11110)	10
M. neurophylla	55200 (40730)	10, 98

a From Queensland (Australia)
b From Sabah

6.5. Chromium

There are some reports of chromium being accumulated by plants, particularly from serpentine soils, which may contain 500-5000 mg/kg Cr. On normal soils, plant Cr is often <1 mg/kg and rarely exceeds 5 mg/kg. Even on serpentine soils, Cr concentrations >50 mg/kg are so uncommon that this may be used as an indicator of soil contamination. Wild [101] reported high concentrations of Cr in ash from leaves of *Dicoma niccolifera* and *Sutera fodina* near a chrome mine in Zimbabwe, probably equivalent to 1500-2400 mg/kg on a dry-weight basis. However, Brooks and Yang [53] found values at least 20 times lower in the same species, and the earlier data may have been affected by windblown ore dust. Unusually high Cr concentrations (1800-7500 mg/kg ash weight, or 180-720 mg/kg dry weight) have also been reported in the epiphytic moss *Aerobryopsis longissima* from ultramafic areas of New Caledonia [102].

6.6. Arsenic

Normal arsenic levels in igneous rocks and in soils lie in the range 1-10 mg/kg. However, much higher soil As concentrations can be found in areas of polymetallic sulfide mineralisation, in areas of geothermal activity, and in areas contaminated through the use of As compounds as horticultural sprays or timber preservation agents. Plant As concentrations are normally of the order of 1 mg/kg, but higher values can be found in mineralised areas, and plant As analysis has been used in several countries as a biogeochemical pathfinder in gold prospecting [103].

Contrary to a large number of recent statements, arsenic hyperaccumulation (based on a 1000 mg/kg dry matter criterion) has been well established for more than 30 years. Warren *et al.* [104], for example, found As in the ash of growing tips of Douglas fir (*Pseudotsuga menziesii*) to be 2500-10,000 mg/kg over soils with 1000-5000 mg/kg As. The highest of these values are almost certainly equivalent to >1000 mg/kg on a dry weight basis.

Studies carried out in the 1970s and 1980s by several groups on the behaviour of aquatic plants in the Waikato river in the centre of the North Island of New Zealand established that three aquatic plants are acting as As hyperaccumulators. Natural geothermal activity, together with the borefield drainage and waste water from the Wairakei geothermal power plant that opened in 1953, combine to raise the As concentration in the river water from about 0.01 mg/L to as much as 0.08-0.09 mg/L before various dilution and sedimentation processes bring about a gradual dilution further downstream. Several adventive aquatic weeds in the river (*Ceratophyllum demersum, Egeria densa, Lagarosiphon major*) act as As hyperaccumulators [105-108]. Depending on the As concentration of the water and on the season and exposure time, As levels in the plant dry matter can range from about 100 mg/kg to 1000-1500 mg/kg. The biological accumulation factor, taken as the plant/substrate concentration ratio, can be as high as 20,000-30,000, e.g. where the plants contain 1400-1500 mg/kg in water with 0.05-0.07 mg/L.

More recent attention has been paid to As accumulation by fern species, particularly those growing in areas of As contamination from waste disposal from timber preservation processes or mining. Arsenic concentrations in *Pteris vittata* were studied by Ma *et al.* [109], who found 3280-4980 mg/kg in plants from soils with 19-1603 mg/kg As. Even from uncontaminated soils with 0.5-7.6 mg/kg, the accumulation was remarkable (12-64 mg/kg). Arsenic hyperaccumulation was also found by Visoottiviseth *et al.* [110] in the fern *Pityrogramma calomelanos*, and several other *Pteris* species have been found to respond with high uptake when the growing medium is spiked with solutions of soluble As compounds such as Na_2HAsO_4 [111]. Although members of the latter group apparently do not show hyperaccumulation in their natural habitats, they could be regarded as facultative hyperaccumulators, in the sense used in section 3, particularly if the same uptake behaviour can be shown to occur on unamended high-As soils. There are indications that a significant number of fern species will prove to possess this capability of As accumulation as a constitutive property [112], accounting for the intense recent activity in testing their phytoremediation potential. At the same time, the prospects for the use of a range of aquatic plants for remediation of As-contaminated waters should not be overlooked.

6.7. Selenium

Selenium is an essential element for animal and human health, with a remarkably narrow range between the levels required to prevent deficiency diseases and those producing symptoms of toxicity. The Se content of soils is usually 0.01-2 mg/kg, but can greatly exceed this range, reaching concentrations of several hundred mg/kg in soils derived from certain Cretaceous shales, or where the soil has become Se-enriched through the disposal of materials such as fly-ash or through the evaporation of irrigation water. Se-rich soils occur naturally in areas of the western part of the United States, and in Ireland, Queensland (Australia), Colombia, and Venezuela. In plant dry matter, Se concentrations are generally below 1 mg/kg, and may even be below 0.01 mg/kg in areas of low-Se soils.

The detailed history of the discovery of Se toxicity to range animals in parts of the USA can be found elsewhere [12, 113], and further aspects of Se uptake by accumulator plants are discussed later in this volume. Notable plant genera in which extreme accumulation of Se can be found include *Astragalus* (Leguminosae), *Stanleya* (Brassicaceae), *Haplopappus* and *Machaeranthera* (Asteraceae).

Because normal Se levels in plants are below 2 mg/kg, a case could be made for considering any plant with more than 100 mg/kg as a hyperaccumulator of this element, especially since levels above 10 mg/kg are considered capable of being toxic to animals. As in the case of Cd hyperaccumulation by *T. caerulescens*, the Se concentrations recorded for many of the Se accumulating plants show a very wide variation, over as much as two orders of magnitude, even within a single locality. Again, further work seems to be needed to establish the relative importance of genotypic variation and the rather wide variations in soil Se concentrations that can occur on even a local scale.

7. THE GENUS *THLASPI*

The Zn-accumulating *Thlaspi* from Germany and Belgium was the subject of occasional reports between 1885 and 1990, but recent years have seen a great upsurge of interest in its metal-accumulating properties. The nomenclature of this species has a complex history, which can be found elsewhere [12, 114], but it is now generally accepted that *T. caerulescens* J. & C. Presl is the legitimate name for the species that occurs sporadically throughout central and western Europe, often at sites of Zn and Pb mining and/or smelting. The name *T. alpestre*, formerly widely used for this species, properly applies to a different plant [115].

Although *T. caerulescens* is often regarded as a plant of metalliferous soils, it does occur on non-metalliferous soils in France, Spain, Scotland, Sweden and some other places in Europe, and is remarkable in its ability to show extreme Zn accumulation even from soils of normal Zn status [26, 62, 63]. This species is not just an accumulator of Zn. It was observed by Reeves and Brooks [26] that it occasionally occurred on serpentine, especially in France and Italy, from which it showed Ni hyperaccumulation. Its wider multielement accumulating abilities were shown in the work of Baker *et al.* [59], who found that Zn, Ni, Co, Mn and Cd could all be accumulated to high concentrations in the leaves when these elements were supplied as low-concentration amendments to standard nutrient solution. Other elements, such as Pb, Cu, Fe, Al and Cr were accumulated strongly in the roots, but not readily translocated to the shoots. The Cd-accumulating ability of *T. caerulescens* has also been borne out by the behaviour of some populations of this species in the field, particularly those on Zn mine and smelter sites in south-central regions of France, and are now attracting attention for their potential for Cd phytoremediation.

Several findings have drawn attention to other *Thlaspi* species as metal accumulators. Rascio [72] showed that the species now listed as *T. cepaeifolium* subsp. *rotundifolium* from zinc-polluted soils near the border of Italy and Austria was also a hyperaccumulator of zinc. Reeves and Brooks [26] observed that not only *T. caerulescens* but a number of other European *Thlaspi* species also occurred on serpentine soils, from which hyperaccumulation of Ni occurred, and that Zn levels above 1000 mg/kg were common in many species, not necessarily from areas of known Zn mineralization. A more complete study of the genus [28, 29] identified high concentrations of Ni and/or Zn in *Thlaspi* species from the Pacific Northwest of the United States, and from Turkey, Cyprus, and Japan. In Turkey, in particular, there are several serpentine-endemic Ni-accumulating *Thlaspi* species, including relatively recent discoveries such as *T. cariense* [31].

Reeves and Baker [86] demonstrated that the ability of the Austrian species *T. goesingense* to accumulate nickel and zinc was an innate property, unrelated to the geochemistry of the area from which the seed originated, giving rise to the notion of a 'constitutional' metal-tolerance and metal-uptake ability. Although much of the focus of recent work has been on the properties of *T. caerulescens*, more detailed study of the closely related *T. brachypetalum* and of other species such as *T. goesingense, T. cepaeifolium, T. alpinum* and *T. ochroleucum* is certainly justified.

The place of the North American *T. montanum* also deserves further investigation. There are strong indications from DNA analysis that this is not identical to the European species of this name [116], which is not a metallophyte. However, two of the American varieties, var. *californicum* and var. *siskiyouense*, are serpentine-endemic Ni hyperaccumulators, a property shared with North American var. *montanum* when it occurs on serpentine [28]. The metal-accumulating abilities of other (non-serpentine) populations of the North American species need further study to help clarify the intra-species relationships and to evaluate their phytoremediation potential.

8. HYPERACCUMULATORS AND PHYTOREMEDIATION

The main attraction of using hyperaccumulators for phytoremediation is the possibility of employing species that remove and concentrate large amounts of a particular element from the soil without significant chemical intervention, other than the application of conventional fertilisers. As pointed out by Salt and Baker [117], it is important that the metal concentration in the harvested plant tissue be greater than that in the soil, to ensure that phytoextraction results in a smaller volume of contaminated material.

If the typical plant dry matter concentrations are known, and an estimate can be made of the annual areal production of biomass, simple calculations can be performed to estimate the annual rate of removal of the contaminant element from the soil. These calculations are often extrapolated to give an estimate of the time needed to 'clean up' the soil to the desired or regulatory level. However, there is now a need for more sophisticated estimation methods, based on two other important pieces of information: (i) the rate of replenishment of the bioavailable metal pool from the contaminated soil; and (ii) the plant concentrations that can be achieved when the levels of soil contamination are lower.

Few hyperaccumulator species have so far been studied agronomically, and it is not known what yields of plant dry matter might be achieved under optimum conditions of climate, nutrition and plant density, particularly since in their natural habitats they are usually growing under adverse physical and nutritional conditions. The interelement selectivity of the metal uptake process is also well established in only a few cases. This is particularly relevant when a plant species is proposed for use in phytoremediation of a substrate that is chemically quite different from the plant's normal habitat.

9. EXPLOITATION AND CONSERVATION OF METAL-ACCUMULATING PLANTS

Many plant species found on serpentine soils, whether hyperaccumulators or not, are rare species [6], often being endemic to a few small areas. Some have been so rarely collected that they can be seen in only one or two herbarium collections worldwide. Several Ni-accumulating species of *Alyssum, Bornmuellera, Thlaspi* and *Centaurea* in Turkey, for example, are known from only two or three sites or collections. Several Cuban hyperaccumulators of Ni are also known from only one or two localities or collected specimens. The hyperaccumulator *Phyllanthus insulae-japen* is known only as seven replicates from a single collection. Efforts are therefore needed to find these species again in their natural habitats, to protect them properly, and to collect seed for further propagation in living collections.

Even some of the more widespread metal hyperaccumulators may be locally under threat: populations of the Zn-accumulating *Thlaspi caerulescens* are dense in some localities, but in danger of being exterminated in others. Populations of *Arabidopsis halleri* can be very small and widely scattered. Because there can be significant variation in metal-accumulating properties among isolated populations of such sporadically distributed species, and intra-population variability as well, it is important to the preserve all populations of such species, and to see that they are maintained in a healthy state.

Metal-tolerant species, and hyperaccumulators in particular, are a valuable and potentially useful biological resource, and concerted efforts are now needed to improve our field knowledge and distribution records of these plants in many parts of the world. This includes tracts of serpentine soils, especially in the tropics, and areas of wastes (e.g. from mining of Cu, Zn, and Pb) in many countries. This information will then be valuable in the formulation of strategies for conserving rare species and threatened populations. These issues have been discussed recently in some detail by Whiting *et al.* [118].

Major threats include the following: (1) mining, (2) agricultural expansion, (3) fire, (4) urban development, (5) recreational development, and (6) the activities of overzealous plant collectors. Specific examples of the effects of these threats on serpentine plants and Ni hyperaccumulators can be found in Kruckeberg [119] and Reeves [6], but they all apply to some degree to hyperaccumulators of other elements, and are discussed in more detail in Reeves and Baker [12].

It is clear that the metal-hyperaccumlating plants represent an unusual and valuable biological resource with great potential for use in a variety of strategies for soil remediation. There is an urgent need for for more exploration of the world's metalliferous soils, so that more accumulator species can be recognised, and the distribution and rarity of these species become better defined. Conservation measures need to be put in place to ensure that rare hyperaccumulator species are not lost through any of the above threats, and there are issues to be resolved about the way in which both scientific and commercially-oriented work can proceed, while promising benefits to the people in whose territory the resource is found in nature.

REFERENCES

1. Forchhammer, J.G. (1855) Über den Einfluß des Kochsalzes auf die Bildung der Mineralien, *Poggendorf's Annal. Phys. Chem.* 91, 60-96.
2. Sachs, J. (1865) *Handbuch der Experimental-Physiologie der Pflanzen. Handbuch der Physiologischen Botanik* (ed. W. Hofmeister), Vol. IV. Engelmann, Leipzig. pp. 153-154.
3. Jaffré, T., Brooks, R.R., Lee, J. and Reeves, R.D. (1976) *Sebertia acuminata*: a hyperaccumulator of nickel from New Caledonia, *Science* 193, 579-580.
4. Jaffré, T. and Schmid, M. (1974) Accumulation du nickel par une Rubiacée de Nouvelle Calédonie, *Psychotria douarrei* (G. Beauvisage) Däniker, *Compt. Rend. Acad. Sci.* (Paris), *Sér. D* 278, 1727-1730.
5. Brooks R.R., Lee, J., Reeves, R.D. and Jaffré, T. (1977) Detection of nickeliferous rocks by analysis of herbarium specimens of indicator plants, *J. Geochem. Explor.* 7, 49-57.
6. Reeves, R.D. (1992) Hyperaccumulation of nickel by serpentine plants, in J. Proctor, A.J.M. Baker and R.D. Reeves (eds.), *The Vegetation of Ultramafic (Serpentine) Soils*, Intercept Ltd., Andover, U.K., pp. 253-277.
7. Malaisse, F., Grégoire, J., Brooks, R.R., Morrison, R.S. and Reeves R.D. (1978) *Aeolanthus biformifolius*: a hyperaccumulator of copper from Zaïre, *Science* 199, 887-888.

8. Brooks, R.R., Reeves, R.D., Morrison, R.S. and Malaisse, F. (1980) Hyperaccumulation of copper and cobalt - a review, *Bull Soc. roy. Bot. Belg.* 113, 166-172.
9. Reeves, R.D. and Brooks, R.R. (1983) Hyperaccumulation of lead and zinc by two metallophytes from a mining area in Central Europe, *Environ. Pollut.* 31, 277-287.
10. Jaffré, T. (1980) Etude écologique du peuplement végétal des sols dérivés de roches ultrabasiques en Nouvelle Calédonie. Paris: Travaux et Documents de l'ORSTOM, 124.
11. Baker, A.J.M. and Brooks, R.R. (1989) Terrestrial higher plants which hyperaccumulate metallic elements - a review of their distribution, ecology and phytochemistry, *Biorecovery* 1, 81-126.
12. Reeves, R.D. and Baker, A.J.M. (2000). Metal-accumulating plants, in I. Raskin and B.D. Ensley (eds.), *Phytoremediation of Toxic Metals: Using Plants to Clean Up the Environment*, Wiley, New York.
13. Minguzzi, C. and Vergnano, O. (1948) Il contenuto di nichel nelle ceneri di *Alyssum bertolonii* Desv., *Atti Soc. Tosc. Sci. Nat., Mem. Serie A* 55, 49-77.
14. Doksopulo, E.P. (1961). Nickel in rocks, soils, water and plants adjacent to the talc deposits of the Chorchanskaya group. Izdatel vo Tbiliskovo Universitet, Tbilisi.
15. Menezes de Sequeira, E. (1969) Toxicity and movement of heavy metals in serpentinitic rocks (North-Eastern Portugal). *Agronomia Lusitana* 30, 115-154.
16. Wild, H. (1970) The vegetation of nickel-bearing soils, *Kirkia* 7 (suppl.), 1-62.
17. Severne, B.C. and Brooks, R.R. (1972) A nickel accumulating plant from Western Australia, *Planta* 103, 91-94.
18. Cole, M.M. (1973) Geobotanical and biogeochemical investigations in the sclerophyllous woodland and scrub associations of the eastern goldfields area of Western Australia, with particular reference to the role of *Hybanthus floribundus* (Lindl.) F. Muell. as nickel indicator and accumulator plant, *J. Appl. Ecol.* 10, 269-320.
19. Brooks, R.R. and Radford, C.C. (1978) Nickel accumulation by European species of the genus *Alyssum*, *Proc. Roy. Soc.* (London) B200, 217-224.
20. Brooks, R.R., Morrison, R.S., Reeves, R.D., Dudley, T.R. and Akman, Y. (1979) Hyperaccumulation of nickel by *Alyssum* Linnaeus (Cruciferae), *Proc. Roy. Soc.* (London) B203, 387-403.
21. Adıgüzel, N. and Reeves, R.D. (2002) A new nickel-accumulating species of *Alyssum* (Cruciferae) from western Turkey, *Edinb. J. Bot.* 59, 215-219.
22. Vergnano Gambi, O. and Gabbrielli, R. (1979) Ecophysiological and geochemical aspects of nickel, chromium and cobalt accumulation in the vegetation of some Italian ophiolitic outcrops, *Ofioliti* 4, 199-208.
23. Reeves, R.D., Brooks, R.R. and Press, J.R. (1980) Nickel accumulation by species of *Peltaria* Jacq. (Cruciferae), *Taxon* 29, 629-633.
24. Vergnano Gambi, O. and Gabbrielli, R. (1981) La composizione minerale della vegetazione degli affioramenti ofiolitici dell'alta Valle d'Ayas, *Rev. Valdôt. d'Hist. Nat.* 35, 51-61.
25. Vergnano Gambi, O., Gabbrielli, R. and Pancaro, L. (1982) Nickel, chromium and cobalt in plants from Italian serpentine areas, *Acta Oecol., Oecol. Plantar.* 3, 291-306.
26. Reeves, R.D. and Brooks, R.R. (1983) European species of *Thlaspi* L. (Cruciferae) as indicators of nickel and zinc, *J. Geochem. Explor.* 18, 275-283.
27. Gabbrielli, R., Pedani, F. and Vergnano Gambi, O. (1987) Ulteriori dati sulla composizione minerale della vegetazione degli affioramente ofiolitici dell'alta Valle d'Ayas, *Rev. Valdôt. d'Hist. Nat.* 41: 99-110.
28. Reeves, R.D., Macfarlane, R.M. and Brooks, R.R. (1983) Accumulation of nickel by western North American genera containing serpentine-tolerant species, *Amer. J. Bot.* 70, 1297-1303.

29. Reeves, R.D. (1988) Nickel and zinc accumulation by species of *Thlaspi* L., *Cochlearia* L., and other genera of the Brassicaceae, *Taxon* 37, 309-318.
30. Reeves, R.D., Brooks, R.R. and Dudley, T.R. (1983) Uptake of nickel by species of *Alyssum*, *Bornmuellera* and other genera of Old World Tribus Alysseae, *Taxon* 32, 184-192.
31. Reeves, R.D., Kruckeberg, A.R., Adıgüzel, N. and Krämer, U. (2001) Studies on the flora of serpentine and other metalliferous areas of western Turkey, *S. Afr. J. Sci.* 97, 513-517.
32. Reeves, R.D., Brooks, R.R. and Macfarlane, R.M. (1981) Nickel uptake by Californian *Streptanthus* and *Caulanthus* with particular reference to the hyperaccumulator *S. polygaloides* Gray (Brassicaceae), *Amer. J. Bot.* 68, 708-712.
33. Brooks, R.R. (1987). *Serpentine and its Vegetation: a Multidisciplinary Approach.* Dioscorides Press, Portland, Oregon.
34. Batianoff, G.N., Reeves, R.D. and Specht, R.L. (1990) *Stackhousia tryonii* Bailey: a nickel-accumulating serpentinite-endemic species of central Queensland, *Austral. J. Bot.* 38, 121-130.
35. Baker, A.J.M., Proctor, J., van Balgooy, M.M.J., and Reeves, R.D. (1992) Hyperaccumulation of nickel by the ultramafic flora of Palawan, Republic of the Philippines, in J. Proctor, A. J. M. Baker and R. D. Reeves (eds.), *The Vegetation of Ultramafic (Serpentine) Soils*, Intercept Ltd., Andover, U.K., pp. 291-304.
36. Rajakaruna, N. and Bohm, B.A. (2002) Serpentine and its vegetation: a preliminary study from Sri Lanka, *J. Appl. Bot.* 76, 20-28.
37. Reeves, R.D. (2003) Tropical hyperaccumulators of metals and their potential for phytoextraction, *Plant Soil* 249, 57-65.
38. Reeves, R.D., Baker, A.J.M., Borhidi, A. and Berazaín, R. (1996) Nickel-accumulating plants from the ancient serpentine soils of Cuba, *New Phytol.* 133, 217-224.
39. Reeves, R.D., Baker, A.J.M., Borhidi, A. and Berazaín, R. (1999) Nickel hyperaccumulation in the serpentine flora of Cuba, *Ann. Bot.* 83, 29-38.
40. Roberts, B.A. (1992) The ecology of serpentine areas, Newfoundland, Canada. in B.A. Roberts and J. Proctor (eds.), *The Ecology of Areas with Serpentinized Rocks – a World View*. Kluwer Academic Publishers, Dordrecht. pp. 75-113.
41. Morrey, D.R., Balkwill, K. and Balkwill, M.-J. (1989) Studies on serpentine flora: preliminary analyses of soils and vegetation associated with serpentine rock formations in the south-eastern Transvaal, *S. Afr. J. Bot.* 55, 171-177.
42. Morrey, D.R., Balkwill, K., Balkwill, M.-J. and Williamson, S. (1992) A review of some studies of the serpentine flora of southern Africa, in J. Proctor, A.J.M. Baker and R.D. Reeves (eds.), *The Vegetation of Ultramafic (Serpentine) Soils*, Intercept Ltd., Andover, U.K., pp. 147-157.
43. Reeves, R.D. and Adıgüzel, N. (2003) Rare plants and nickel accumulators from Turkish serpentine soils, with special reference to *Centaurea* species, *Turk. J. Bot.* (in press).
44. Brooks, R.R., Reeves, R.D. and Baker, A.J.M. (1992) The serpentine vegetation of Goiás State, Brazil, in J. Proctor, A.J.M. Baker and R.D. Reeves (eds.), *The Vegetation of Ultramafic (Serpentine) Soils*, Intercept Ltd., Andover, U.K, pp. 67-81.
45. Vergnano Gambi, O., Brooks, R.R. and Radford, C.C. (1979) L'accumulo di nichel nelle specie italiane del genere *Alyssum*, *Webbia* 33, 269-277.
46. Dudley, T.R. (1986) A new nickelophilous species of *Alyssum* (Cruciferae) from Portugal, *Alyssum pintodasilvae* T.R. Dudley, sp.nov., *Feddes Rep.* 97, 135-138.
47. Dudley, T.R. (1986) A nickel hyperaccumulating species of *Alyssum* from Spain: *Alyssum malacitanum* (Rivas Goday) T.R. Dudley, comb. et stat. nov., *Feddes Rep.* 97, 139-142.

48. Berazain Iturralde, R. (1981) Sobre el endemismo de la florula serpentinicola de Lomas de Galindo, Canasi, Habana. *Rev. Jard. Bot. Nac.* (Cuba*)* 2, 29-59.
49. Jaffré, T., Brooks, R.R. and Trow, J.M. (1979) Hyperaccumulation of nickel by *Geissois* species, *Plant Soil* 51, 157-162.
50. Homer, F.A., Reeves, R.D., Brooks, R.R. and Baker, A.J.M. (1991) Characterization of the nickel-rich extract from the nickel hyperaccumulator *Dichapetalum gelonioides*, *Phytochem.* 30, 2141-2145.
51. Kersten, W.J., Brooks, R.R., Reeves, R.D. and Jaffré, T. (1979) Nickel uptake by New Caledonian species of *Phyllanthus*, *Taxon* 28, 529-534.
52. Schmid, M. (1991) *Phyllanthus.* in *Flore de la Nouvelle Calédonie et Dépendances. 17. Euphorbiacées.* Museum National d'Histoire Naturelle, Paris. pp. 39-40.
53. Brooks, R.R. and Yang, X.H. (1984) Elemental levels and relationships in the endemic serpentine flora of the Great Dyke, Zimbabwe, and their significance as controlling factors for this flora, *Taxon* 33, 392-399.
54. Jaffré, T., Kersten, W.J., Brooks, R.R. and Reeves, R.D. (1979) Nickel uptake by the Flacourtiaceae of New Caledonia, *Proc. Roy. Soc.* (London) B205, 385-394.
55. Baker, A.J.M., Brooks, R.R. and Kersten, W.J. (1985) Accumulation of nickel by *Psychotria* species from the Pacific Basin, *Taxon* 34, 89-95.
56. Wither, E.D. and Brooks, R.R. (1977) Hyperaccumulation of nickel by some plants of South-East Asia, *J. Geochem. Explor.* 8, 579-583.
57. Brooks, R.R., Reeves, R.D., Baker, A.J.M., Rizzo, J.A. and Diaz Ferreira, H. (1990) The Brazilian serpentine plant expedition (BRASPEX), 1988. *Nat. Geog. Res.* 6, 205-219.
58. Brooks, R.R. and Wither, E.D. (1977) Nickel accumulation by *Rinorea bengalensis* (Wall.) O.K., *J. Geochem. Explor*. 7, 295-300.
59. Baker, A.J.M., Reeves, R.D. and Hajar, A.S.M. (1994) Heavy metal accumulation and tolerance in British populations of the metallophyte *Thlaspi caerulescens* J. & C. Presl (Brassicaceae), *New Phytol.* 127, 61-68.
60. Blaylock, M.J., Salt, D.E., Dushenkov, S., Zakharova, O., Gussman, C., Kapulnik, Y., Ensley, B.D. and Raskin, I. (1997) Enhanced accumulation of Pb in Indian mustard by soil-applied chelating agents, *Environ. Sci. Technol.* 31, 860-865.
61. Robinson, B.H., Leblanc, M., Petit, D., Brooks, R.R., Kirkman, J.H. and Gregg, P.E.H. (1998) The potential of *Thlaspi caerulescens* for phytoremediation of contaminated soils, *Plant Soil* 203, 47-56.
62. Escarré, J., Lefebvre, C., Gruber, W., Leblanc, M., Lepart, J., Rivière, Y. and Delay, B. (2000) Zinc and cadmium accumulation by *Thlaspi caerulescens* from metalliferous and nonmetalliferous sites in the Mediterranean area: implications for phytoremediation, *New Phytol.* 145, 429-437.
63. Reeves, R.D., Schwartz, C., Morel, J.L. and Edmondson, J. (2001) Distribution and metal-accumulating behavior of *Thlaspi caerulescens* and associated metallophytes in France. *Int. J. Phytoremed.* 3, 145-172.
64. Ernst, W.H.O. (1968) Das Violetum calaminariae westfalicum, eine Schwermetallpflanzengesellschaft bei Blankenrode in Westfalen, *Mitt. Flor.-soziol. Arbeitsgem.* 13, 263-268.
65. Baumann, A. (1885) Das verhalten von zinksalzen gegen pflanzen und im boden, *Die Landwirts. Vers.-Stat.* 31, 1-53.
66. Linstow, O. von (1924). Die natürliche Anreicherung der Metallsalzen und anderen anorganischen Verbindungen in den Pflanzen, *Feddes Rep.* 31, 1-151.
67. Ernst, W.H.O. (1966). Ökologisch-soziologische Untersuchungen an Schwermetallpflanzen-gesellschaften Südfrankreichs und des östlichen Harzvorlandes, *Flora* (Jena) B156, 301-318.
68. Shimwell, D.W. and Laurie, A.E. (1972) Lead and zinc contamination of vegetation in the southern Pennines, *Environ. Pollut.* 3, 291-301.

69. Denaeyer-De Smet, S. and Duvigneaud, P. (1974) Accumulation de métaux lourds dans divers écosystèmes terrestres pollués par les retombées d'origine industrielle, *Bull. Soc. roy. Bot. Belg.* 107, 147-156.
70. Ernst, W.H.O. (1974). *Schwermetallvegetation der Erde*, Fischer, Stuttgart.
71. Kelepertsis, A.E. and Bibou, A. (1991) Heavy metal contamination of soils at old mining sites on Thasos Island, Greece, *Environ. Geochem. Health* 13, 23-28.
72. Rascio, N. (1977) Metal accumulation by some plants growing on zinc-mine deposits, *Oikos* 29, 250-253.
73. Johnston, W.R. and Proctor J. (1977) A comparative study of metal levels in plants from two contrasting lead-mine sites, *Plant Soil* 46, 251-257.
74. Barry, S.A.S. and Clark, S.C. (1978) Problems of interpreting the relationship between the amounts of lead and zinc in plants and soil on metalliferous wastes, *New Phytol.* 81, 773-783.
75. Cole, M.M., Provan, D.M.J. and Tooms, J.S. (1968) Geobotany, biogeochemistry and geochemistry in the Bulman-Waimuna Springs area, Northern Territory, Australia, *Trans. Inst. Min. Met. Sec. B* 77, 81-104.
76. Williams, S.T., McNeilly, T. and Wellington, E.M.H. (1977) The decomposition of vegetation growing on metal mine waste, *Soil Biol. Biochem.* 9, 271-275.
77. Deram, A. and Petit, D. (1997) Ecology of bioaccumulation in *Arrhenatherum elatius* L. (Poaceae) populations - applications of phytoremediation of zinc, lead and cadmium contaminated soils, *J. Exper. Bot.* 48, Spec. Suppl., p. 98.
78. Dykeman, W.R. and De Sousa, A.S. (1966) Natural mechanisms of copper tolerance in a copper swamp, *Can. J. Bot.* 44, 871-878.
79. Blissett, A.H. (1966) Copper tolerant plants from the Ukaparinga copper mine, Williamstown. *Quart. Geol. Notes, Geol. Surv. S. Austral.* 18, 1-3.
80. Beeson, K.C., Lazar, V.A. and Boyce, S.G. (1955) Some plant accumulators of the micronutrient elements, *Ecology* 36, 155-156.
81. Kubota, J., Lazar, V.A. and Beeson, K.C. (1960) The study of cobalt status of soils in Arkansas and Louisiana using the black gum as the indicator plant, *Soil Sci. Proc.* 24, 527-528.
82. Brooks, R.R., McCleave, J.A. and Schofield, E.K. (1977) Cobalt and nickel uptake by the Nyssaceae, *Taxon* 26, 197-201.
83. Duvigneaud, P. (1959) Plantes cobaltophytes dans le Haut Katanga, *Bull. Soc. roy. Bot. Belg.* 91, 111-134.
84. Morrison, R.S. (1980) *Aspects of the accumulation of cobalt, copper and nickel by plants.* PhD Thesis, Massey University, NZ.
85. Brooks, R.R., Naidu, S.D., Malaisse, F. and Lee, J. (1987) The elemental content of metallophytes from the copper/cobalt deposits of Central Africa, *Bull. Soc. roy. Bot. Belg.* 119, 179-191.
86. Reeves, R.D. and Baker, A.J.M. (1984) Studies on metal uptake by plants from serpentine and non-serpentine populations of *Thlaspi goesingense* Hálacsy (Cruciferae), *New Phyt.* 98, 191-204.
87. Homer, F.A., Morrison, R.S., Brooks, R.R., Clemens, J. and Reeves, R.D. (1994) Comparative studies of nickel, cobalt and copper uptake by some nickel hyperaccumulators of the genus *Alyssum*, *Plant Soil* 138, 195-205.
88. Duvigneaud, P. and Denaeyer-De Smet, S. (1963) Cuivre et végétation au Katanga, *Bull. Soc. roy. Bot. Belg.* 96, 93-231.
89. Malaisse, F., Grégoire, J., Brooks, R.R., Morrison, R.S. and Reeves, R.D. (1979) Copper and cobalt in vegetation of Fungurume, Shaba Province, Zaïre, *Oikos* 33, 472-478.
90. Malaisse, F., Brooks, R.R. and Baker, A.J.M. (1994) Diversity of vegetation communities in relation to soil heavy metal content at the Shinkolobwe copper/cobalt/uranium mineralization, Upper Shaba, Zaïre, *Belg. J. Bot.* 127, 3-16.

91. Malaisse, F. and Grégoire, J. (1978) Contribution á la phytogéochimie de la Mine de l'Étoile (Shaba, Zaïre), *Bull. Soc. roy. Bot. Belg.* 111, 252-260.
92. Brooks, R.R., Morrison, R.S., Reeves, R.D. and Malaisse, F. (1978) Copper and cobalt in African species of *Aeolanthus* Mart. (Plectranthinae, Labiatae), *Plant Soil* 50: 503-507.
93. Brooks, R.R. (1977) Copper and cobalt uptake by *Haumaniastrum* species, *Plant Soil* 48, 541-544.
94. Paton, A. and Brooks, R.R. (1995) A re-evaluation of *Haumaniastrum* species as geobotanical indicators of copper and cobalt, *J. Geochem. Explor.* 56, 37-45.
95. Denaeyer-De Smet, S. (1966) Note sur un accumulateur de manganèse, *Vaccinium myrtillus, Bull. Soc. roy. Bot. Belg.* 99, 331-343.
96. Bidwill, S.D. (2000) *Hyperaccumulation of metals in Australian native plants*. PhD Thesis, University of Melbourne, Australia.
97. Jaffré, T. (1977) Accumulation du manganèse par les espèces associées aux terrains ultrabasiques de Nouvelle Calédonie, *Compt. Rend. Acad. Sci.* (Paris), *Sér. D* 284, 1573-1575.
98. Jaffré, T. (1979). Accumulation du manganèse par les Proteacées de Nouvelle Calédonie. *Compt. Rend. Acad. Sci.* (Paris), *Sér. D* 289, 425-428.
99. Brooks, R.R., Trow, J.M., Veillon, J.-M. and Jaffré, T. (1981) Studies on manganese-accumulating *Alyxia* from New Caledonia, *Taxon* 30, 420-423.
100. Proctor, J., Phillipps, C., Duff, G.K., Heaney, A. and Robertson, F.M. (1989). Ecological studies on Gunung Silam, a small ultrabasic mountain in Sabah, Malaysia. II. Some forest processes, *J. Ecol.* 77, 317-331.
101. Wild, H. (1974). Indigenous plants and chromium in Rhodesia, *Kirkia* 9, 233-241.
102. Lee, J., Brooks, R.R., Reeves, R.D. and Jaffré, T. (1977) Chromium-accumulating bryophyte from New Caledonia, *Bryologist* 80, 203-205.
103. Brooks, R.R. (1972) *Geobotany and Biogeochemistry in Mineral Exploration*. Harper & Row, New York.
104. Warren, H.V., Delavault, R.E. and Barakso, J. (1964) The role of arsenic as a pathfinder in biogeochemical prospecting, *Econ. Geol.* 59, 1381-1389.
105. Lancaster, R.J., Coup, M.R. and Hughes, J.W. (1971) Toxicity of arsenic present in lakeweed. *N.Z. Vet. J.* 19, 141-145.
106. Aggett, J. and Aspell, A.C. (1980) Arsenic from geothermal sources in the Waikato catchment, *N.Z. J. Sci.* 23, 77-82.
107. Liddle, J.R. (1982) *Arsenic and other elements of geothermal origin in the Taupo Volcanic Zone*. PhD Thesis, Massey University, New Zealand.
108. Reeves, R.D. and Liddle, J.R. (1986) Dispersal of arsenic from geothermal sources of the central North Island, in M.J. Baker (ed.) *Trace Elements in the Eighties*, NZ Trace Element Group, Palmerston North, NZ.
109. Ma, L.Q., Komar, K.M., Tu, C., Zhang, W.H., Cai, Y. and Kennelley, E.D. (2001) A fern that hyperaccumulates arsenic, *Nature* 409, 579.
110. Vittoottiviseth, P., Francesconi, K. and Sridokchan, W. (2002). The potential of Thai indigenous plant species for the phytoremediation of arsenic contaminated land, *Environ. Pollut.* 118, 453-461.
111. Zhao, F.J., Dunham, S.J. and McGrath, S.P. (2002) Arsenic hyperaccumulation by different fern species, *New Phytol.* 156, 27-31.
112. Meharg, A. (2002) Arsenic and old plants, *New Phytol.* 156, 1-4.
113. Rosenfeld, I. and O.A. Beath. (1964). *Selenium - Geobotany, Biochemistry, Toxicity and Nutrition*. Academic Press, New York.
114. Gutermann, W. (1975) Notulae nomenclaturales 1-18, *Phyton* (Austria) 17, 31-50.
115. Akeroyd, J.R. (1993) *Thlaspi*, in *Flora Europaea*, Vol. 1, 2nd ed., Cambridge University Press, Cambridge. pp. 384-388.

116. Mummenhoff, K. and Koch, M. (1994) Chloroplast DNA restriction site variation and phylogenetic relationships in the genus *Thlaspi* sensu lato (Brassicaceae), *Syst. Bot.* 19, 73-88.
117. Salt, D.E. and Baker, A.J.M. (2000) Phytoremediation of metals, in J. Klein (ed.) *Biotechnology* (2nd ed.), *Vol. 11b Environmental Processes II – Soil Decontamination.* Wiley, New York. pp. 386-397.
118. Whiting, S.N., Reeves, R.D. and Baker, A.J.M. (2002) Mining, metallophytes and land reclamation: conservation of biodiversity, *Mining Environ. Management* 10 (2), 11-16.
119. Kruckeberg, A.R. (1992) Serpentine biota of western North America, in J. Proctor, A. J. M. Baker and R.D. Reeves (eds.), *The Vegetation of Ultramafic (Serpentine) Soils*, Intercept Ltd., Andover, U.K., pp. 19-33.

CHAPTER 3

PLANT ACCUMULATION CAPACITY FOR POTENTIALLY TOXIC ELEMENTS

P. TLUSTOŠ, D. PAVLÍKOVÁ,
J. SZÁKOVÁ, AND J. BALÍK
Czech University of Agriculture in Prague,
165 21 Prague – Suchdol, Czech Republic

Abstract

Arsenic and lead always showed limited mobility. Their uptake by plants was low and if element was taken up by plant roots usually stayed there without substantial movement. The highest content was found in thin secondary roots on the opposite seeds were well protected. Arsenic mainly showed the lowest content in plant tissues among all elements tested. Lead showed similar properties, its content was usually higher but transfer coefficients (TC) showed low similar values for both elements. Cadmium and zinc were more mobile and easily taken by plants. Secondary roots contained usually their highest levels, but their translocation to aboveground biomass was rapid and in some cases bulk roots contained lower element contents than leaves and stems. Importance of zinc for plants was confirmed by its elevated content in seeds, while seeds and edible parts of vegetables were well protected from Cd accumulation.

1. INTRODUCTION

Plants covering almost all solid inurbane surface of Earth can give different use not just in the way of foods for humans and animals but they can be also used in other different ways as building materials, for production of energy mainly heat and for many industrial purposes. In last decade another very important role of plants was introduced. Plant can also help to protect and clean our environment. We can talk about new phenomena and widen our scope of interest.

We always need to think about top quality of produced biomass in case of food. The effort trying to decrease content of all risk compounds and elements in food for human population and in fodder materials for animals is never ending and we need to deep our knowledge to establish the goals we can meet in next decade.

J.-L. Morel et al. (eds.), Phytoremediation of Metal-Contaminated Soils, 53–84.

On the other side results showing ability of plants to accumulate potentially toxic elements in high amounts giving new expectations to use plants for improvement quality of surrounding environment. The idea is completely opposite to the first one because we are looking for crops accumulating high amounts of adverse compounds or helping to decompose them in plant surrounding environment.

Natural sensitivity or tolerance of plants to accumulate metals is substantially affected by plant species and genotypes. Plant can be divided into three groups (Fig. 1):

Excluders - there are plants insensitive for uptake and accumulation of potentially toxic elements. Mainly monocotyledon grasses belong into this group (sudangrass, fescue).

Indicators - majority of agricultural plants whose content of elements more or less linearly responding to increasing available content of trace elements in soil (wheat, oats, maize).

Accumulators - plants accumulating higher contents of elements in their tissues according to their increase in the soil. Accumulators include many species of *Brassicaceae* (mustard), and *Compositae* (lettuce, spinach) families. There are extreme accumulators (hyperaccumulators) that can even prosper on contaminated soils and accumulate extremely high contents of trace elements (Baker, 1987).

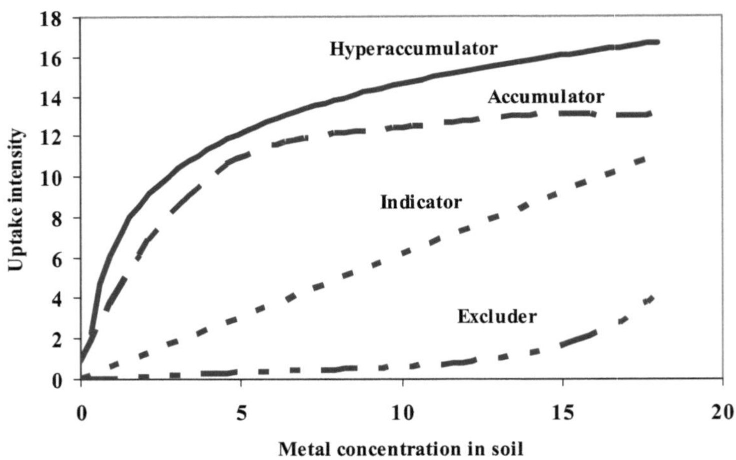

Fig. 1. Relative uptake and bioaccumulation potential among plant species (Adriano, 2001)

Not just content of elements but also the mass of dry matter produced or consumed plays an important role either in intake or offtake of total amount of elements. Generally we can find at least five groups of plants artificially or naturally grown in soil and subsequently divided into many subgroups. We usually talk about agricultural and horticultural crops, trees and bushes, weeds and naturally grown wild plants. This groups mainly show specific use of plants implying many plant species in each ones. In this context, zinc accumulating *Brassica spp.* seems to be more effective for removing this element from the polluted soil compared to zinc hyperaccumulator *Thlaspi caerulescens* producing ten time less amount of shoot biomass (Ebbs et al., 1997).

Not just differences in accumulation of elements in plants and their yield of biomass affect total amount in biomass, but also specific properties and mobility of each element can significantly affect their uptake and distribution in plant tissues. Kabata-Pendias and Pendias (1986) simplified relationship between accumulation of individual elements in plant tissues and soil solution (Fig. 2). The content of Cd grew the fastest in plant tissues with increased Cd concentration in soil solution, Zn and Cu showed much lower response and Pb was the least active among all of investigated elements. Alloway (1990) showed that Cr, Pb and Hg were the least translocated, Ni, Co and Cu intermediate, and Cd, Mn, Mo and Zn were readily transported to the plant tops.

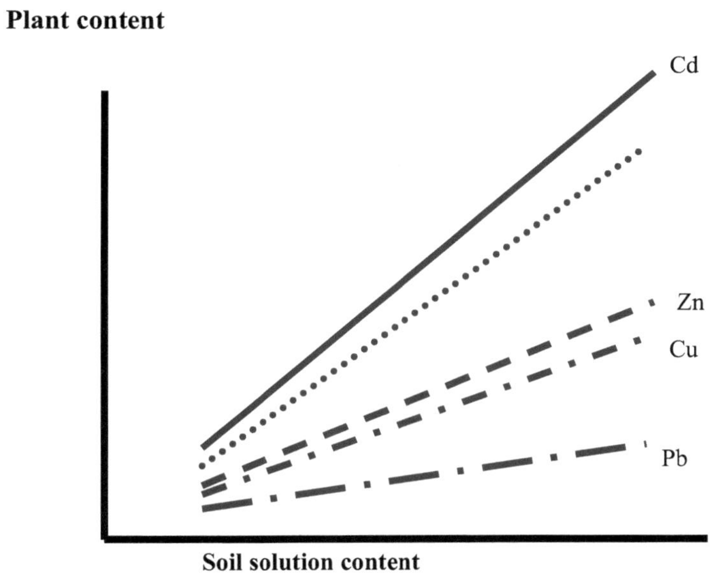

Fig. 2. The uptake of heavy metals by plants with relation to their concentration in soil solution (Kabata-Pendias, Pendias, 1986)

Mobility of trace elements in plant tissues and their total content in plants did not correspond with the element content in soil solution and their changes. The content of individual elements can differ by two or by three (for slow mobile ones) orders in the same plant specie. Kabata-Pendias and Pendias (1986) summarized mean contents of trace elements in plants leaves of series of plant species (Table 1). They divided content of elements into three groups: deficient contents in case of micronutrients, common contents of elements and elevated ones. Among twelve elements the highest contents were usually determined for Zn and Mn and the lowest for Cd and Cr.

Accumulation of elements in plants did not depend just on the plant specie and element behaviors. There are also high differences in total soil element content that can be dramatically changed by soil contamination. Soil properties (pH, CEC, clay, oxide and organic matter contents are the most important) play also important role in the release of individual elements.

Table 1. Approximate concentration of trace elements in mature leaf tissues generalized for various speci es (ppm DW*) (Kabata-Pendias, Pendias, 1986).

Element	Approximate concentrations (ppm DW)		
	Deficient	Sufficient or normal	Excessive or toxic
As	-	1-1.7	5-20
Be	-	<1-7	10-50
Cd	-	0.05-0.2	5-30
Co	-	0.02-1	15-50
Cr	-	0.1-0.5	5-30
Cu	2-5	5-30	20-100
Hg	-	-	1-3
Mn	15-25	20-300	300-500
Mo	0.1-0.3	0.2-1	10-50
Ni	-	0.1-5	10-100
Pb	-	5-10	30-300
Zn	10-20	27-150	100-400

*DW – expressed in dry matter

Among attempts to eliminate different soil contents **transfer coefficient (TC)** was released. TC was gained by dividing of total element content in plant tissues by total soil content (Kiekens and Camerlych, 1982). Detailed distribution of grain crops and vegetables according to their TC is presented in table 2 (Lubben and Sauerbeck, 1991). All elements showed at least one order difference between the highest and the lowest transfer coefficient. TC confirmed a different mobility of trace elements. Plants usually introduced higher TC for Cd, except for maize and peas, and low TC in the case of Cr and mainly Pb, as well. There were also other parameters introduced. Among them, **remediation factor (RC)** describing portion of element taken off by above ground biomass to total amount of element present in surface layer of soil (Vyslouzilova et al., 2003).

Table 2. Margins of TC values of trace metals from vegetables and grains (Lubben, Sauerbeck 1991).

Cd	Zn	Ni	Cu	Pb	Cr
0.01-0.1	0.1-0.2	0.01-0.1	0.01-0.05	0.001-0.005	0.001-0.005
Maize	Maize	Maize	Maize	Maize	Maize
Pea	Pea	Lettuce	Pea	Oats	Pea
			Carrot	Wheat	Oats
				Pea	Wheat
				Beans	Beans
					Carrot
0.1-0.5	0.2-0.5	0.1-0.2	0.05-0.1	0.005-0.01	0.005-0.01
Lettuce	Oats	Wheat	Beans	Carrot	Lettuce
Oats	Carrot		Oats		
Wheat	Wheat		Lettuce		
Carrot	Beans		Wheat		
Beans	Lettuce		Spinach		
0.5-1.0	0.5-1.0	0.2-0.5	0.1-0.2	0.01-0.02	0.01-0.02
		Carrot	Radish	Lettuce	Spinach
		Pea	Lettuce	Radish	
1.0-3.0	1.0-3.0	0.5-1.0		0.02-0.05	0.02-0.03
Radish	Lettuce	Radish		Spinach	Radish
Lettuce	Spinach	Oats		Radish	
Spinach	Radish	Lettuce			
		Beans			
		Spinach			

2. DIFFERENT CAPACITY OF AGRICULTURAL CROPS AND VEGETABLES TO ACCUMULATE ELEMENTS IN PLANT TISSUES

The precise test of differences in element uptake has to be done under standard conditions for all tested species and at the same soil. The pot experiment presented here was done at experimental site of our institute with the same soil clayloam Chernozem (loc. Suchdol) with pH_{KCl} 7.0 and CEC = 255 $mval.kg^{-1}$. Soil contained 0.23 ppm Cd, 16.5 ppm As and 96.6 ppm Zn. None of the tested elements showed their elevated total contents and presented results more or less presented values of real situation at arable soils in Czech Republic.

Understanding of different accumulation patterns for tested plants cannot be clear without determination of biomass yield. Pot experiments (5 kg of soil in each pot) allowed us to determine yield of roots, too. Figure. 3 shows yield distribution among five parts of plants. Differences among plant species cannot allow us to divide all five species in each plant. We can nicely see that total dry biomass production fluctuated from less than 20 g up to 130 g per pot. The highest biomass production always showed grains and the lowest vegetables. Roots represented significant portion of whole plant biomass. The lowest root portion was determined in case of tobacco about 18%, twice higher portion of roots was found at wheat and barley treatments and the highest 77% was determined for carrot, but majority of it was formed by consumable part. Seeds have always formed smaller portion of above ground biomass then leaves and stems.

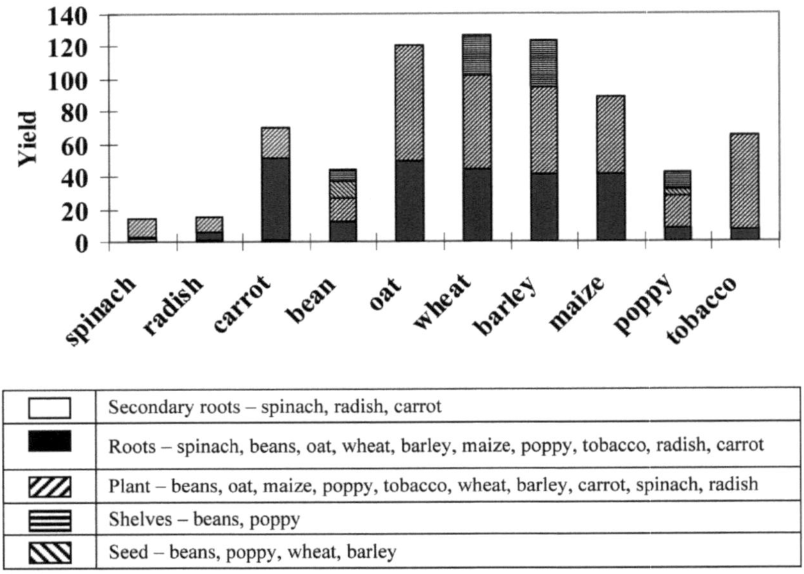

Fig. 3. Mean yield of total biomass at different plants (g of dry matter per pot)

Arsenic accumulation at different plant parts is introduced in (Table 3). Separation of secondary roots at three plant species showed significantly higher As accumulation than other plant parts. Differences in As content were not determined even if plants originated from different families. Only two plant species (oats and beans) showed similar As accumulation in bulk roots and in leaves and stems. Bean plants belong to the most sensitive plants for arsenic amendment in soilless culture (Carbonell-Barrachina et al., 1997, Lario et al., 2002) easily translocating this element into the above ground biomass. The rest showed always significantly higher As content in bulk roots than in stems and leaves. Specific protection effect was found in carrot roots, where As content was the lowest among parts analyzed. Radish accumulated significantly higher As contents in roots and leaves without difference for plant part. Similar results also found Merry et al. (1986). Plants tried to protect generative organs, therefore As seed content was lowest among all tested parts and especially As content in bean seeds was extremely low.

Table 3. The content of As in individual parts of plants ($mg.kg^{-1}$).

Parts of plants	Spinach	Radish	Carrot	Bean	Oat	Wheat	Barley	Maize
Sec. roots	8.896	8.602	6.270					
Roots	0.696	1.008	0.110	1.072	0.233	3.691	2.414	2.450
Plant	0.166	0.904	0.401	1.085	0.218	0.570	0.247	0.200
Shelves				0.113				
Seed				0.008		0.130	0.150	

As_{total} = 16.5 $mg.kg^{-1}$
As_{avail} = 0.05 $mg.kg^{-1}$

Affinity of planted species for Cd is displayed in (Table 4). The same as arsenic content the cadmium content in soil was relatively low and portion of available Cd was about 2% from the total one. Cd is usually introduced as the most mobile element but its content in secondary roots was the highest one among all tested plant parts. Secondary roots of carrots accumulated the highest content of Cd, its content in spinach was lower and in radish the lowest. Bulk roots analysed in other seven plants have lower Cd content, with the lowest one in grain roots and the highest one in poppy and bean roots. Above ground parts showed two different patterns.

Plants with low root content showed usually higher Cd levels in plant leaves and stems and on the contrary higher Cd in roots limited its accumulation in plant and leaves tissues. Among seven tested species only tobacco gave an opposite response and accumulated four fold higher Cd in above ground biomass (1.25 mg $Cd.kg^{-1}$) than in bulk roots. Tobacco Cd content in above ground biomass was the highest among all tested plant species. If we check the ratio between consumable roots and leaves we can see again good protection of edible parts of both vegetables and significantly lower Cd content in roots. Keefer et al. (1986) investigated proportion of Cd content in roots and leaves of radish and carrots and found higher Cd content in leaves of both plants, as well. Cobb et al. (2000) also found lower accumulation of metals in radish roots compared to leaves. Ratio between Cd content in roots and straw of grains was different and higher Cd was always found in roots. The results were confirmed by Lubben and Sauerbeck (1991) in the case of wheat. The highest ratio (about ten-fold greater content in roots) was determined in bean and poppy plants. Seeds and shelves accumulated the lowest Cd content among all plant parts. Cd content

Table 4. The content of Cd in individual parts of plants ($mg.kg^{-1}$).

Parts of plants	Spinach	Radish	Carrot	Bean	Oat	Wheat	Barley	Maize	Poppy	Tobacco
Sec. roots	1.422	0.710	1.942							
Roots	0.161	0.131	0.111	0.662	0.574	0.276	0.311	0.370	0.667	0.43
Plant	0.489	0.353	0.463	0.072	0.195	0.162	0.117	0.052	0.04	1.24
Shelves				0.002					0.07	
Seed				0.014		0.075	0.040		0.104	

$Cd_{total} = 0.23\ mg.kg^{-1}$ $Cd_{avail.} = 0.004\ mg.kg^{-1}$

in grains (0.04 ppm) was about twice lower than Cd content in wheat grain. Results well corresponded with results of Chaudrí et al. (1995) that determined mean Cd content (0.042 ppm) in great number of wheat grain samples.

Zinc content in soil used for experiment was slightly higher than total mean Zn content determined in agricultural soils of Czech Republic. Higher total content was partially eliminated by very low Zn availability for plants (0.12%) and Zn plant contents are comparable with other authors grown plants on Zn unpolluted soil. In agreement with other elements the highest Zn content was also determined in secondary roots with the maximum 292 ppm in carrots fine roots (Table 5). Zinc differences in secondary roots of three plant species well corresponded with Cd content. Bulk roots of other plants showed the highest variability among all tested plants. While Zn content in roots of bean and wheat was very close to content of Zn in fine roots of radish and spinach (more than 100 ppm), Zn content in tobacco roots was almost one order lower. Very low Zn content was also found in edible parts of radish and carrots the same as in their leaves. Much higher differences between roots and shoots Zn content in radish determined Davies (1992) but Keefer et al. (1986) presented similar ratio like we found. Both plants have taken up low amount of Zn compare to other plants. Third vegetable - spinach - showed higher affinity to Zn and especially its accumulation in leaves was high compare to other crops and exceeded 100 ppm. The ratio between accumulation of Zn in leaves, stems and seeds was different from other two elements. Higher zinc content was always found in seeds than in the rest of above ground biomass. Our results did not correspond with findings of Schaller and Diez (1991) introducing higher Zn content in vegetative parts of plants than in seeds. Similar Zn contents in straw and grain were presented by Lubben and Sauerbeck (1991a). They explained higher content of Zn in grain by its accumulation in grain shelves. These results well corresponded with plant need for zinc (Marschner, 1995), Zn content was the highest in seeds among three investigated elements. Plants protected their reproductive organs from toxic elements and content of As as well as Cd was generally the lowest there, but content of Zn was always higher than in vegetative parts of plants.

Transfer coefficient (TC) was used for comparison of element behavior with partial elimination of soil element content. TC published in the literature are usually specify for plant specie and not for plant part, but differences can be significant and did not meet values presented in table 2. Even if we omit very specific fine roots we can get almost one order high differences among plant parts of the same species or to be almost identical. Good examples are TC of As, Cd and Zn determined for bean seeds and for bean plants. TC_{As} for seed was 0.0005 and for plants 0.064, TC_{Cd} for seeds was 0.05 and for plants 0.26, and TC_{Zn} for seeds was 0.78 and for plants 0.50. These results confirmed that only precise description of plant part can be used for comparison of results from different sources. Our TC results fluctuated from 0.0005 for bean seeds to 0.52 for fine roots of spinach in case of As. TC for Cd were determined from 0.05 for bean shelves to 6.28 for fine roots of carrot and TC for Zn from 0.11 for poppy shelves to 4.26 for fine carrot roots again.

Comparison of our TC_{As} values determined for main parts of crops with marginal values of Kloke et al. (1984) show our TC usually lower and in case of carrot, beans, and grain of wheat and barley even lower than their minimum values 0.01. Similar results were also revealed for TC_{Cd} and TC_{Zn} in few cases. Lubben and Sauerbeck (1991) showed marginal values of TC_{Cd}. Our values usually correspond with their lower marginal values. Lower TC_{Cd} than ours determined for wheat, spinach and maize Smilde et al. (1992). The highest fluctuation between our results and published by other authors was for TC_{Zn}. TC_{Zn} for spinach leaves 1.74 and barley grain 0.48 corresponded with values presented in table 2. Values for radish 0.19 and carrot 0.15 roots were lower and for beans 0.78 and wheat grain 0.83 higher than values introduced by Lubben and Sauerbeck (1991).

Omitting the specificity of edible root parts of carrot and radish plants we can simplify the relationship in element accumulation of arsenic, cadmium and zinc in plants as following:

As: Secondary roots > primary roots > leaves and stems > seeds
Cd: Secondary roots > primary roots > leaves and stems > seeds
Zn: Secondary roots > primary roots > seeds > leaves and stems

TC values are not only affected by plant species, plant part and element origin but there are also affected by total content of element in soil and element mobility. Table 6 shows differences in TC_{Cd} determined in main parts of five plant species at three soils substantially differed in their total content of Cd. TC differences except bean seeds were similar among plant species as well as among soils used for planting of individual species. Growing soil Cd content, about twice higher at Luvisol than Fluvisol, led usually to similar decrease of TC at all crops. The exception was only found in spinach leaves. Spinach was able to take proportionally higher amount of Cd into plant biomass to soil Cd content probably due to its capacity to mobilize labile Cd forms in soil. One and half order higher Cd content at Chernozems than at other two soils could be accepted by growing plants and TC of Cd was the lowest for all plants at this soil.

As we mentioned early total plant content is easily determined and can be limited by legislation, but total intake or offtake play the most important role in the process to determine input or output of investigated element. Fig. 4 shows the total As offtake by tested plants. The highest one was introduced by group of grains (wheat and barley the highest, and maize slightly lower) and the lowest by group of vegetables (spinach, radish, carrot). The comparison of total maximum As offtake with total amount of As in pot was used

Table 5. The content of Zn in individual parts of plants ($mg.kg^{-1}$).

Parts of plants	Spinach	Radish	Carrot	Bean	Oat	Wheat	Barley	Maize	Poppy	Tobacco
Sec. roots	113.5	100.2	292.3							
Roots	51.3	12.9	10.1	146.9	60.7	113.2	41.9	37.5	60.4	16.2
Plant	119.7	17.0	16.7	34.5	42.5	24.2	20.7	25.4	9.3	31.5
Shelves				30.7					7.9	
Seed				53.8		57.1	33.3		47.5	

$Zn_{total} = 96.9\ mg.kg^{-1}$
$Zn_{avail.} = 0.12\ mg.kg^{-1}$

Table 6. TC values of Cd in main parts of plants.

Soil		Plants									
Type	Cd content (ppm)	Oats	Spinach		Radish		Carrot		Beans		
			Leaves	Roots	Leaves	Roots	Leaves	Roots	Plant	Shelves	Seeds
Fluvisol	0.36	0.60	3.32	3.09	3.32	1.48	1.90	0.96	0.320	0.100	0.140
Chernozem	17.56	0.15	1.50	1.62	1.65	0.69	0.45	0.25	0.048	0.010	0.005
Luvisol	0.60	0.26	3.59	1.69	1.91	0.48	1.01	0.41	0.095	0.038	0.037

to determine the As portion taken from the soil. The maximum content was very low 0.15%, showing only negligible part of arsenic removed from the soil, annually. Low arsenic mobility and its accumulation in roots led to substantial reduction of As offtake by above ground biomass or harvestable parts. Only 30% of total offtake was formed by wheat aboveground biomass and 15% by barley, two crops with the highest total As offtake.

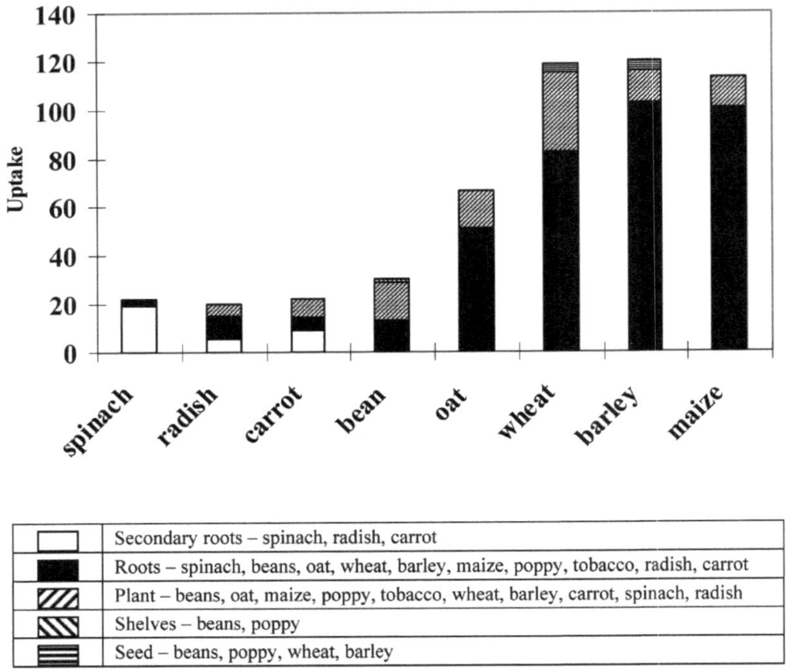

Fig. 4. Mean uptake of As by different parts of plants (μg per pot)

Total offtake of Cd by same crops was few fold lower than in case of As (Fig. 5) which corresponds with lower soil Cd content. Highest Cd offtake was found by tobacco, almost twice higher than the second one, by oats. Extremely low Cd uptake was reported in the case of vegetables again, mainly by radish and beans. Also Cd offtake by poppy was low. All three mentioned plant species introduced medium mean Cd content and biomass yield bellow average. Due to several fold lower Cd content in soils, portion of Cd taken off from the soil was substantially higher with maximum 4.9% found for tobacco. Portion of Cd taken by oats was also reasonable 2.7% and for other grains approximately twice lower. Higher Cd mobility led to translocation Cd in higher extent. In case of tobacco 95% of Cd was in harvestable portion of plant, for other plants about 35% of Cd was accumulated there. Mench et al. (1989) and Macek et al. (2002) confirmed high Cd offtake by harvestable portion of tobacco (about 80%).

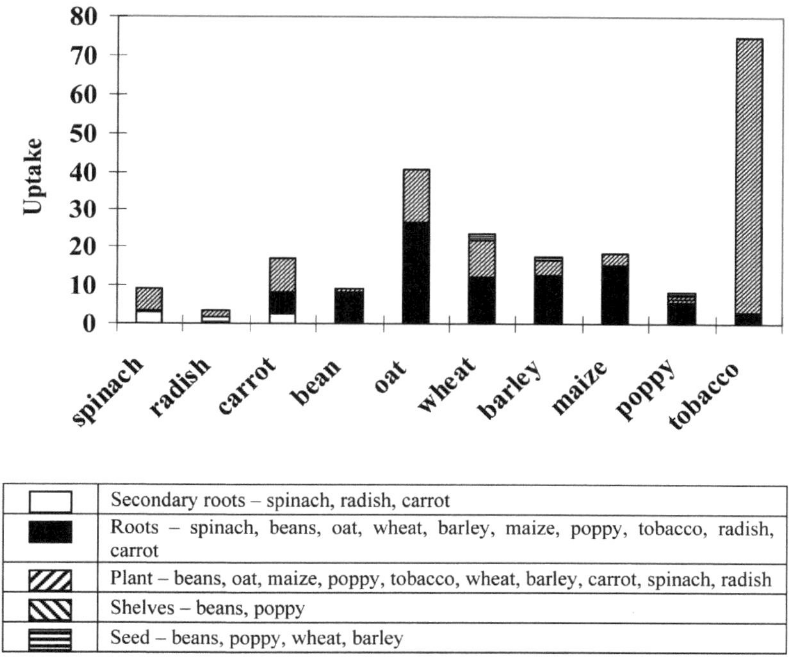

Fig. 5. Mean uptake of Cd by different parts of plants (μg per pot)

The highest total offtake was observed for Zn, which correlates with its highest content in soil among all three studied elements (Fig. 6). The highest offtake was determined for wheat and lower for other grain crops. Very low Zn offtake was determined at vegetables, especially at radish, where total offtake caused only 3% of it by wheat. Surprisingly low Zn amount was also found at tobacco, confirming its selectivity to accumulate mainly Cd. Portion of Zn taken off by wheat to total Zn amount in soil was 2.5% and by oats slightly lower than 2%. Zinc introduced also higher mobility mainly due to its higher content in seeds. Above ground wheat biomass accumulated one third of total zinc and oat biomass roughly half of zinc taken by plants.

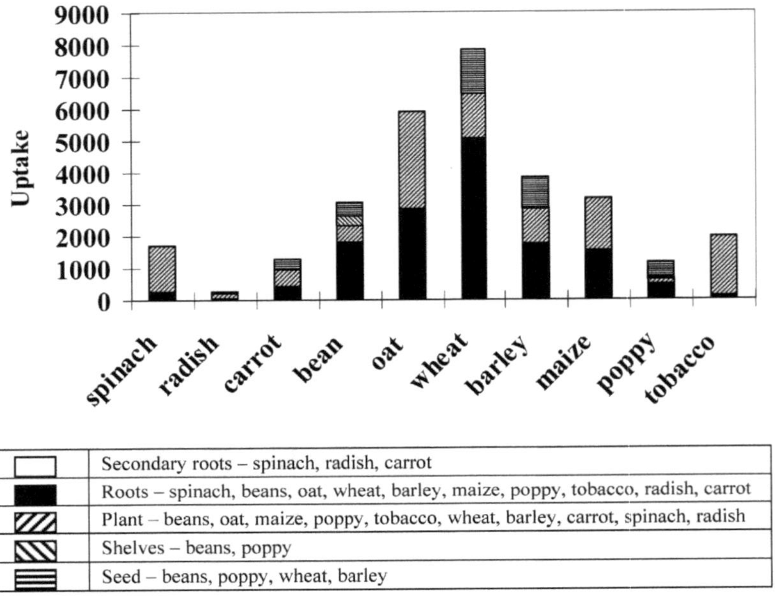

Fig. 6. Mean uptake of Zn by different parts of plants (µg per pot)

3. THE EFFECT OF SOIL METAL CONTENT AND CROP ROTATION ON ELEMENT ACCUMULATION IN PLANT BIOMASS

Soil element content can differ substantially on agricultural soils due to geogenic and anthropogenic origin. Many soils with elevated metal contents are in use for agricultural and horticultural production affecting element accumulation in plant tissues. Fig. 7 shows accumulation of Cd in above ground biomass of three common agricultural crops. Growing contents of Cd at both Luvisols led to accumulation more Cd in green biomass of both tested plants (oat, maize) as well as in barley straw and in lesser extend in barley grain. Twice higher soil Cd content did not make significant differences at any of analyzed plant part, confirming the role of soil properties for element accumulation. Third soil containing twenty fold higher Cd amount than the second one. Long-term contamination of Chernozem did not cause any yield reduction but significantly affected Cd accumulation in all plant parts. The highest increase of Cd content was determined in maize biomass almost linearly responding to Cd soil content increase. Cd accumulation in oat biomass as well as in barley straw less responded to elevated Cd content in Chernozems and Cd content in both parts of plants was significantly lower compare to maize. The lowest response determined in grains at both Luvisols and also confirmed at polluted Chernozem. Barley grain showed well protective mechanism and Cd content in it was only seven fold higher, compare to twenty fold differences in soils.

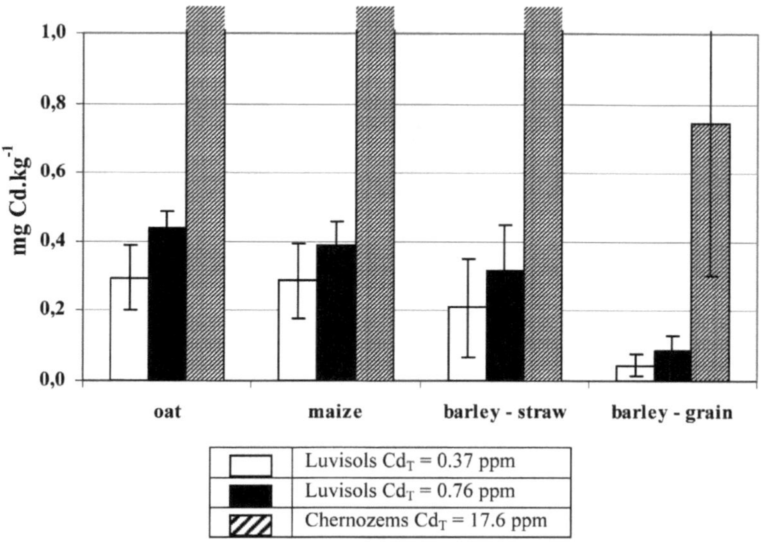

Fig. 7. Content of Cd in aboveground crop biomass

Total element content affects metal accumulation in plants, but availability of elements is also affected by soil properties. Száková et al. (2000) showed high fluctuation of Cd availability from 0.38 to 25.3% of total content mainly depended on soil pH and sorption properties. Results in Fig. 8 demonstrated different Zn content in plant biomass grown at soils with different total Zn content. Individual plants contained almost identical Zn content determined at specific soils, except barley grain. Luvisol 1 with lower soil Zn content usually allowed to plants to accumulate the same amount of Zn as Chernozem with twice higher Zn content. The proportion was same at green parts as well in barley grains. On the other site same Zn plant content was determined at both Luvisols twice differing in soil Zn content. Good answer of found pattern can give available portion of element. Our analyses has shown the lowest available Zn portion at Chernozem (0.05%) higher at Luvisol 1 (0.09%), with twice lower total Zn content, and significantly higher availability at Luvisol 2 (0.72%). Closer relationship between available Cd content in three soils and its accumulation in vegetable plant species also confirmed Tlustos et al. (1998).

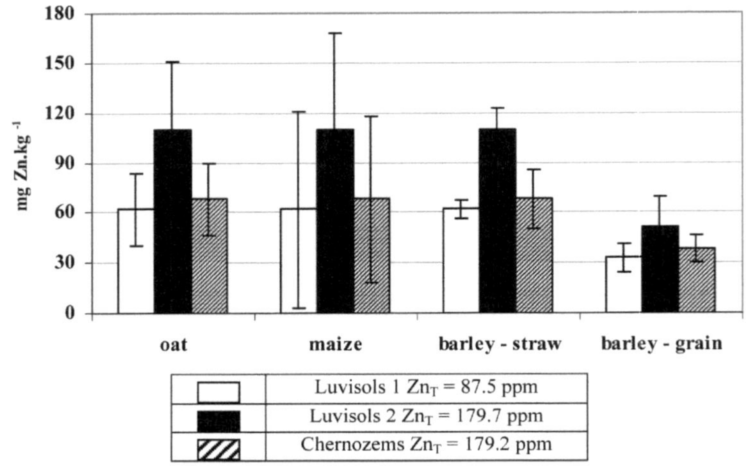

Fig. 8. Content of Zn in aboveground crop biomass

Forecrop can also affect the element accumulation in directly growing crops due to specific accumulation capacity of individual plants, different amount of roots and plant residues leaving on the field after the harvest and their decomposition, and due to an release of exudates affecting mobility of bound metals in the soil. The cumulative uptake of elements was tested at two soils substantially differed in total Cd content. Four crops were grown in different sequences at both soils at each year and after four years the experiment was evaluated (Fig. 9). Results showed differences in total Cd offtake at both soils. The highest offtake was found at treatment with sequence maize and oat as two first crops at both soils. Only slightly lower Cd amount in above ground biomass was found at similar oat and maize sequence (4 and 2 rotation). Other two crops (poppy and barley) accumulated significantly lower Cd amount at first two years (1 and 3 rotation). Especially barley cumulated different Cd amount at the first and other years. First year low accumulation was caused by low ability of barley to release soil Cd from firm bounds. Roots with low Cd content did not supply subsequent poppy plants by available Cd and its accumulation by poppy was also very low.

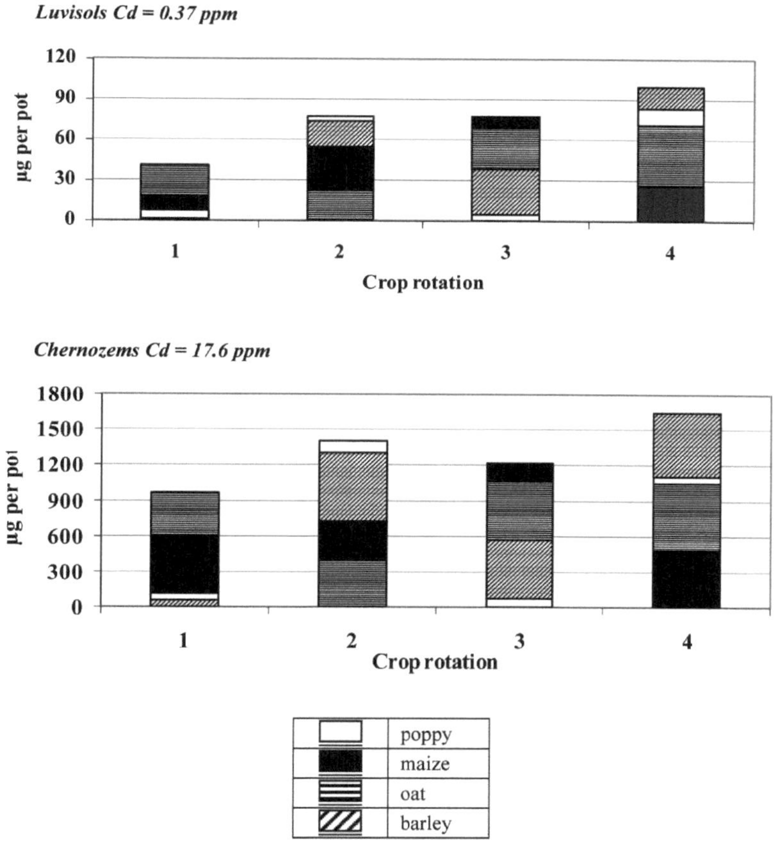

Fig. 9. Cumulative Cd uptake (µg per pot) by crops in different plant rotation

Higher capacity of oat and maize to release Cd from soil first year meaningfully helped to start total high Cd offtake at treatments 2 and 4. Higher root biomass of both crops also captured higher amount of Cd and released it into soil solution after root decomposition during the growth of subsequent plants. Plants usually contained lower Cd amount at the second half of experiments except those accumulating high contents of Cd from decomposed root biomass.

4. THE TOXIC ELEMENT ACCUMULATION IN FODDER CROPS

Grasses of many families, *Fabaceae* plants and other species are usually planted to feed livestock or other animals, to make nice green color in parks and gardens, to cover soil surface and protect it from the erosion and other damages. Many from the mentioned groups of plants can develop very thick layer of fine roots with high capacity to excavate elements enable to produce high amount of fresh or dry biomass. Very specific plants belong to *Fabaceae* family having ability to bind nitrogen from the air. Gerard et al. (2000) studied the Cd accumulation in ryegrass at three polluted soils with Cd content from 0.6 up to 25 ppm and found its content in plant biomass from 0.2 to 2.29 ppm Cd at the first harvest and lower mean contents at second and the third ones. Content of Cd in biomass grew with the elevating content of Cd in the soil. Low affinity to accumulate Cd in plant tissues introduced also red fescue. Shoot tissues contained from 2 to 4 ppm Cd planted at soil with 164 ppm Cd (Li et al., 2000).

In our experiment four different plant species were tested for their ability to accumulate elements in plant tissues at control unpolluted and two polluted soils. One polluted soil Cambisol came from smelter site and second one was made artificially by adding of As, Cd and Pb into control Chernozem. Among tested plants, elements and soils significant differences were appeared. Arsenic accumulation was always the highest at mild polluted Cambisol (1.82-2.78 ppm As) showing significantly higher differences at both *Fabaceae* species than at others (Fig. 10). Among all four species mallow (*Malva verticillata)* showed higher ability to accumulate As at all three tested soils. High binding capacity of Chernozem sorbed added As and level of arsenic in plant tissues was at these treatments same or even lower than at controlled soil of the same type and origin. Another explanation for it is the As precipitation after an application into soil. All plant species has confirmed generally low affinity to accumulate As in their plant tissues and very weak response to elevated As contents.

Better responses of plants were found for Cd. The highest Cd contents in plant tissues were determined at spiked Chernozem with significantly higher differences between safflower (*Carthamus tintorius*) and mallow at one site and both clovers at the other. Differences in plant Cd contents reached up to one order between the most cumulative safflower and the least red clover (*Trifolium pratense*), the pattern was similar at all three tested soils (Fig. 10). The highest Cd safflower content 23.9 ppm Cd was determined at spiked soil with addition of 40 ppm of Cd, and the lower one 20.6 ppm Cd at Cambisol from the vicinity of smelter. Safflower showed a linear accumulation response between control Chernozem and Cambisol, but another increase of soil Cd content did not substantially change Cd content in plants. Also

Tlustos et al. (1998) found good response of five vegetable species for elevated soil Cd contents up to one order, but higher soil contents did not meaningfully affect it plant contents.

Lead is usually very immobile in the soil and its transfer coefficients are much lower compare to Cd and Zn (Adriano, 2001). Also our results with tested plants confirmed their low Pb content compare to lead amount in soil. The differences in Pb accumulation in tested plants among soils showed similar patterns as As and Cd. Lead accumulation was the highest at polluted Cambisols, similar like as arsenic one. Spiking of Chernozem by lead salt led to a growth of its content in plant tissues compare to control soil, but almost twice lower Pb content in Cambisol introduced the highest plant levels in all tested plants. Suitable soil properties caused firm lead bounds not decomposable by plants. The highest Pb accumulation potential showed mallow at all three soils. Lower and similar Pb contents were determined in white sweet clover (*Melilotus alba*) and safflower and the mean lowest in red clover.

Mallow and safflower introduced usually higher accumulation potential for all three elements than white sweet clover. Red clover contained always the least contaminants from the soil.

5. THE ELEMENT ACCUMULATION IN FAST GROWING TREES

Replacement of agricultural crops by energy crops producing biomass as renewable source of energy is becoming the reality in many developed countries through the world. Non-food or foodstuff biomass production can be also suitable for phytoextaction of contaminants from the soils especially when suitable accumulation of elements as well biomass production can occur. Among many plant species mainly willows and poplars have been tested. Willows have a capacity for efficient uptake of macro- and micro-nutrients and have high evapotranspiration, which reflected in their high productivity (Perttu and Kowalik, 1997). Willows also meet requirements of metal tolerance and high concentrations of heavy metals (Dickinson et al., 1994, Greger and Landberg, 1995). Under temperate climate of Central Europe annual yield of dry biomass can reach about 10 tons per ha (Weger and Havlickova, 2002).

The differences in trace element accumulation in above ground willow biomass were tested in pot experiments with seven willow clones at two different soils. Sandy Fluvisol containing more than 3700 ppm of total Zn introduced its phytotoxicity and the yield of all clones was significantly reduced there, while Cambisol contained more than one order lower amount of Zn and willow biomass yield was not affected. Results of element accumulation at Cambisol (total soil element contents bellow Figs.) are

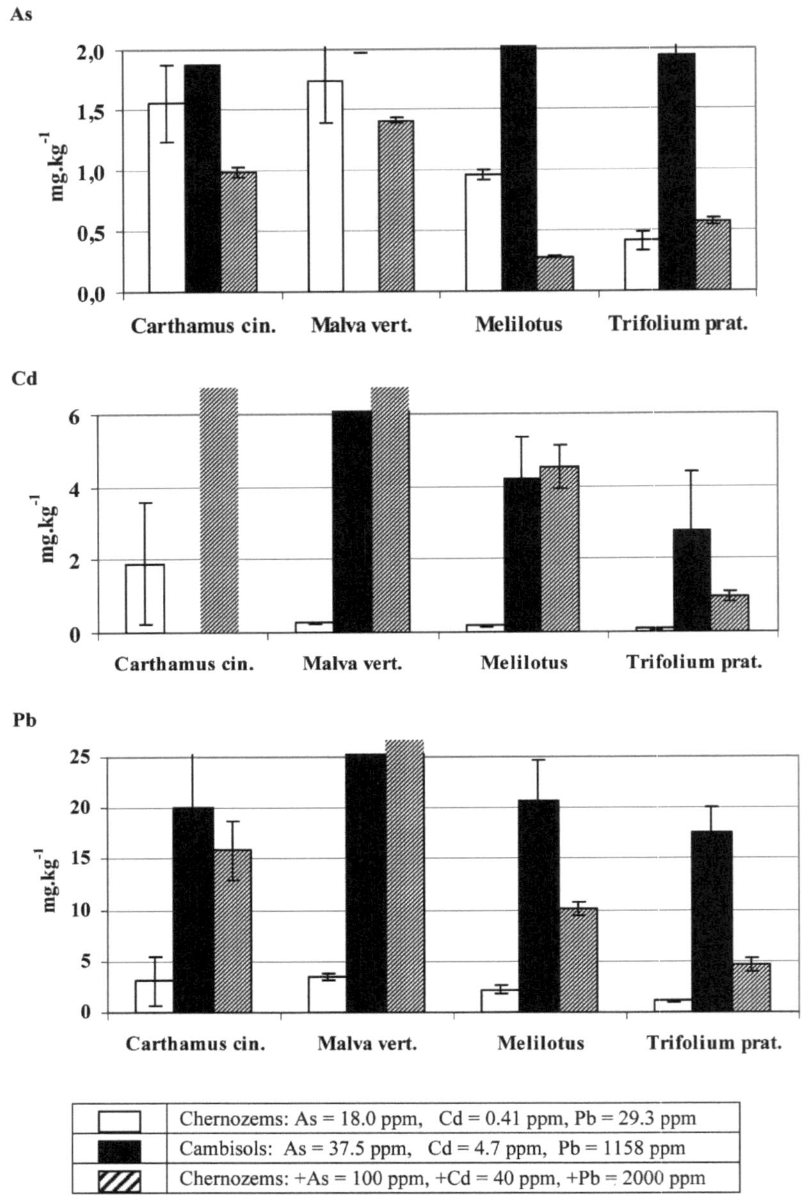
As
2,0
1,5
1,0
0,5
0,0
mg.kg-1
Carthamus cin.
Malva vert.
Melilotus
Trifolium prat.
Cd
6
4
2
0
mg.kg-1
Carthamus cin.
Malva vert.
Melilotus
Trifolium prat.
Pb
25
20
15
10
5
0
mg.kg-1
Carthamus cin.
Malva vert.
Melilotus
Trifolium prat.
Chernozems: As = 18.0 ppm, Cd = 0.41 ppm, Pb = 29.3 ppm
Cambisols: As = 37.5 ppm, Cd = 4.7 ppm, Pb = 1158 ppm
Chernozems: +As = 100 ppm, +Cd = 40 ppm, +Pb = 2000 ppm

species

displayed in Figures 11 and 12. The As level in all clones was generally low, fluctuating between 0.41-2.35 ppm of As in leaves and 0.32-0.75 ppm of As in twigs. While twig As content did not significantly differ among clones with exception of clone 1 (*Salix smithiana*), leaves showed higher differences and especially both clones (1 and 2) of *Salix smithiana* accumulated significantly higher As content. Cadmium again confirmed its mobility and Cd content in above ground biomass was generally one order higher than As content reaching from 38 to 65 ppm of Cd in leaves and mainly significantly lower from 19 to 40 ppm in twigs. Among all tested species *Salix smithiana* (clones 1,2) and *Salixes dasyclados* and *rubens* (clones 6,7) showed the best accumulation capacity. Taking soil content in account (1158 ppm of Pb) the lowest mobility was determined for Pb. Biomass Pb contents reached only 6.3 ppm in leaves and 10.8 ppm in twigs (Fig. 12). Lead was only the element accumulating higher content in twigs than in leaves. Clones 5 and 6, *Salixes pyramidilis* and *dasyclados* showed the highest Pb content. High Zinc mobility and accumulation potential was also found at willows. Zn mobility increased Zn accumulation in leaves. Leaves Zn content was always higher than soil content and reached up to 572 ppm of Zn in *Salix rubens* (clone 7). Twigs Zn content was usually three to five fold lower with the maximum 126 ppm at clone 7, again. Nissen and Leep (1997) also found evident accumulation of Zn in eight willow clones, Zn contents decreased in order leaves > bark > twigs > wood. Element mobility also affected TC by both plant parts. Due to lower element accumulation in willow wood than leaves TC of leaves was almost higher TC of twigs, except Pb. Maximum TC was found at Cd 18.9 for leaves and 7.1 for twigs. On the other hand TC of Pb at leaves as well at twigs did not exceed 0.008. The highest remediation factors (RC) were determined at best clone for Cd 12.5% and for Zn 17%. RC for other two elements As and Pb were almost negligible.

The uptake of elevated elements in soil at smelter site was also tested at willow species grown close to Litavka river. Willows can grow at this site, however, Czech Republic tested high productive willow clones suffered at this soil during pot experiments. Landberg and Greger (1996) found that clones from polluted areas had higher metal accumulation in their roots but the transport those metals to the shoots was more restricted than in clones from unpolluted areas. Our results provided at extremely contaminated site with As = 81, Cd = 45, Pb = 1320 and Zn = 3197 ppm showed that *Salix fragilis* grown at polluted Litavka embankment accumulated usually lower element contents with different distribution than willows grown in pots at less polluted site. Table 7 showed element distribution into main willow parts.

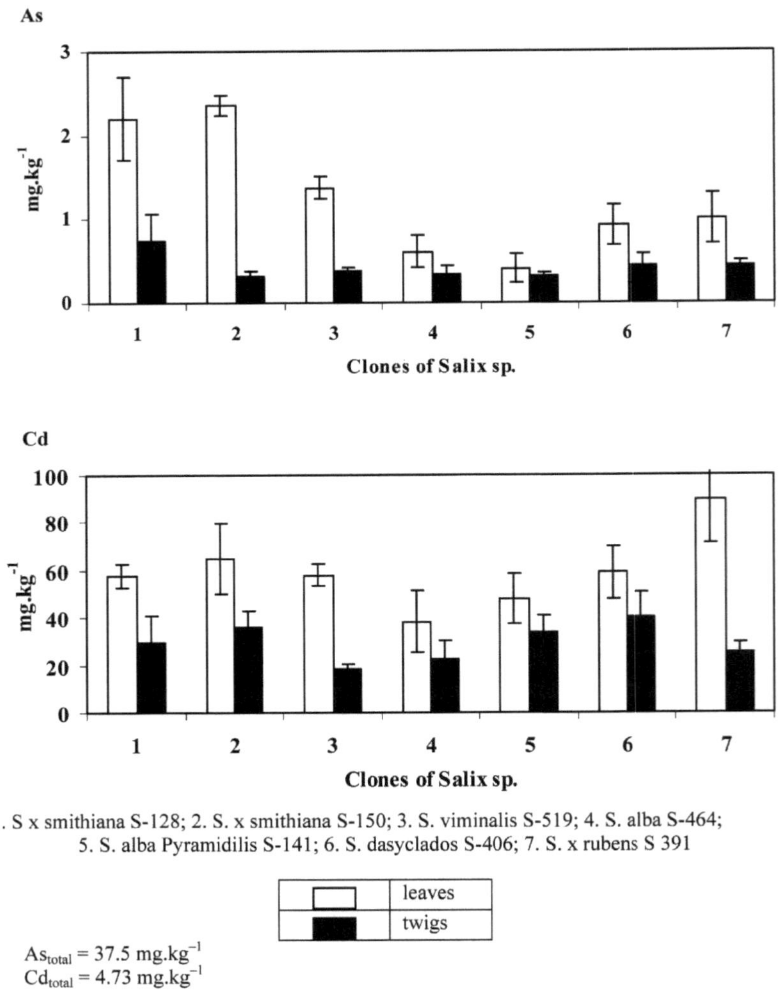

Fig. 11. Concentration of As and Cd (mg.kg^{-1} d.wt.) in leaves and twigs in seven clones of *Salix sp.*

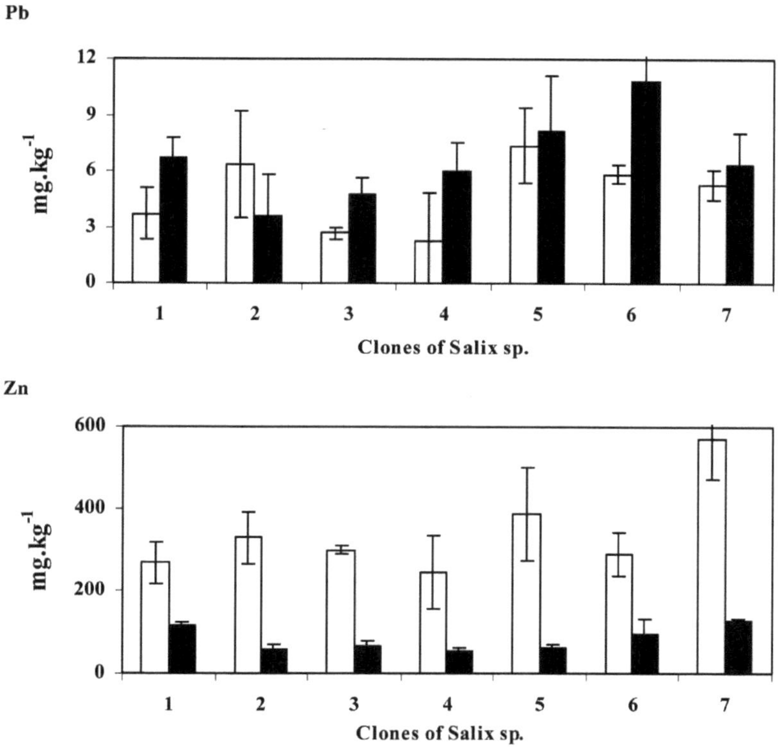

1. S x smithiana S-128; 2. S. x smithiana S-150; 3. S. viminalis S-519; 4. S. alba S-464; 5. S. alba Pyramidilis S-141; 6. S. dasyclados S-406; 7. S. x rubens S 391

□	leaves
■	twigs

Pb_{total} = 1158 mg.kg^{-1}
Zn_{total} = 180 mg.kg^{-1}

Fig. 12. Concentration of Pb and Zn (mg.kg^{-1} d.wt.) in leaves and twigs in seven clones of *Salix sp.*

Table 7. Content of toxic elements ($mg.kg^{-1}$) in different parts of willow tree from Litavka river bank.

	Part	Element			
		As	Cd	Pb	Zn
Salix fragilis	leaves	0.156	24.4	1.89	1196
	twigs	0.543	29.7	3.82	824
	roots	3.770	28.2	56.87	696

Low mobility of As and Pb determined at previous experiment was also found at about thirty years old willow tree. Both elements were accumulated in much higher extend in roots, than in twigs and the lowest content was determined in leaves. Cd and Zn were translocated into above ground biomass in much higher extent. Cd content was similar at all investigated parts and Zn content grew from roots through twigs to leaves. Lower element contents, except Zn, can be caused by older trees, inserting roots into deeper soil layers containing less elements. Higher Zn plant content was caused by roughly three fold higher Zn content in surface soil surrounding analyzed tree. Other willow varieties found at embankment of Litavka river with similar contents of trace elements has not shown high differences in element accumulation (tables 8 and 9). As content was almost identical in two varieties and even higher in leaves of other two varieties. Also Pb very immobile in plants always showed at least slightly higher Pb content in leaves than in twigs. Higher mobility of elements can be explained by lower difference of young twigs and young leaves at old willow trees than at one year old cuttings and also by different properties of both soils. Accumulation of mobile elements introduced the same pattern as in pot experiment. Cd and Zn contents were always higher in plant leaves than in twigs.

Table 8. Content of toxic elements ($mg.kg^{-1}$) in different plant clones of willows – leaves.

Variety	As	Cd	Pb	Zn
Salix caprea	0.704	22.0	19.1	1233
Salix fragilis	0.101	22.6	1.43	1744
Salix purpurea	0.217	29.9	5.86	1613
Salix viminalis	0.715	14.1	9.89	1198

Table 9. Content of toxic elements ($mg.kg^{-1}$) in different plant clones of willows – twigs.

Variety	As	Cd	Pb	Zn
Salix caprea	0.339	13.1	18.42	542
Salix fragilis	0.150	14.9	0.47	650
Salix purpurea	0.215	11.2	1.12	237
Salix viminalis	0.380	9.8	3.51	644

6. THE ACCUMULATION OF POTENTIALLY TOXIC ELEMENTS IN WEEDS AND OTHER PLANTS

Weed plants have extremely good ability to survive among other crops and to suppress their growth. Weed species can also develop great amount of biomass without close relation to climate and soil conditions. Accumulation capacity of weeds is not usually well known. Daniel et al. (1997) tested many weed species for their heavy metal accumulation along the roads and found high metal concentrations in weed biomass. Among range of weeds tested they also suggested *Artemisia vulgaris* and *Lathyrus pratensis* as suitable bioindicators of element contamination.

Seven weed species were tested for their accumulation of Cd and Zn at site with their soil background total content (Cd 0.37 ppm and Zn 98 ppm) and at plots with twice higher metal content derived from inorganic salts applied before soil plough. Figure 13 described Cd accumulation in roots and above ground weed biomass of seven species. Species distribution was not homogeneous at all plots, therefore the same species can not be compared at both treatments. The addition of Cd into soil led to its increase in all tested weeds. Highest increase of Cd content introduced aboveground biomass of *Artemisia vulgaris and Lathyrus tuberosus* with about four fold higher Cd accumulation. *Artemisia vulgaris* and *Taraxacum officionale* accumulated the highest Cd contents at natural as well at abundant soils. Rebele et al. (1993) analyzed *Artemisia vulgaris* at smelter site and they found Cd content substantially higher from 0.75 to 12.4 ppm. Analyses of roots confirmed high Cd mobility and especially at low Cd soil contents majority of species cumulated more metal in green part than in roots. Contrary results introducing higher element contents in roots found Szarek-Lukasevka and Niklinska (2002) testing Cd accumulation in *Plantago lanceolata* at calamine site. Large root surface with many active spots probably led to higher Cd accumulation in roots of *Elytrigia repens* compare to other plants.

The application of Zn salt into soil showed the same response as Cd and led to Zn increase in all tested weed species (Fig. 14). Zn content was usually doubled in green plant at treated soil, only in *Artemisia vulgaris* grew Zn content several times up to 237.5 ppm, the highest content among all species tested. Rebele et al. (1993) found 215-2301 ppm Zn in green matter of *Artemisia vulgaris*, which corresponded with several times higher soil Zn content at smelter site. Zinc also introduced high mobility and transport in plants. Higher Zn content was determined in aboveground biomass in all cases only the exception of *Elytrigia repens* was found as for cadmium.

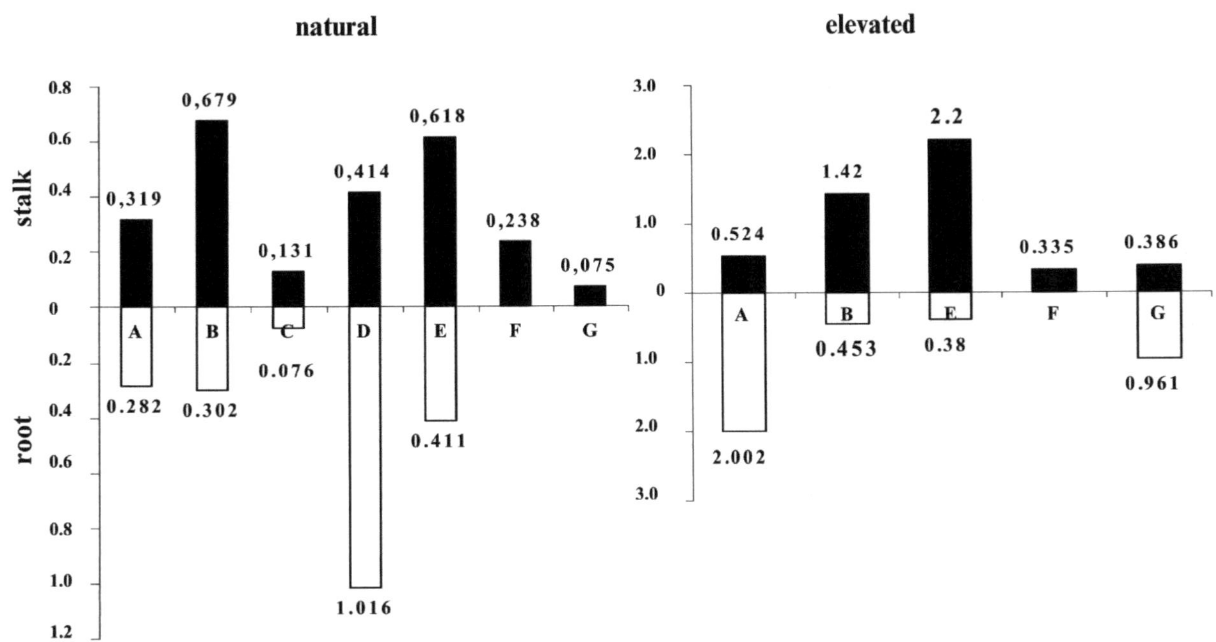

Fig. 13. The accumulation of Cd in weeds (mg Cd . kg^{-1})

A. Plantago lanceolata; B. Taraxacum officinale; C. Chenopodium hybridum;
D. Elytrigia repens; E. Artemisia vulgaris; F. Convolvulus arvensis; G. Lathyrus tuberosus
(Cd content = .0.37 mg.kg^{-1} + 0.37 mg.kg^{-1}

The highest accumulation potential among all plant species was introduced by hyperaccumulators, plants which have ability to take up enormously high amount of trace elements (Brooks, 1998). Many of species mainly from *Brassicaceae* family have ability to accumulate Cd up to 0.2% (Brown et al., 1994) and Zn and Ni in the content more than 1% (Brooks, 1998). Extraordinary accumulation potential can give these plants chance to be used for phytoextraction of metals from contaminated sites. Unfortunately an advantage of high accumulation is suppressed by low biomass yield.

High contamination of soil can significantly limit plant biodiversity. Investigation of Litavka embankments site contaminated with approximately As 89 ppm, Cd 43 ppm, Pb 2583 ppm and Zn 3312 ppm showed very restricted plant population. Besides few species of *Poaceae* family only very few dicotyledonous species were found there. Among them only *Arabidopsis halleri* and *Rumex acetosa* covered larger spots at the area. *Arabidopsis halleri* showed its capacity to take up metals (Table 10). Cd and Zn confirmed their mobility and higher content of both elements was found in green biomass than in roots. Minimum and maximum Cd and Zn contents in green parts of plants at three spots analyzed four times during vegetation always reached higher values than mean soil contents of both elements. TC_{Cd} fluctuated between 1.1-4.4 and TC_{Zn} between 1.3-5.0, confirming great capacity of plant to take up both elements even at high soil contamination. Other two contaminants were accumulated in lesser extent, and their mobility was significantly lower. Lead accumulation in roots was about one order higher than in aboveground biomass. *A. halleri* has capacity to take up Pb,.but its mobility is very low and translocation by xylem can not occur in large extent. Arsenic was not taken in reasonable amounts by tested plant. Unfortunately, among all the plant species, only a few fern species like *Pteris vittata* (Tu et al., 2002) and *Pityrogramma calomelanos* (Francesconi et al., 2002) demonstrated hypeaccumulating potential for this element.

Table 10. The accumulation of toxic elements in leaves and in roots of Arabidopsis *halleri.*

Leaf content	As	Cd	Pb	Zn
minimum	0.07	48.0	10.2	4370
maximum	4.47	184	255	16634
Root content				
minimum	0.95	32.3	358	2381
maximum	6.21	99.4	1924	82802

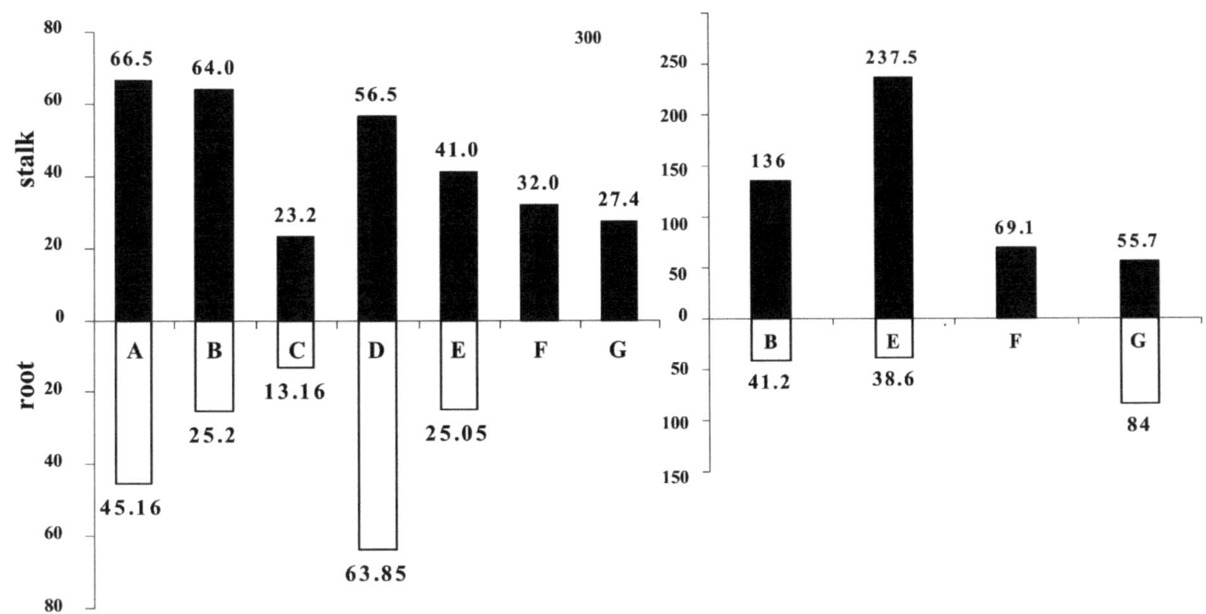

Fig. 14. The accumulation of Zn in weeds (mg Zn . kg^{-1})

A. A. Plantago lanceolata; B. Taraxacum officinale; C. Chenopodium hybridum;
D. D. Elytrigia repens; E. Artemisia vulgaris; F. Convolvulus arvensis; G. Lathyrus tuberosus
Zn Zn content = 98 mg.kg¯1 + 110 mg.kg¯1

Rumex acetosa found at same spots cumulated significantly lower contents in green tissues like in roots of all investigated elements (Table 11). *Rumex acetosa* introduced not just lower capacity to take elements but also to transfer them into aboveground tissues. Usually high Cd and Zn mobilities caused higher contents of both elements in green parts was not found for this specie and Cd, Zn, like Pb, and As were found in greater extent in roots. Mean Zn contents in leaves were approximately twice lower than in roots, Cd three, As four and Pb eight fold lower. The least affinity showed *Rumex acetosa* for Pb, and mean contents in roots and leaves were about eight fold lower than in *Arabidopsis halleri*. TC by leaves did not reach 1.0 for any element studied and was highest for Zn 0.2-0.6 lower for Cd 0.01-0.3 and very low for As and Pb 0.001-0.02. *Rumex acetosa* can be ranked among plants with the ability to grow under adverse soil conditions excluding element accumulation especially Pb and Cd in their green tissues.

Table 11. The accumulation of toxic elements in leaves and roots of *Rumex acetosa*.

Leaf content	As	Cd	Pb	Zn
minimum	0.10	0.4	3.0	845
maximum	1.67	13.4	67.0	2521
Root content				
minimum	0.85	2.8	24.0	1156
maximum	7.18	44.1	453	5980

Determination of accumulation capacity of potentially toxic elements, mainly playing negative role in soil contamination of Czech Republic, showed high different among species and plant parts.

7. SUMMARY

Determination of accumulation capacity of potentially toxic elements playing negative role in soil contamination of Czech Republic showed high differences among tested elements, plant species and their individual parts. Arsenic and lead always showed limited mobility. Their uptake by plants was low and if element was taken up by plant roots usually stayed there without substantial movement. The highest content was found in thin secondary roots on the opposite seeds were well protected. Arsenic mainly showed the lowest content in plant tissues among all elements tested. Lead showed similar properties, its content was usually higher but transfer coefficients (TC) showed low similar values for both elements. Cadmium and zinc were more mobile and easily taken by plants. Secondary roots contained usually their highest levels, but their translocation to aboveground biomass was rapid and in some cases bulk roots contained lower element contents than leaves and stems. Importance of zinc for plants was confirmed by its elevated content in seeds, while seeds and edible parts of vegetables were well protected from Cd accumulation.

Increased soil element contents differently affected plant element accumulation. Growth of As content in soil did not provide any significant changes in plant element content, the increase of soil Pb level usually changed plant lead content, soil properties played important role in behavior of both elements. Increased soil contents of mobile Cd and Zn led to higher accumulation of both elements by all species tested here. Mild growth of soil element content resulted in direct response of plants in accumulation of both elements. Plant depleted their Cd and Zn accumulation potential under high soil element contents and the increment of soil element content was higher than the growth of plant content. Extremely high Zn soil contents reduced plant development, elevated contents of other elements did not show any phytotoxicity effect. Among tested species significant differences in element accumulation were found. Consumable parts of vegetables as well seeds of cereals were well protected and the accumulation of toxic elements did not occur there. Different development and accumulation capacity of roots for heavy metals did not affect direct crop but played role in planting of subsequent crop, due to root decay and element release. The highest accumulation capacity of Cd and Zn was confirmed for *Arabidopsis halleri* grown at contaminated site. Promising results mainly for Cd introduced three clones of high productive willows, but higher amount of Cd and Zn was accumulated in leaves than in twigs. Other species did not accumulate sufficient amount of elements in their green tissues and some of them like *Trifolium pratense* or *Rumex acetosa* showed low affinity to take up elements.

More investigation especially at contaminated sites has to be done with either agricultural or horticultural plants and also with other suitable species. Some sites are medium populated, people have their own gardens there and they did not want to leave these places. We have to develop strategy what can people grow on the garden to meet required criteria and how we can steadily, environmentally friendly clean polluted areas.

REFERENCES

1. Alloway, B.J. (1990): Heavy metals in soil. J. Wiley & Sons, New York, 331 pp.
2. Adriano, D.C. (2001): Trace elements in terrestrial environments. Springer-Verlag, New York, 866 pp.
3. Baker, A.J.M. (1987): Metal tolerance. New Phytol., 160 (1): 93-111.
4. Brooks, R.R. (1998): Plants that hyperaccumulate heavy metals. CAB Intern., UK, 357 pp.
5. Brown, S.L., Chaney, R.L., Angle, J.S., Baker, A.J.M. (1994): Phytoremediation potential of *Thlaspi caerulescens* and bladder campion for zinc- and cadmium-contaminated soil. J. Environ. Qual., 23: 1151-1157.

6. Carbonell-Barrachina, A.A., Burlo, F., Burgos-Hernandez, A., Lopez, E., Mataix, J. (1997): The influence of arsenite concentration on arsenic accumulation in tomato and bean plants. Sci. Hortic. 71: 167-176.
7. Chaudri, A.M., Zhao, F.J., McGrath, S.P., Crosland, A.R. (1995): The cadmium content of British wheat grain. J. Environ. Qual., 24: 850-855.
8. Cobb, G.P., Sands, K., Waters, M., Wixson, B.G., Dorward-king, E. (2000): Accumulation of heavy metals by vegetables grown in mine wastes. Environ. Toxicol. Chem., 19 (3): 600-607.
9. Daniel, P., Kovacs, B., Prokisch, J., Gyori, Z. (1997): Heavy metal dispersion detected in soils and plants alongside roads in Hungary. Chem. Speciation Bioavailability 9 (3): 83-93.
10. Davies, B.E. (1992): Uptake of cadmium, copper, lead and zinc from contamined soils by radish (*Raphanus sativus* L.). Water, Air, Soil Pollut., 63: 331-342.
11. Dickinson, N.M., Punshon, T., Hodkinson, R.B., Lepp, N.W. (1994): Metal tolerance and accumulation in willows. Willow vegetation filters for municipal wastewaters and sludges. A biological purification system. Proc. of a study tour, conference and workshop in Sweden, Section of Short Rotation Forestry, Uppsala: 121-127.
12. Ebbs, S.D., Lasat, M.M., Brady, D.J., Cornish, J., Gordon, R., Kochian, L.V. (1997): Phytoextraction of cadmium and zinc from a contaminated soil. J. Environ. Qual., 26: 1424-1430.
13. Francesconi, K., Visootiviseth, P., Sridokchan, W., Gössler, W. (2002): Arsenic species in an arsenic hyperaccumulating fern, Pityrogramma calomelanos: a potential phytoremediator of arsenic – contaminated soils. Sci. Total Environ., 284: 27-35.
14. Gérard, E., Echevarria, G., Sterckeman, T., Morel, J.L. (2000): Cadmium availability to three species varying in cadmium accumulation pattern. J. Environ. Qual., 29: 1117-1123.
15. Greger, M., Landberg, T. (1995): Use of willow clones with high Cd accumulating properties in phytoremediation of agricultural soils with elevated Cd levels. Proc. of the 3rd Int. Conference on the Biogeochemistry of Trace Elements. Contaminated Soils, Paris: 505-511.
16. Kabata-Pendias, A., Pendias, H. (1986): Trace elements in soils and plants. CRC Press, Inc., Boca Raton, Florida, 315 pp.
17. Keefer, R.F., Singh, R.N., Horvath, D.J. (1986): Chemical composition of vegetable grown on an agricultural soil amended with sewage sludge. J. Environ. Qual., 15 (2): 146-152.
18. Kiekens, L., Camerlyck, R. (1982): Transfer characteristics for uptake of heavy metals by plants. Landwirtschaftliche Forschung SH39: 255-261.
19. Kloke, A., Sauerbeck, D.R., Vetter, H. (1984): The contamination of soils and plants with heavy metals and the transport of heavy metals with terrestrial food chains. In Changing metal cycles and human health (ed. Nriagu, J.O.), Springer-Verlag, Berlin: 113-141.
20. Landberg, T., Greger, M. (1996): Differences in uptake and tolerance to heavy metals in Salix from unpolluted and polluted areas. Appl. Geochem., 11: 175-180.
21. Lario, Y., Burlo, F., Aracil, P., Martinez-Romero, D., Castillo, S., Valero, D., Carbonell-Barrachina, A.A. (2002): Methylarsonic and dimethylarsinic acids toxicity and total arsenic accumulation in edible bush beans, *Phaseolus vulgaris*. Food Add. Contam., 19: 417-426.
22. Li, Yin-Ming, Chaney, R.L., Siebielec, G., Kerschner, B.A. (2000): Response of four turfgrass cultivars to limestone and biosolids-compost amendment of a zinc and cadmium contaminated soil at Palmerton, Pelnnsylvania. J. Environ. Qual., 29: 1440-1447.
23. Lubben, K.R., Sauerbeck, D.R. (1991): Transferfaktoren und Transferkoeffizienten für den Schwermetallübergang Boden-Pflanze. In Auswirkungen von Siedlungsabfällen auf Böden, Bodenorganismen und Pflanzen (eds. Sauerbeck, D.R., Lübben, S.), Forschungszentrum Jülich GmbH, Berichte aus der ökologischen Forschung, 6: 180-233.
24. Lubben, K.R., Sauerbeck, D.R. (1991a): The uptake and distribution of heavy metals by spring wheat. Water, Air, Soil Pollut., 57-58: 239-247.

25. Macek T., Macková M., Pavlíková D., Száková J., Truksa M., Singh-Cundy A., Kotrba P., Yancey N. and Scouten W.H., 2002. Accumulation of cadmium by transgenic tobacco. Acta Biotechnol., 22 (2):101-106.
26. Marschner H. (1995): Mineral nutrition of higher plants. 2nd edition, Academic Press London, 889 pp.
27. Mench M., Tancogne J., Gomez A., Juste C. (1989): Cadmium bioavailability to *Nicotiana tabacum* L., *Nicotiana rusticana* L., and *Zea mays* L. grown in soil amended or not amended with cadmium nitrate. Biol. Fert. Soils, 8: 48-53.
28. Merry, R.H., Tiller, K.G., Alston, A.M. (1986): The effects of soil contamination with copper, lead and arsenic on the growth and composition of plants. II. Effects of source of contamination, varying soil pH, and prior waterloggin. Plant Soil, 95: 255-269.
29. Nissen, L.R., Lepp, N.W. (1997): Baseline concentrations of copper and zinc in shoot tissued of a range of *Salix* species. Biomass Bioenerg., 12 (2): 115-120.
30. Perttu, K.L., Kowalik, P.J. (1997): Salix vegetation filters for purification of waters and soils. Biomass Bioenerg., 12 (1): 9-19.
31. Rebele, F., Surma, A., Kuznik, C., Bornkamm, R., Brej, T. (1993): Heavy-metal contamination of spontaneous vegetation and soil around the copper smelter Legnica. Acta Soc. Bot. Pol., 62 (1-2): 53-57.
32. Schaller, A., Diez, Th. (1991): Pflanzenspezifische Aspekte der Schwermetallaufnahme und vergleich mit dem Richt - und Grenzwerten für Lebens – und Futter Mittel. In Mitteilungen Dt. Bodenkundl. Gesellsch.: 92-125.
33. Smilde, K.W., van Luit, B., van Driel, W. (1992): The extraction by soil and absorption by plants of applied zinc and cadmium. Plant Soil, 143: 233-238.
34. Száková, J., Tlustoš, P., Balík, J., Pavlíková, D., Balíková, M., 2000: Efficiency of extractants to release As, Cd, and Zn from main soil compartments, Analusis, 28 (9): 808-812.
35. Szarek-Lukaszewska, G., Niklinska, M. (2002): Concentration of alkaline and heavy metals in *Biscutella laevigata* L. and *Plantago lanceolata* L. growing on calamine spoils (S. Poland) Acta Biol. Cracov., Ser. Bot., 44: 29-38.
36. Tlustoš, P., Pavlíková, D., Balík, J., Száková, J., Hanč, A., Balíková, M. (1998). The accumulation of arsenic and cadmium in plants and their distribution. Rostl. Výr., 44 (10): 463-470.
37. Tu, C., Ma, L.Q., Bondada, B. (2002): Arsenic accumulation in the hyperaccumulator Chinese brake and its utilization potential for phytoremediation. J. Environ. Qual. 31: 1671-1675.
38. Vysloužilová, M., Tlustoš, P., Száková, J., Pavlíková D. (2003): As, Cd, Pb and Zn uptake by different *Salix spp.* grown at soils enriched by high loads of these elements. Plant, Soil, Environ. 49 (5): in press.
39. Weger, J., Havlickova, K. (2002): The first results of the selection of woody species for short rotation coppices in the transitional oceanic-continental climate of the Czech Republic. Proc. of the 12th European Conference and Technology Exhibition on Biomass for Energy, Industry and Climate Protection, Amsterdam (In print).

CHAPTER 4

GENETICALLY MODIFIED PLANTS WITH IMPROVED PROPERTIES FOR PHYTOREMEDIATION PURPOSES

T. MACEK[a,b], K. FRANCOVA[b], M. SURA[a] and M. MACKOVÁ* [a, b]
[a]*Department of Natural Products, Institute of Organic Chemistry and Biochemistry, Academy of Sciences of the Czech Republic, Flemingovo n. 2, 166 10 Prague 6, Czech Republic*
[b]*Department of Biochemistry and Microbiology, Faculty of Food and Biochemical Technology, ICT Prague, Technická 5, 166 28 Prague 6, Czech Republic*

Abstract

Genetically modified plants should be considered as a potential tool for decontamination of environmental pollutants. Transgenic plant technology showed many obvious advantages over conventional plant breeding approaches for crop improvement. Recent discoveries allowed the engineering of new transgenic plants generating desirable products, such as enzymes, polymers and vaccines. Among new approaches, the use of transgenic plants specifically tailored for the bioremediation of organic pollutants and heavy metals have recently occurred. This paper gives an overview of the results of present research in phytoremediation of the polluted environment using transgenic plants and their improved properties. Preparation and properties of genetically modified plants changing mercury into less toxic forms, plants with improved abilities of Cd accumulation, degrading explosives (TNT), PCBs or others oxidising TCE (trichloroethylene) are described, as well as approaches to change cooperation of organisms living in the rhizosphere (bacteria, mycorrhizal fungi) in order to improve their remediation potential. Advantages and disadvantages of the use of GM plants are discussed.

1. INTRODUCTION

1.1. What is a Transgenic Plant, and How it can be Prepared

Transgenic plants used in everyday practice are nowadays represented especially by agricultural crops with increased resistance against insect pests, viral infections or herbicides (Koprowski and Yusibov, 2001). As the next generation of genetically modified (GM) crops were introduced plants

J.-L. Morel et al. (eds.), Phytoremediation of Metal-Contaminated Soils, 85–108.

that can serve as a source of food products with increased nutritive value (e.g. with better composition and content of unsaturated fatty acids, vitamins etc.). Further, here belong plants with increased resistance against extreme weather conditions, or changes in their ability of ripening. In addition to these advantages, exploited in agricultural practice (wheat, corn, soy, rape, lemon, etc.) were prepared also GM plants able to produce pharmaceutically important products like antibodies or enzymes (Mason and Arntzen, 1995).

In the nineties, the technology of GM plants slowly shifted towards a new field, the use of plants as heterologous expression systems, among others for production of antigens of mammalian pathogens. Different medically important antigens were already expressed in transgenic plants, e.g. the surface antigen of hepatitis B (Mason and Arntzen, 1995) or rabbit virus glycoprotein (McGarvey, et al., 1995). As can be seen from the previous paragraph, genetic manipulations are finding increased use also in plant biotechnology. Different transgenes influence the yield and qualities of harvested crops, genes regulate ripening, composition of fats and the quality of seed proteins. The studies of tissue specific promoters will allow genetic engineering to create new categories of GM plants producing the required products (Poirier, et al., 1992).

Not less important and in the literature often discussed is the exploitation of GM plants in the removal of environmental contaminants by using transgenic plants with increased ability to accumulate or metabolise different xenobiotics and pollutants from soil, sediments and water (Macek, et al., 1997; Macek, et al., 1996; Francová, et al., 2001; Rugh, et al., 1996).

In last decade, the increasing use of transgenic plants caused large discussions. This is especially concerning agricultural crops grown on large scale as source of human food. Fear of GM crops induced an increased interest of politicians, and in many countries as well as in the Czech Republic, new laws were prepared and approved on handling of genetically modified organisms and products.

Unfortunately, most of the arguments against the use of GMO, especially those from activists of different green movements, are missing scientific background. That is the reason why even sound arguments are often not accepted. On the other hand, there is no dispute that carefulness is necessary in handling, cultivation and especially exploitation of genetically modified organisms and that it is a must to evaluate the risks connected with the release of GMOs to the environment or to the market. Concerning food production, the regulations are strict enough in many countries, including the obligatory labelling of any food product containing transgenic DNA in levels above some limit. The detection methods for analysis of the presence of transgenes are developing and improving rapidly.

1.2. Methods of Preparation of Transgenic Plants

The revolution in plant biotechnology launched two main findings. First was the unique ability of any single plant cell to regenerate into a whole plant. Of similar importance was the description of the until now most suitable vector for gene transfer into plants, the Ti plasmid of soil bacterium *Agrobacterium tumefaciens*, and the plasmid Ri of *A. rhizogenes*. The exploitation of the agrobacterial gene transfer techniques is one of the main approaches for introduction of foreign genes into the plant genome (Barrett, et al., 1997), but even this already widespread technology has some inborn limitations. Agrobacterial plasmids as vectors are well suited for transformation of dicotyledonous plants, but cases of successful use on monocotyledonous plants (corn, rice, wheat) were already also described (Ondrej, et al., 1999). An other technique using "natural" gene transfer is the use of viral vectors, but these are leading only to transient, not stable transformation.

In cases, where the above technique is inapplicable, other methods of transfer have to be exploited. Some were developed after finding of methods allowing regeneration of a whole plant from callus culture or from a single protoplast. These techniques include direct DNA transfer (Gelvin, 1998), electroporation, DNA vacuum infiltration, different chemical modifications, microinjection or ballistic methods (Maliga, 1990).

1.2.1. Transformation of Plant Cells by Bacteria of the Genus Agrobacterium

This method of genetic information transfer into plant cells is based on the natural ability of soil bacteria *Agrobacterium tumefaciens* and *A. rhizogenes* to attack higher plants and introduce into them their specific genes. These genes are localised into a part of a large plasmid known as Ti (tumour inducing) or Ri (rhizome inducing) (Ondrej, et al., 1999), and are able to induce morphological changes in attacked plants. The part of Ti or Ri plasmids, which is transferred into plants, is called T-DNA. Plasmid Ti has size usually between 150 000– 200 000 base pairs, what represents about 3% of the *A. tumefaciens* chromosome. Plasmid Ti contains two main parts:

- T-DNA, which enters the plant cell, but has no genes for its own integration,
- virulence region, which contains genes necessary for functions allowing transfer of T-DNA to plant cells and their integration into the plant genome.

For the function of the plasmid in bacterial cells and for the interaction between bacteria and plant cells are also necessary other parts of the genome, some of which are localised also in the bacterial chromosome.

T-DNA of wild-type Agrobacteria introduces into the plant genome two groups of genes (Ondrej, et al., 1999):

- Genes for biosynthesis of plant hormones, auxins and cytokinins. These are responsible for changes in tissue morphology and differentiation of plant cells, causing their growth as undifferentiated tumours or so called hairy roots.

- Genes for synthesis of tumour specific metabolites, so called opines. Opines serve as a source of carbon, nitrogen and energy for that type of bacteria, which induced the transformation. At present the natural wild-type strains of Agrobacterium are no more used for the genetic manipulations, these were replaced by strains with modified Ti plasmids. The genes responsible for plant hormone production, that cause the formation of tumours, and genes coding opine formation, were removed, being not necessary in constructed GM plants. In the same time genes serving as selective markers were introduced into the original T-DNA. Among the most exploited marker genes belong bacterial β-glucuronidase (GUS) and luciferase genes, and recently also the gene coding the green fluorescent protein (GFP). Because the big size of the natural Ti-plasmid cause problems in handling, the T-DNA with border sequences was introduced into smaller plasmids derived from *pBR 322*, which are nowadays commercially available. Within the T-DNA region of these newly prepared plasmid unique restriction sites are present, allowing introduction of constructs prepared for transfer into plant genome.

Genetic construct, intended to be transferred into plants, has to contain proper promoter region (most often constitutive promoter from cauliflower mosaic virus CaMV 35S, promoter controlling ubiquinone expression in corn, or other organ or tissue specific promoters active in plants), sequence coding the required gene and termination signal. At present much effort is devoted to isolation of novel promoters inducible e.g. by some environmental effect (like the presence of heavy metals).

1.2.2. Other Methods used in Plant Molecular Genetics

Protoplasts are a useful tool for gene manipulations, because under specific conditions fusion can be achieved between cells of closely related species, as well as very different plant species and thus allow formation of somatic hybrids. Protoplasts can be obtained from plant tissues or from callus or suspension cultured cells (Old and Primrose, 1994). After fusion and transfer of protoplasts to a nutrient medium containing cytokinins and auxins in proper ratio, callus tissue can develop. This can be achieved roughly after 5 to 10 days after formation of a new cell wall and beginning of cell division, than cell differentiation occurs and embryo is formed. Such an embryo can develop to a fully functional plant with changed genetic equipment.

Other useful tools are ballistic methods, allowing transfer of genes into cells and into whole intact plants. These methods are based on use of accelerated heavy metal particles (gold or wolfram) covered by genetic material (Maliga, 1990). The relatively new methods are used for species where failed the use *Agrobacterium*. The approach can be used with success also in other applications, including direct transformation of organelle genome or fast introduction of a genetic information into cells without tissue damage, e.g. into growth meristeme (Old and Primrose, 1994).

Another known method of gene transfer is microinjection, where DNA is transferred into the plant cell under microscopic control (Meyers, 1995). DNA can be placed into defined subcellular compartments of single protoplasts, isolated cells, but also into particular cells of a multicellular structure. This method shows the highest effectivity (yield in tens of percents) and can be used for plant tissues, too (Ondrej, et al., 1999). The disadvantage is, that during a single experiment only a low number of cells can be affected. Most of laboratories working in this field prefers from the above approaches the Agrobacterial transformation due to high effectivity, low copy number, intact transgenes, and ballistic methods (allows transformation independent on target plant species).

2. FIELDS OF GM PLANT APPLICATION

2.1. Agriculture and Food Industry

As shown above, despite the controversial attitude of public, largest exploitation found the GM plants in agriculture and in food production. The largest problem that could be addressed by the wider use of transgenic plants is the steadily increasing food consumption. A recent study showed that the world population increased within last 40 years by 90%, while the food production increased only by 25%. An other problem represent the annual crop losses due to attack by insects or other pests (up to 40% of harvest), while the amount of agrochemicals used (pesticides, insecticides or herbicides) cannot be for ecological reasons further increased. The main aim of genetic modifications is to achieve formation or organisms resistant to herbicides, pests, bacteria, fungal or viral infections, extreme weather conditions, and organisms with increased nutritive value of produced food. From practical applications, as the most important can be considered the introduction of resistance against herbicides and insect pests. In the USA the soy (Roundup Ready), cotton and canola resistant to the herbicide glyphosate introduced into market and since 1996 their percentage of total harvest is rapidly increasing.

Plants resistant against insects represent the highest percentage of all plant GMOs (Ondrej, et al., 1999). These plants contain in their genome genes for toxins derived from bacterium *Bacillus thuringiensis*. Earlier extracts of these bacteria were sprayed over the crops, what often lacked the expected effect. The experience collected lead to the construction of GMOs containing genes coding the formation of a toxin, the so-called bt toxin. Such plants are able to survive attack of insect pests (e.g. corn, tobacco). Some GM plants, at present approved, grown and distributed to the market, are shown in Table 1.

Interesting examples of GMOs are transgenic tomatoes. Their genome has been modified in such a way to achieve delayed ripening and longer storage. First example is the tomato "Flavr Savr" (Ondrej, et al., 1999; Flavell, 1995), containing transgene leading to inhibition or suppression of the function of gene for the enzyme polygalacturonase (PG), which catalyses the hydrolysis of α-1, 4 bond of the polygalacturonic acid, a component of the cell wall of tomatoes (Stilwell, 1997). PG is synthesised during ripening and by decomposing pectin in the middle lamellae of fruit cell walls causes softening of the fruits. The fruits with strongly reduced PG activity (under 10%) changes of fruit characteristics were achieved. Fruits were substantially larger, less susceptible to bacterial and fungal diseases; the juice was thicker and contained more dry mass. Against expectations, the tendency to softening during ripening did not change considerably.

Table 1. Examples of genetically modified plants already used in agricultural practice and their properties.

GM plant	Properties
Soy	increased production of oleic acid, resistant against insect pests and herbicides
Corn	resistant against insect pests and herbicides
Papaya	resistant against viruses
Potatoes	resistant against insect pests and herbicides
Cotton	resistant against insect pests and herbicides
Rape	increased amount of lauric acid, resistant against herbicides
Tomato	delayed ripening

Another type is tomatoes with transgene preventing expression of a plant gene responsible for ethylene formation (a gaseous plant hormone triggering among others also the cascade of processes leading to fruit ripening. The GM corn, resistant against phosphinothricine and insects, contains in comparison with classical varieties three additional genes. By direct transformation, using protoplasts were into the plant chromosome introduced:

a) genes for resistance to herbicide phosphinothricine (trade names: Basta, Liberty, Finale, Radicale) – that is a new type of herbicide, blocking glutamine synthetase, a key enzyme of nitrogen metabolism,
b) gene for δ-endotoxin from *Bacillus thuringiensis*, which brings resistance against insect pests,

c) gene for β-lactamase, responsible for the resistance to ampiciline – this gene allows selection of modified plants.

2.2. Production of Pharmaceutically Important Compounds

The field of preparation of pharmaceutically important compounds gains far lower negative publicity than the use of GMOs in agriculture for food and feed production. This is probably due to the fact, that studies leading to such GMOs are performed mostly in laboratory scale. Here belong production of mammalian antigens, which after consumption of the proper plant might lead to induction of immune reaction and function as a vaccine. The expression of immunoglobulin and their fragments has been also tested (Ma, et al., 1995). Transgenic potatoes were used for successful expression and confirmation of functioning (in mice) of a vaccine against surface antigen of hepatitis B, the same vaccine has been expressed later also in tobacco. Among others were also expressed e.g. enterotoxin (LT-B) from *E. coli*, or capside protein of the Norwalk virus, glycoprotein B of the cytomegalovirus, subunit B of cholera toxin and some others (Ma, et al., 1995).

2.3. Preparation of Transgenic Plants for Phytoremediation Purposes

Within the last two centuries the industrial production, mining industry and different urban activities caused environmental contamination on a large scale. The agricultural practices, especially those of the twentieth century are also responsible for the widespread contamination of soil, sediments and water, not speaking about wars and different conflicts responsible for production, use and storage of explosives and chemical warfare agents (Macek, et al., 1998). The air contamination by volatile toxic compounds is also an important problem.

One possibility of removal of contaminants from soil represents the physico-chemical methods. These techniques are unfortunately highly economically demanding and often further destroy the environment. For this reason in last decade much of the research was oriented towards the use of biological remediation methods. One of them is the use of plants, phytoremediation, which can be defined as the use of green plants for transfer, accumulation and removal of pollutants from the environment, or at least reduction of their spreading (Cunningham, et al., 1995; Macek, et al., 2000). It is an approach that can be used for removal of both inorganic and organic pollutants present in the soil, water and air (Salt, et al., 1995). The application of phytoremediation technologies represented in 1999 the market value of 35 million of US dollars, while it was expected that this value should increase ten-times within the next five years.

It is generally accepted, that phytoremediation of soil can be best used on large areas, where contamination reaches maximum into 5 meters of depth. In general the use of plants in remediation can be divided into some main areas (Macek, et al., 2000; Chaney, et al., 1997).

Phytoextraction (sometimes called phytoaccumulation) is based on the ability of plants to collect some toxic compounds from soil and accumulate them in the above-ground parts. This method is mostly connected with the removal of trace elements, heavy metals and radioactive compounds. Most of plants is able to accumulate heavy metals present in soil or water. The plant biomass containing heavy metals should be than harvested, reduced in volume e.g. by composting, low temperature ashing or other methods, stored on special places or used for reutilisation of the metal.

Phytotransformation is the use of plants for conversion of organic pollutants.

Phytostimulation is representing the stimulation of microbial degradation by the effect of plant exudates.

Rhizofiltration is the exploitation of plant roots for absorption or adsorption of pollutants, especially metals, from water. From many tested plants, e.g. hydroponically grown sunflowers proved to be very effective in removal of radionuclides from surface waters around Chernobyl (Meagher, 2000).

Phytostabilisation is a method that uses plants to reduce the mobility and availability of the pollutants in the environment.

Phytovolatilisation is uptake and transfer of some pollutants into gas phase by plants.

Use of plants for the removal of pollutants from the air – represents the ability of plants to accumulate and metabolise toxic compounds from the air.

One of the main tasks of basic research in the field of phytoremediation can be defined as the selection of plant species most suitable for phytoremediation purposes (Macek, et al., 2002). Not all plant species are equally well suited to metabolise or accumulate pollutants. Requirement for plants removing heavy metals is to grow fast in the contaminated environment, to be resistant, able to accumulate toxic metals and transfer cations or oxyanions into the harvestable (aboveground) parts, or transform them into less-toxic forms (Kraemer and Chardonnens, 2001). Especially useful are plants able to remove more than one pollutant, because contamination is usually caused by a mixture of more toxic compounds together. During remediation of organic pollutants, it would be optimal, if the plants or plants together with a microbial consortium (Ryslava, et al., 2003) would be able to mineralise the compound fully to non-toxic products.

It is clear that pollutants can be removed using a range of physico-chemical and/or biochemical processes, sorption, transport and translocation, accumulation or transformation. Many elemental pollutants enter to plants by basic transport systems designated for nutrient uptake. A number of xenobiotics is than stored in vacuoles as protection against their toxic effects.

Especially during the accumulation of heavy metals, but also during removal of organic contaminants the main limiting factor in large-scale exploitation of plants is the long time necessary for soil decontamination. For this reason, much effort is devoted to preparation of plants specifically tailored for phytoremediation purposes by breeding or genetic manipulations.

Recently more informations appeared in literature about possibilities to increase the expression of the genes already present, or introduction of bacterial or mammalian genes into plants, in order to increase the natural ability of plants to cope with xenobiotics (Macek, et al., 2002). Especially in the case of metal uptake the basic research studies are reaching quite deep. Complex interactions of transport and chelating activities control the rates of metal uptake and storage. In recent years, several key steps have been identified at the molecular level, enabling us to initiate transgenic approaches to engineer the transition metal content of plants (Clemens, et al., 2002). Clemens et al. in this excellent review summarise the determinants of metal accumulation, mobilisation, uptake, sequestration, xylem transport, discussing the role of different families of transporter proteins. Many problems are still unclear, the transport and storage forms of transition metals are largely unknown and many system have yet to be analysed. Despite the unsolved questions, different attempts were already done to prepare transgenic plants with improved properties for phytoremediation purposes. The use of GM plants did not reach yet application on large scale, but the start of field experiments is just the question of time.

Table 2 shows some examples of transgenes and transgenic plants prepared for increase of accumulation, resistance and transport of some inorganic pollutants. Table 3 summarises examples of GM plants characterised by increased ability to transform organic pollutants.

The effect of expression of the discussed transgenes alone, and its increased effectivity in practical scale for phytoremediation of contaminated fields was not much discussed yet, most experiments that should show the properties of improved plants were performed in laboratory scale with agar-grown plants or in hydropony (see Table 2, 3). On real contaminated soil, only a few transgenic plants have been tested (Grichko, et al., 2000; Karenlampi, et al., 2000; Bizily, et al., 1999; Macek, et al., 2002).

Table 2. Examples of transgenic plants with increased accumulation, resistance or increased transport of heavy metals (modified from Kraemer and Chardonnens, 2001).

Transgene	Product	Source	Transgenic plant	Aim/effect	Cultivation medium
mer A (Macek, et al., 1996)	Hg (II) reductase	G^- bacteria	*Liriodendron tulipifera* *Nicotiana tabacum*	increased Hg accumulation + evaporation	agar, soil
mer A *mer B* (Meagher, 2000)	Hg (II) reductase lyase	G^- bacteria	*Arabidopsis thaliana*	increased Hg tolerance + evaporation	agar, solution
APS1	ATP sulphurylase	*A. thaliana*	*Brassica juncea*	doubled Se accumulation	hydropony
MT-I	MT[a]	mouse	*N. tabacum*	20-times more resistant to Cd	agar
CUP1	MT[a]	*S. cerevisiae*	*B. oleracea*	16-times more resistant to Cd	hydropony
HisCUP (Macek, et al., 1998, 2002)	polyHis-MT[a]	*S. cerevisiae*	*N. tabacum*	accumulation of Cd 190%	agar, soil
gsh2	GSH-synthase	*E. coli*	*B. juncea*	accumulation of Cd 125%	hydropony
gsh1	γ-Glu-Cys synthase	*E. coli*	*B. juncea*	accumulation of Cd 190%	hydropony
NtCBP4	channel for M+[b]	*N. tabacum*	*N. tabacum*	tolerance to Ni, accumulation of Pb 200%	hydropony
ZAT1	transport of Zn	*A. thaliana*	*A. thaliana*	increased tolerance to Zn	hydropony
ACC (Grichko, et al., 2000)	ACC[c] deaminase	*bacteria*	*Lycopersicon esculentum*	increased resistance, accumulation of Cd, Co, Cu, Ni, Pb, Zn	soil
FRO2 (Karenlampi, et al., 2000)	reductase of Fe(III) chelate	*A. thaliana* *S. cerevisiae*	*A. thaliana* *N. tabacum*	reduction of Fe(III) to Fe (II) on root surface	hydropony

[a]MT = metallothionein, [b]M+ = cation, [c]ACC = aminocyclopropane-carboxylic acid

Table 3. Examples of GM plants with increased ability to transform organic pollutants.

Transgene	Product	Source	Transgenic plant	Aim/effect	Cultivation medium
PETNred (French, et al., 1999)	pentaerythritol tetranitrate reductase	*Enterobacter cloacae* PB2	*N. tabacum*	degradation of explosives, ability to grow on them	agar
cyp2E1 (Kellner, et al., 1997)	dehalogenase cytP450 2E1	mammalian	*N. tabacum*	640 x higher TCE oxidation debromination of EDB	hydropony
bph C (Francova, et al., 2003)	dihydroxy-biphenyl-dioxygenase	*Comamonas testosteroni*	*N. tabacum*	degradation of PCB hydroxyderiva-tives, increase of resistance	agar

2.4. Examples of Gm Plants Tailored for Phytoremediation

2.4.1. Phytoremediation of Mercury-Contaminated Environment

A special and interesting approach was chosen for removal of mercury from contaminated environment. In former times mercury used to be widely exploited in chemical industry, paper or mining industries. Due to limited knowledge of the toxicity of mercury and its potential effect on the environment the wastewaters from production facilities were often released into nearby water resources. This caused e.g. two wide intoxication cases in Japan in the sixties.

Clinical research shows that organic derivatives like the cation of methylmercury (CH_3Hg^+), are the most often species of Hg accumulating in fish and the consumers of fish, causing neurodegenerative symptoms (Heaton, et al., 1998). Methylmercury belongs to relatively rare wastes, but it can be a product of microbial conversion of inorganic mercury. The cation CH_3Hg^+ is in sediments bound in complexes with organic and inorganic ligands less strongly than the salt of Hg(II) and for this reason it has better biological availability (Rugh, et al., 1998). In addition, it is more soluble in lipids, what adds to its toxicity.

Due to the high price of excavation and storage of dangerous sediments relatively few places contaminated by mercury has been remediated.

Gram-negative bacteria growing in Hg-contaminated soil developed a resistance towards mercury, which is genetically coded by the *mer* operon (Bizily, et al., 2000). This operon contains about 5 to 6 genes coding for sensitive regulatory proteins, transport proteins, organomercury lyase, and a Hg-reductase. Within this operon, genes *merA* a *merB* are present, coding for enzymes catalysing the conversion of mercury. *MerB* is coding organomercury lyase, which is responsible for the cleavage of of the carbon-Hg bond in many organic compounds of mercury, e.g. in methylmercury acetate or phenylmercury. The product of this reaction, salt of Hg(II), is the substrate of the enzyme called reductase of Hg ions, coded by *merA*. This enzyme is a soluble, NADPH-dependent, disulphide oxidoreductase containing FAD, which catalyses the reduction of Hg(II) to the less toxic Hg(0) (see Scheme 1).

Scheme 1

$$R\text{-}CH_2\text{-}Hg^+ + H^+ \rightarrow R\text{-}CH_3 + Hg^{2+} \qquad \textit{merB}$$

$$Hg^{2+}\ NADPH \rightarrow Hg^0 + NADP^+\ H^+ \qquad \textit{merA}$$

These reactions decrease the relative toxicity of the metal and allow its evaporation from the system. The use of bacteria is not effective on large contaminated areas and for this reason approaches based on exploitation of GM plants have been designed for remediation.

During preparation of such plants, the expression of *merA* from bacterial transposon Tn21 in plants was the first step, as discussed by Meagher and coworkers (Macek, et al., 1996; Bizily, et al., 2000). Even using a very effective plant expression system, no *merA*-protein has been detected. The original bacterial *merA* is rich in GC (67%) in comparison with plant genes, and in addition the *merA* has many CpG dinucleotide sequences, which are potential sites of DNA methylation followed by gene silencing. For this reason, Rugh et al. (1998) constructed a mutant *merA* with a modified coding sequence (*merA9*) and introduced it into *Arabidopsis thaliana* plant. The seeds germinated and seedlings were able to grow in media containing up to 100 μM of mercury ($HgCl_2$). Transgenic plants released between two to three times more elemental mercury Hg^0 than the control plants. The plants proved to be resistant to a toxic concentration of compounds Au^{3+}. To test the effectivity of using transgenic plants for phytoremediation of Hg contaminated soil the yellow poplar has been chosen (*Liriodendron tulipifera*) due to its promising biological and structural characteristics (Rugh, et al., 1998). It was proved that transgenic poplar is releasing ten times more elemental Hg^0 in comparison with then control plant. Although this system was not tested under field conditions, the results clearly show that genetic engineering is able to increase the ability of plants to remove undesired compounds from the soil. Because the volatilisation, release of mercury into air is for ecological reasons undesirable, further plants were prepared that are able to accumulate the mercury instead of releasing it (Bizily, et al., 2000).

Volatilisation of selenium was chosen as the aim of research in California, this time with expected positive effect of evaporated selenium (Berken, et al., 2002).

2.4.2. Increased Accumulation of Heavy Metals

Among other natural mechanisms allowing plants to grow in heavy metal contaminated environment especially many studies dealt recently with the phytochelatins. In plant kingdom it is a general feature to complex heavy metals by phytochelatins. These peptides are formed by phytochelatin synthase from glutathione, when plant are challenged by Cd or other heavy metals. Much effort to explanation of their formation has been devoted by Zenk and co-workers (e.g. Chasaigne, et al., 2001; Oven, et al., 2001).

Different species of plants have been described to be able to accumulate substantial amounts of heavy metals (Salt, 1995). These so-called hyper-accumulators are able to accumulate many times a higher amount of heavy metals than do other species growing under the same conditions at the same site. Hyperaccumulators, found usually at places where the metal ore with high metal content is surfacing (e.g. Congo, New Caledonia), are unfortunately characterised by a relatively low biomass formation and that is why they are not well suited for phytoremediation purposes. The mechanisms involved in hyperaccumulation are not yet fully understood, but already attempts were made to increase the accumulation of heavy metals in plants forming high amounts of biomass (Kraemer and Chardonnens, 2001; Clemens, et al., 2002). Morel et al. (1999) have discussed some important aspects of plant - soil interaction with respect to hyperaccumulation. Many data are gradually collected on the role of peptides and proteins in metal accumulation (Kotrba, et al., 1999; Mejare and Bulow, 2001).

With the aim to improve metal uptake transgenic plants were prepared with increased formation of glutathione synthase or phytochelatin synthase, what in both cases led to increased Cd accumulation (Karenlampi, et al., 2000). Transgenic plants bearing foreign genes for proteins transporting metal across membranes have been also prepared, as summarised in recent review (Kraemer and Chardonnens, 2001), see Table 2. In some laboratories were into plants introduced genes coding different types of metallothionein (mammalian, yeast, insect and human), this effort led mostly to increase of the resistance towards some heavy metals (Karenlampi, et al., 2000; Macek, et al., 1996; Macek, et al., 1997), but not yet to an increase of accumulation (Liu, et al., 2000).

In order to increase the metal-binding capacity we studied the possibility of introducing an additional metal/binding domain with high affinity towards heavy metals into the implemented protein (Macek, et al., 2002). In this way by combining the gene *CUP1* for yeast metallothionein from *Saccharomyces cerevisiae* and the gene for the histidine anchor from the commercial plasmid pTrcHis (Invitrogene) a construct has been prepared and introduced by *A. tumefaciens* into tobacco (Macek, et al., 1997; Macek, et al., 1996). The comparison of genetically modified lines of tobacco with control plants showed an increase of resistance and good growth in cadmium contaminated soil, together with increased accumulation of cadmium during cultivation in hydropony and in sand with cadmium (190% of the control), as described in Macek et al. (2002).

The importance of improvement of metal uptake by breeding or genetic modification can be illustrated by the fact, that simultaneously appeared more important and interesting reviews on engineering GM plants suitable for metal accumulation, e.g. in Trends in Plant Science (Clemens, et al., 2002), and in Critical Reviews in Plant Science (Pilon-Smith and Pilon, 2002). These works cover practically all aspects of metal phytoremediation, especially also the basic research questions and biological mechanisms (Lasat, 2000), often omitted in earlier papers. Reserves are not only in better understanding, possibility of modifications and improvements of normal plant mechanisms themselves, but also in other ways to improve the overall yield of the remediation process. Improved bioavailability of metals caused by changes of exudate, increased excretion of organic acids, or cooperation with rhizospheric microorganisms might be itself target of genetic engineering, or increase the performance of transgenic plants. Our very early screening of different plant species for production of phytosiderofores (Rudolph, et al., 1985) showed on the example of nicotianamine, that their presence is ubiquitous in higher plants, might be increased or alternatively produced by proper bacteria.

2.4.3. Biodegradation of Explosives

Another problem of interest, often discussed in literature, is the contamination of the environment by explosives. The most widespread are 2,4,6-trinitrotoluene (TNT), hexahydro-1,3,5-trinitro-1,3,5-triazine (RDX), oktahydro-1,3,5,7-tetranitro-1,3,5,7-tetraazocine (HMX) and their aminoderivatives (Rudolph, 1985). Studies showed that plants are able to transform TNT without the help of microorganisms. Most of the research was targeted onto TNT, but it is very rarely present at contaminated sites as a single contaminant alone.

Some experiments dealt with the abilities of water plants *Myriophyllum spicatum*, *Myriophyllum aquaticum* and of hairy root cultures of *Catharanthus roseus* to absorb and transform TNT (Rivera, et al., 1998). It was proved that these plants are able to transform TNT fast and especially without accumulation of aminonitrotoluenes or azoxy dimers. Due to this fact most of the studies were aimed at establishing the time interval necessary for an effective transformation of TNT and identification of its side products (Rudolph, et al., 1985). In parallel, also the kinetic parameters of the degradation of aminodinitrotoluene (ADNT–2 isomers) to diaminonitrotoluene (DANT–2 isomers) and other products, probably triaminotoluene (TAT) have been established.

Bruce with co-workers studied the degradation of explosive residues by soil bacteria *Enterobacter cloacae* PB2 and proved its ability to utilise nitrates of explosives like pentaerythritoltetranitrate (PETN) and glyceroltrinitrate (GTN) as the sole nitrogen source for their growth, French et al. (1999). Denitrification is effective due to the NADPH-dependent PETN-reductase. Studies reveal that PETN-reductase is able to degrade aromatic explosives as 2,4,6-trinitrotoluene, the most ubiquitous and recalcitrant pollutant of the military grounds. Already for a longer time period it is known that plant cells are able to denitrify GTN. It is expected that denitrification is performed by a glutathione-dependent mechanism, similarly to the mammalian cells. Transgenic tobacco (*Nicotiana tabacum* cv. xanthi) bearing the bacterial PETN-reductase was in comparison with control plants able to germinate and grow on media containing 1 mM GTN or 0,05 mM TNT. Even if no TNT PETN reductase products have been identified, it is obvious that these compounds are less inhibitory for the germinating seed than are the products of the normal TNT metabolism in plants (aminodinitrotoluenes and diaminodinitrotoluenes), which are more toxic than the starting compound itself.

From the above it follows the conclusion that the expression of PETN reductase in plants might lead to detoxification of TNT residues. It can be also expected, that the expression of PETN reductase in plants may increase their resistance towards nitroaromatic herbicides, what could be of great importance for the agriculture.

2.4.4. The Exploitation of Cytochrome P450 for Bioremediation

Cytochrome P450 is known for its ability to participate in oxidation of a wide range of compounds, including different xenobiotics (Newman, et al., 1998). Cytochrome P450 is present in microorganisms, as well as in plants and animals, including mammals, and is involved in many types of chemical transformations, like aliphatic hydroxylation, epoxidation, dealkylation, dehalogenation and other different mechanisms leading to inactivation and detoxification. Many of these reactions are a key to successful bioremediation.

Microbial P450 are especially useful for biodegradation purposes, because these enzymes are ubiquitous and have a strong reduction and oxidation potential. Recent studies proved the ability of cytochromes P450 in natural conditions to degrade compounds like atrazine, thiocarbamate herbicides and haloalkanes.

The ability of cytochrome P450 cyp 2E1 to oxidise trichloroethylene (TCE), tetrachloromethane (TCM), dibromoethylene (DBE) and some other chlorinated compounds has been described (Kellner, et al., 1997). TCE is oxidised by cyp 2E1 to chloral, which is further metabolised to trichloroethanol, trichloroacetic acid, and partly to carbon dioxide, chloride ion and water. This knowledge led to introduction of the gene for cyp 2E1 into tobacco and poplar. To ensure the activity and stability of the enzyme, together with this gene were introduced genes for oxidoreductase and cytochrome B5. These three proteins form a complex in cellular microsomes. The experiments with hybrid poplars as axenic tumor cultures of poplar cells, and in greenhouse and field studies already showed, that this species can metabolise TCE and other compounds (Gordon, et al., 1998), thus allowing optimism concerning also the GM plants.

2.4.5. PCBs and GM Remediation

Polychlorinated biphenyls represent another important global problem. Even now, 35 years after their production was banned in civilised countries, and 18 years after finished in former Czechoslovakia, these persistent compounds still represent a danger for biological diversity and for man. Being taken up by plants, PCBs can enter food chain and accumulate especially in fat tissues. The ability of plants to metabolise them has been already proved, and is further studied (Mackova, et al., 2001). In our laboratories, we have a quite complex approach to PCB bioremediation (Kucerova, et al., 2000; Demnerova, et al., 2002), which includes studies on enzymes involved (Chroma, et al., 2003), products formed and attempts to introduce bacterial genes from the bph operon (Francova, et al., 2003).

Rice plants have been prepared expressing chlorocatechol dioxygenase from *Ralstonia eutropha* (Shimizu, et al., 2002). Callus cultures of this GM plant were able to convert 3-chlorocatechol to 3-chloromuconate.

Worldwide many laboratories deal with the performance of organisms present in the rhizosphere of plants in contaminated soil. Studies were performed to improve the rhizospheric cooperation between plants and microorganisms in PCB degradation (Brazil, et al., 1995), by construction of a rhizosphere Pseudomonad with potential to degrade PCBs and detection of *bph* gene expression in the rhizosphere.

2.4.6. Other Approaches to Improve the Effectivity of Phytoremediation Process

Genetically modified symbiotic bacteria. Here we should mention construction of a novel system, based on symbiosis between leguminous plants and genetically engineered rhizobia (Sriprang, et al., 2002). The authors introduced gene for human metallothionein MTL4 into rhizobium strain forming nodules on *Astragalus* plant roots. The symbionts increased

Cd accumulation in nodules up to two-fold, but no significant increase in Cd accumulation was noted.

Studies on GM organisms involved recently also other approaches to genetic manipulation of the properties of phytoremediation system, e.g. growth promotion for better biomass formation. Glick and co-workers (Saleh and Glick, 2001; Ma, et al., 2002) manipulated for this reason plant growth promoting bacteria *Enterobacter cloacae* using genes from *Pseudomonas fluorescens*, in order to facilitate the growth of plants. Factors tested included indoleacetic acid production, antibiotic production, 1-aminocyclopropane-1-carboxylic acid deaminase activity (important for decreasing the root growth inhibiting ethylene production), and siderophore production. Overproduction of the genes in bacteria inoculated into rhizosphere of canola resulted in significant increase in root and shoot elongation, thus increasing promises towards exploitation in phytoremediation.

Mycorrhizal symbiosis represents another very important factor influencing the behaviour of plants in contaminated soil. Over 80% of land vascular plants can establish symbiotic associations with arbuscular mycorrhizal fungi (AMF) (Jolicoeur, 2002). Such a symbiosis has an important effect on nutrient uptake by plants (e.g. phosphorus), especially under starvation stress. It is expected that establishing symbiosis of plants and AMF in contaminated soil will support the plant growth and survival rate, still an open question is if root colonisation will lead to increased accumulation trace elements.

One of the approaches to evaluate the effect of mycorrhizal fungi on phytoremediation process, in case of cadmium accumulation, represents inoculation of transgenic tobacco characterised by increased Cd uptake, by mycorrhizal fungus (Vosatka and Janouskova, 2003). Such pot experiments were performed in sand with nutrients and Cd, and in real Cd contaminated soil sterilised by gamma-irradiation, using control tobacco and transgenic tobacco with a polyhistidine chain in fusion with yeast metallothionein (Macek, et al., 2002), with and without colonisation by *Glomus intraradices*. While phosphorus uptake was much increased in the presence of root colonisation by *G. intraradices*, both in transgenic and non-transgenic tobacco, Cd uptake and accumulation in above-ground parts was suppressed by the presence of AMF *G. intraradices* in this model system (Vosatka and Janouskova, 2003).

Roots and hairy roots, formed as the result of plant infection by *Agrobacterium rhizogenes*, are an additional example of innovative approach in screening for plants suitable for phytoremediation, and for detailed studies of xenobiotics metabolism. The excellent monography of Doran (Doran, 1997) covers all aspects of studies and exploitation of hairy roots. In last years more reviews followed, but application in phytoremediation studies has been discussed only in some of them (Pletsch, et al., 1999; Gleba, et al.,

1999). The exploitation of hairy root cultures in phytoremediation studies is known more than a decade; we used them for cadmium uptake studies (Macek, et al., 1994; Macek, et al., 1997) and for PCB degradation studies (Mackova, et al., 1997).

Recently very interesting results have been obtained using hairy root cultures derived from hyperaccumulator plants; a range of *Agrobacterium rhizogenes* transformed hairy roots was tested and compared with hairy roots of closely related non-hyperaccumulator plant species. Using long-term hairy root cultures the authors on demonstrated *Alyssum* species that hyperaccumulation does not necessarily depend on the presence of shoots or root-shoot transfer (Nedelkoska and Doran, 2002). A choice of available unique hairy root cultures of different hyperaccumulating species and their non-hyperaccumulator relatives allows important comparison and selection of traits responsible for high metal accumulation.

The importance of plant roots for successful performance of phytoremediation has been many times advocated by Fletcher and co-workers. The authors also discuss the potential use of mycorrhizal fungi, effect of plant compounds on PCB-degrading bacteria, stimulation of these bacteria by compounds released from plant roots (Donnely, et al., 1994; Fletcher, et al., 1995). Recent review of Gleba and co-workers (Gleba, et al., 1999) is cowering specific aspects of plant roots exploitation in phytoremediation, their use in hydropony, and their genetic manipulation for release of compounds into exudate.

Secondary plant metabolites and their role in phytoremediation are discussed in (Singer, et al., 2003). Concerning the metabolism of organic xenobiotic compounds, the whole process is based on mechanism originally developed to cope with different allelopathic compound of natural origin. The content of root exudate (lactate, acetate, oxalate, succinate, fumarate, malate, citrate, isocitrate, aconitate, sugars, amino acids) and compounds released from dying and decaying roots (Fletcher, et al., 1995) represent a waste array of molecules that might be involved directly in metal uptake, and stimulation of bacteria. Plant secondary metabolites are discussed especially with respect to their possibility to stimulate microbial degradation of PAHs and PCBs (Singer, et al., 2003). Biosynthesis of such compounds could be also a suitable target for genetic manipulation.

3. ADVANTAGES AND DISADVANTAGES OF GMO USE

There are many reasons for preparation and exploitation of transgenic plants. On the first place, it is the possibility of increasing yields of agricultural crops without needing larger areas, and preparation of varieties more resistant against external damage, including pests. Further it is the increase in nutritive value of food derived from plants, production of novel compounds (monoclonal antibodies, vaccines, enzymes, proteins like insulin or interferon), and remediation of the soil contaminated by different pollutants and xenobiotics (Koprowski and Yusibov, 2001; Ondrej, et al., 1999; van der Lelie, et al., 2001). The removal of contaminants is the basic requirement for preventing toxic compounds entering the food chain, what is the most important factor in maintaining human health and biological diversity. From this overview, it can be clear that some transgenic plants already found practical applications and supporters. The opposers of GMO usually claim the following disadvantages:

1) Due to the complexity of the plant genome, it is impossible to predict all changes that might happen and their outcomes.
2) The possibility of horizontal gene transfer cannot be totally excluded as well as the formation of superweeds resistant to insect pests and viruses.
3) The need of application of increasing amounts of herbicides due to the formation of new plant varieties resistant towards usual herbicides.
4) Development of allergic reactions to GMO containing products.
5) Replacement of the rich endemic flora by monocultures, loss of biodiversity.
6) Induction of unwanted resistance to antibiotics in man, animals, eventually microorganisms.
7) Effort of many GMO-producing companies to push forward their interests over all the negative effect, which might arise from the introduction of GMOs.

Most of the above fears and arguments can be controverted after appropriate analysis of the problem and detailed discussion. The advantages of GMOs (Macek, et al., 2000; Macek, et al., 2002; Schnoor, et al., 1995) can be summarised as follows:

1) Increase of the growth and production of usable plant product.
2) Increase of the quality including nutritive and storage properties.
3) Increase of the ability to adapt to specific environmental conditions causing stress, including environments contaminated by inorganic and organic pollutants.
4) Increased resistance to diseases and plant pests.
5) Lower consumption of agrochemicals and reduction of improper approaches used in agricultural practice that endanger the environment. This fact is on the other hand endangering the profits of companies producing herbicides and pesticides.
6) Production of novel compounds.
7) Exploitation of new raw materials.
8) The possibility of much more effective phytoremediation on larger scale.

4. CONCLUSIONS

Due to different accidents and non-careful production technologies, the contamination of the environment is increasing faster than it can be removed. For this reason new remediation techniques are needed, approaches effective but economically acceptable so that they might be exploited on a large enough scale (van der Lelie, et al., 2001). Classical chemical and physical methods cannot be used for economical reasons e.g. for remediation of large areas of agricultural land, from where the toxic compounds are entering directly the food raw materials. The use of plants and their root system in removal of pollutants and xenobiotics from the soil, water, sediments and the air, phytoremediation and rhizoremediation, is one of the approaches that might support the solution of this problem. Most aspects of exploitation of plants for heavy metal removal from contaminated environment are discussed in other chapters of this volume. We tried to concentrate on the hopes and promises connected with the introduction of genetically modified organisms into phytoremediation. According to the recent results, the application of transgenic plants might represent an important part of the mentioned techniques (Raskin, 1996; Bock, et al., 2002). Other discussed approaches might increase the effectivity of the process, and probably concerted action of more methods together will allow to exploit the real potential of transgenic plants, e.g. by increasing of bioavailability of the metal in case of lead using biodegradable chelates (Kos and Lestan, 2003). Only the decreasing of the amount of environmental contamination can prevent the toxic compounds entering the food chain, help to conserve the biological diversity (which is surely more endangered by the contaminated environment than by the use of genetically modified plants) (Macek, et al., 2002), and this way contribute to the sustainable development.

5. ACKNOWLEDGEMENT

The authors are thankful for support of their research by grant 526/02/0293 of the Grant Agency of the Czech Republic, research projects J19/98:2232500003 and Z4-055-905, and EU 5th FW grant QLK3-CT-2001-00101.

REFERENCES

1. Barrett, C., Cobb, E., McNicol, R. and Lyon, G. (1997) A risk assessment study of plant genetic transformation using *Agrobacterium* and implications for analysis of transgenic plants, *Plant Cell Tissue Organ Cult.* 47, 135-144.

2. Berken, A., Mulholland, M.M., LeDuc, D.L. and Terry, N. (2002) Genetic engineering of plants to enhance selenium phytoremediation, *Crit. Reviews Plant Sci.* 21 (6), 567-572.
3. Bizily, S., Rugh, C. and Meagher, R. (2000) Phytodetoxification of hazardous organomercurials by genetically engineered plants, *Nat. Biotech.* 18, 213-217.
4. Bizily, S., Rugh, C., Summers, A.O. and Meagher, R. (1999) Phytoremediation of methylmercury pollution: *merB* expression in *Arabidopsis thaliana* confers resistance to organomercurials, *Proc. Natl. Acad. Sci. USA*, 96, 6808-6813.
5. Bock, C., Kolb, M., Bokern, M., Harms, H., Mackova, M., Chroma L., Macek, T., Hughes, J., Just, C. and Schnoor, J. (2002) Advances in phytoremediation: Phytotransformation, in D. Reible and K. Demnerova, (eds.), *Inovative Approaches to the On-Site Assessment and Remediation of Contaminated Soils*, Kluwer Academic Publishers, Dordrecht, pp. 115-140.
6. Brazil, G.M., Kenefick, L., Callahan, M., Haro, A., de Lorenzo, V., Dowling, D.N. and O'Gara, F. (1995) Construction of rhizosphere Pseudomonad with potential to degrade polychlorinated biphenyls and detection of *bph* gene expression in the rhizosphere, *Appl. Environ. Microbiol.* 61, 5, 1946-1952.
7. Chaney, R.L., Malik, M., Li, Y.M., Brown, S.L., Angle, J.S. and Baker, A.J.M. (1997) Phytoremediation of soil metals, *Curr. Opin. Biotechnol.* 8, 279-284.
8. Chasaigne, H., Vacchina, V., Kutchan, T., Zenk, M.H. (2001) Identification of phytochelatin-related peptides in maize seedlings exposed to cadmium and obtained enzymatically *in vitro*, *Phytochemistry* 56, 7, 657-668.
9. Chroma, L., Moeder, M., Kucerova, P., Macek, T., Mackova, M. (2003) Plant enzymes in metabolism of polychlorinated biphenyls, *Fresenius Environmental Bull.* 12, 2, in press.
10. Clemens, S., Palmgren, M., Kraemer, U. (2002) A long way ahead: understanding and engineering plant metal accumulation, *Trends in Plant Sciences* 7, 7, 309-315.
11. Cunningham, S., Berti, W. and Huang, J. (1995) Phytoremediation of contaminated soils, *TIBTECH* 13, 393-397.
12. Demnerova, K., Mackova, M., Pazlarova, J., Vosahlikova, M., Novakova, H., Jindrova, E., Ryslava, E., Macek, T., Vrchotova, N., Brenner, V., Pavlu, L., Totevova, S., Kristoffer, T., Focht, D.D., Fava, F., Di Gioia, D., Marchetti, L., Fletcher, J.S., Leigh, M.B., Kucerova, P., Stiborova, H., Mateju, V., Sobotka, M., Kastánek, F., Kastánek, P. and Kasak, L. (2002) PCB - approaches to the removal from the environment, in D. Reible and K. Demnerova, (eds.), *Inovative Approaches to the On-Site Assessment and Remediation of Contaminated Soils*, Kluwer Academic Publishers, Dordrecht, pp. 217-263.
13. Donnely, P.K., Hedge, R.S., Fletcher, J.S. (1994) Growth of PCB-degrading bacteria on compounds from photosynthetic plants, *Chemosphere* 28, 984-988.
14. Doran, P.M. (ed.) (1997) *Hairy roots: Culture and Application*, Harwood Academic Publishers, London.
15. Flavell, R.B. (1995) Plant biotechnology R & D - the next ten years, *TIBTECH* 13, 313-323.
16. Fletcher, J.S., Donnely, P.K., Hedge, R.S. (1995) Biostimulation of PCB-degrading bacteria by compounds released from plant roots, in: R.E. Hinchee, D.B. Anderson and R.E. Hoeppel (eds.), *Bioremediation of recalcitrant organics*, Batelle Press, Columbus, pp. 131-136.
17. Francová, K., Macek, T., Demnerova, K. and Mackova, M. (2001) Transgenic plants – potential tool for the decontamination of the environment, *Chemické Listy* 95, 630-637.
18. Francova, K., Sura, M., Macek, T., Szekeres, M., Bancos, S., Demnerova, K., Sylvestre, M. and Mackova, M. (2003) Generation of plants carrying bacterial enzymes for degradation of polychlorinated biphenyls, *Fresenius Environmental Bull.*, 12, in press.

19. French, C.E., Rosser, S.J., Davies, G.J., Nicklin, S. and Bruce, N.C. (1999) Biodegradation of explosives by transgenic plants expressing pentaerythritol tetranitrate reductase, *Nat. Biotechnol.* 17, 491-494.
20. Gelvin, S.B. (1998) The introduction and expression of transgenes in plants, *Curr. Opinion Biotechnol.* 9, 227-235.
21. Gleba, D., Borisjuk, N.V., Borisjuk, L.G., Kneer, R., Poulev, A., Skarzhinskaya, M., Dushenkov, S., Logendra, S., Gleba, Y.Y., Raskin, I. (1999) Use of plant roots for phytoremediation and molecular farming, *Proc. Natl. Acad. Sci. USA* 96, 5973-5977.
22. Gordon, M., Choe, N., Duffy, J., Ekuan, G., Heilman, P., Muiznieks, I., Ruszaj, M., Shurtleff, B.B., Strand, S., Wilmroth, J. and Newman, L. (1998) Phytoremediation of trichloroethylene with hybrid poplars, *Environ. Health Perspect.* 106 (Suppl. 4), 1001-1004.
23. Grichko, V.P., Filby, B. and Glick, B.R. (2000) Increased ability of transgenic plants expressing the bacterial enzyme ACC deaminase to accumulate Cd, Co, Cu, Ni, Pb, and Zn, *J. Biotechnol* 81, 45-54.
24. Heaton, A.C.P., Rugh, C.L., Wang, N. and Meagher, R.B. (1998) Phytoremediation of mercury and methylmercury polluted soil using genetically engineered plants, *J. Soil Contamin.* 7, 497-509.
25. Jolicoeur, M., Bouchard-Marchand, E., Becard, G., Perrier, M. (2002) Regulation of mycorrhizal symbiosis: development of a structured nutritional dual model, *Ecological Modelling* 158, 121-142.
26. Karenlampi, S., Schat, H., Vangronsveld, J., Verkleij, J.A.C., van der Lelie, D., Mergeay, M. and Tervahauta, A.I. (2000) Genetic engineering in the improvement of plants for phytoremediation of metal polluted soils, *Environ. Pollut.* 107, 225-231.
27. Kellner, D.G., Maves, S.A. and Sligar, S.G. (1997) Engineering cytochrome P450s for bioremediation, *Curr. Opinion Biotechnol.* 8, 274-282.
28. Koprowski, H. and Yusibov, V. (2001) The green revolution: plants as heterologous expression vectors, *Vaccine* 19, 2735-2740.
29. Kos, B., Lestan, D. (2003) Induced phytoextraction/soil washing of lead using biodegradable chelate and permeable barriers, Environmental Sci. Technol.
30. Kotrba, P., Macek, T., Ruml, T. (1999) Heavy-metal binding peptides and proteins in plants. A review, *Coll. Czech Chem. Commun.* 64, 1057-1086.
31. Kraemer, U. and Chardonnens, A. (2001) The use of transgenic plants in the bioremediation of soils contaminated with trace elements, *Appl. Microbiol. Biotechnol.* 55, 661-672.
32. Kucerova, P., Mackova, M., Chroma, L., Burkhard, J., Triska, J., Demnerová, K. and Macek, T. (2000) Metabolism of polychlorinated biphenyls by *Solanum nigrum* hairy root clone SNC-9O and analysis of transformation products, *Plant and Soil*, 225, 109-115.
33. Lasat, M.M. (2000) Phytoextraction of toxic metals: A review of biological mechanisms, *Journal of Environmental Quality* 31, 109-125.
34. Liu, J.R., Suh, M.Ch., Choi, D. (2000) Phytoremediation of cadmium contamination: Overexpression of metallothionein in transgenic tobacco plants, *Bundesgesundheitsbl-Gesundheitsforsch-Gesundheitsschutz* 2, 126-130.
35. Ma, J.K., Hiatt, A., Hein, M., Vine, N.D., Wang, F., Stabila, P., van Dolleweerd, C., Mostov, K. and Lehner, T. (1995) Generation and assembly of secretory antibodies in plants, *Science* 268, 716-719.
36. Ma, W., Penrose, D.M., Glick, B.R. (2002) Strategies used by rhizobia to lower plant ethylene levels and increase nodulation, *Can. J. Microbiol.* 48, 947-954.
37. Macek T., Macková M., Pavlíková D., Száková J.,. Truksa M, Cundy A.S., Kotrba P., Yancey N., Scouten W.H. (2002) Accumulation of cadmium by transgenic tobacco, *Acta Biotechnol.* 22, 2, 101-106.
38. Macek, T., Kas, J. and Mackova, M. (2000) Exploitation of plants for the removal of organics in environmental remediation, *Biotechnol. Advances* 18, 23-34.

39. Macek, T., Kotrba, P., Ruml, T., Skacel, F., Mackova, M. (1997) Accumulation of cadmium by hairy root cultures, in: P.M. Doran, (ed.) *Hairy roots: Culture and Application*, Harwood Academic Publishers, London, pp. 133-138.
40. Macek, T., Kotrba, P., Suchova, M., Skacel, F., Demnerova, K., Ruml, T. (1994) Accumulation of cadmium by hairy root cultures, *Biotechnol. Lett.* 16, 6, 621-624.
41. Macek, T., Macková, M., Burkhard, J. and Demnerová, K. (1998) Introduction of green plants for the control of metals and organics in environmental remediation, in F.W. Holm (ed), *Effluents from Alternative Demilitarization Technologies*, NATO PS Series 1, 12, Kluwer Academic Publishers, Dordrecht, Boston, London, pp. 71-85.
42. Macek, T., Mackova, M., Kotrba, P., Truksa, M., Singh-Cundy, A., Scouten, W.H. and Yancey, N. (1997) Attempts to prepare transgenic tobacco with higher capacity to accumulate heavy metals containing yeast metallothionein combined with a polyhistidine, in H. Verachtert and W. Verstraete (eds.), *Environmental Biotechnology*, Technological Institute Gent, pp. 263-266.
43. Macek, T., Mackova, M., Kucerova, P., Chroma, L., Burkhard, J. and Demnerova, K. (2002) Phytoremediation, in M. Hofman and J. Anne (series eds.), *Focus on Biotechnology*, Vol. 3, Kluwer Academic Publishers, Dordrecht, pp 115-137.
44. Macek, T., Mackova, M., Truksa, M., Singh-Cundy, A., Kotrba, P., Yancey, N. and Scouten, W.H. (1996) Preparation of transgenic tobacco with a yeast metallothionein combined with a polyhistidine tail, *Chem. Listy* 90, 690.
45. Mackova, M., Chroma, L., Kucerova, P., Burkhard, J., Demnerova, K. and Macek, T. (2001) Some aspects of PCBs metabolism by horseradish cells, *International Journal of Phytoremediation* 3 (4), 401-414.
46. Mackova, M., Macek, T., Kucerova, P., Burkhard, J., Pazlarova, J., Demnerova, K. (1997) Degradation of polychlorinated biphenyls by hairy root culture of *Solanum nigrum*, *Biotechnol. Lett.* 19, 8, 787-790.
47. Maliga, P. (1990) Transformation of *N. tabacum* by *Agrobacterium* and particle gun, *Molecular and Developmental Biology of Plants*, Cold Spring Harbor Course.
48. Mason, H. and Arntzen, C. (1995) Transgenic plants as vaccine production systems, *Science* 268, 714-716.
49. McGarvey, B., Hammond, J., Dienet, M., Hooper, D., Fu, Z., Dietzschold, B., Koprowski, H. and Michaels, F. (1995) Expression of the rabies virus glycoprotein in transgenic tomatoes, *Bio/Technology* 13, 1484-1489.
50. Meagher, R.B. (2000) Phytoremediation of toxic elemental and organic pollutants, *Current. Opinion Plant Biol.* 3, 153-162.
51. Mejare, M. and Bulow, L. (2001) Metal-binding proteins and peptides in bioremediation and phytoremediation of heavy metals, *Trends in Biotechnology* 19, 2, 67-73.
52. Meyers, R.A. (1995) *Molecular biology and biotechnology*, Verlag Chemie Publishers, Inc., New York.
53. Morel, J.L., Chaineau, C.H., Schiavon, M., Lichtfouse, E. (1999) The role of plants in the remediation of contaminated soils, in: Ph. Baveye et al. (eds.), *Bioavailability of organic xenobiotics in the environment*, Kluwer Academic Publishers, Dordrecht, pp. 429-449.
54. Nedelkoska, T.V., Doran, P.M. (2002) Hyperaccumulation of nickel by hairy root of Alyssum species: Comparison with whole regenerated plants, *Biotechnology Progress* 17, 4, 752-759.
55. Newman, L.A., Doty, S.L., Gery, K.L., Heilman, P.E., Muiznieks, I., Shang, T.Q., Siemieniec, S.T., Strand, S.E., Wang, X., Wilson, A.M. and Gordon, M.P. (1998) Phytoremediation of organic contaminants: a review of phytoremediation research at the University of Washington, *J. Soil Contamin.* 7, 531-542.
56. Old, R.W. and Primrose, S.B. (1994) *Gene transfer to plants. Principles of Gene Manipulation*, Blackwell Scientific Publications.
57. Ondrej, M., Drobnik, J., Gartland, K.M.A. and Gartland, J.S. (1999) *Gene engineering*, (in Czech, textbook of TEMPUS PHARE programme), ICT Prague.

58. Oven, M., Raith, K., Neubert, R.H.H., Kutchan, T.M., Zenk, M.H. (2001) Homo-phytochelatins are synthesized in response to cadmium in Azuki beans, *Plant Physiol.* 126, 3, 1275-1280.
59. Pilon-Smith, E. and Pilon, M. (2002) Phytoremediation of metals using transgenic plants, *Crit. Reviews Plant Sci.* 21 (5), 439-456.
60. Pletsch, M., Santos de Araujo, B., Charlwood, B.V. (1999) Novel biotechnological approaches in environmental remediation research, *Biotechnology Advances* 17, 679-687.
61. Poirier, Y., Dennis, D., Klomparens, K. and Somervill, C. (1992) Targeting of the polyhydroxybutyrate biosynthetic pathway to the plastids of *Arabidopsis thaliana* results in high levels of polymer accumulation, *Science 256*, 520-522.
62. Raskin, I., (1996) Plant genetic engineering may help with environmental cleanup, *Proc. Natl. Acad. Sci. USA* 93, 3164-3166.
63. Rivera, R., Medina, V.F., Larson, S.L. and McCutcheon, S.C. (1998) Phytotreatment of TNT/contaminated groundwater, *J. Soil Contamin.* 7, 511-529.
64. Rudolph, A., Becker, R., Scholz, G., Procházka, Z, Toman, J., Macek, T., Herout, V. (1985) The occurence of the amino acid nicotianamine in plants and microorganisms. A reinvestigation, *Biochem. Physiol. Pflanzen* 180, 557-563.
65. Rugh, C.L., Wilde, H.D., Stack, N.M., Thompson, D.M., Summers, A.O. and Meagher, R. B. (1996) Mercuric ion reduction and resistance in transgenic *Arabidopsis thaliana* plants expressing a modified bacterial *merA* gene, *Proc. Natl. Acad. Sci.* 93, 3182-3187.
66. Rugh, C., Senecoff, J., Meagher, R. and Merkle, S. (1998) Develoment of transgenic yellow poplar for mercury phytoremediation, *Nat. Biotech.* 16, 925-928.
67. Ryslava, E., Krejcik, Z., Macek, T., Novakova, H., Demnerova, K., Mackova, M. (2003) Study of PCB degradation in real contaminated soil, *Fresenius Environmental Bull.*, 12, in press.
68. Saleh, S.S. and Glick, B.R. (2001) Involvement of *gac*S and *rpo*S in enhancement of the plant growth-promoting capabilities of *Enterobacter cloacae* CAL2 and UW4, *Can. J. Microbiol.* 47, 698-705.
69. Salt, D., Blaylock, M., Kumar, N., Duschenkov, V., Ensley, B. and Chet I. (1995) Phytoremediation: A novel strategy for the removal of toxic metals from the environment using plants, *Biotechnology* 13, 468-474.
70. Schnoor, J.L., Licht, L.A., McCutcheon, S.C., Wolfe, N.L., Carreira, L.H. (1995) Phytoremediation of organic contaminants, *Environ. Sci. Technol.* 29, 318-323.
71. Shimizu, M., Kimura, T., Koyama, T., Suzuki, K., Ogawa, N., Miyashita, K., Sakka, K., Ohmiya, K. (2002) Molecular breeding of transgenic rice plants expresssing a bacterial chlorocatechol dioxygenase gene, *Appl. Environm. Microbiol.* 68, 8, 4061-4066.
72. Singer, A., Crowley, D.E., Thompson, I.P. (2003) Secondary plant metabolites in phytoremediation and biotransformation, *TRENDS in Biotechnol.* 21, 3, 123-130. Saleh, S.S. and Glick, B.R. (2001) Involvement of *gac*S and *rpo*S in enhancement of the plant growth-promoting capabilities of *Enterobacter cloacae* CAL2 and UW4, *Can. J. Microbiol.* 47, 698-705.
73. Sommerville, C. (2000) The Genetically Modified Organism Conflict, *Plant Physiology* 123, 1201-1202.
74. Sriprang, R., Hayashi, M., Yamashita, M., Ono, H., Saeki, K., Murooka, Y. (2002) A novel bioremediation system for heavy metals using the symbiosis between leguminous plant and genetically engineered rhizobia, *Journal of Biotechnology* 99, 279-293.
75. Stilwell, D. (1997) Public acceptance of genetically modified foods, *The Biochemist*, 20-23.
76. van der Lelie, D., Schwitzguebel, J.-P., Glass, D.J., Vangronsveld, J., Baker, A. (2001) Assessing phytoremediation´s progress in the US and Europe, *Environmental Science and Technol.* 12, 446A-452A.
77. Vosatka, M. and Janouskova, M. (2003) personal communication.

CHAPTER 5

PHYTOSTABILISATION OF METAL-CONTAMINATED SITES

M. MENCH[1], J. VANGRONSVELD[2], N. LEPP[3], P. BLEEKER[4], A. RUTTENS[2] AND W. GEEBELEN[2]

[1] *INRA Agronomy, Centre Bordeaux-Aquitaine, 71 avenue E. Bourlaux, BP81, F-33883 Villenave d'Ornon, France.*
[2] *Limburgs Universitair Centrum, Environmental Biology, Universitaire Campus, Building D, B-3590 Diepenbeek, Belgium.*
[3] *Liverpool John Moores University, Byrom Street, Liverpool, L3 3AF, United Kingdom.*
[4] *Department of Ecology and Physiology of Plants, Vrije Universiteit, De Boelelaan 1087, 1081 HV Amsterdam, The Netherlands.*

Abstract

The oldest field experiments confirm that phytostabilisation is an unobtrusively and cost-effective technique for treating metal-contaminated soils. University research teams having relatively few means have mostly established trials, but lack funds to deliver crucial long-term assessments. Soil amendments can initiate a healthy and diverse ecosystem at a contaminated site. Inputs of soil amendment are usually economically reasonable and environmentally benign. Selection of additives in batch experiments is necessary and can avoid increasing contaminant toxicity to living organisms, but may not forecast some long-term effects. Materials found to be promising under batch experiment and/or glasshouse conditions can be less effective in the field. Decline of effectiveness can occur over time. Failures and side effects such as induced deficiencies can arise and tend to be site specific. To avoid failures, a thorough evaluation of materials must be conducted, in particular in field lysimeters and/or in plots for each site, in combination with or without the choice of biological material.

1. INTRODUCTION

There are an estimated 750,000 contaminated sites in Europe, the majority of which are contaminated with mixtures of organic and inorganic pollutants. The US Environmental Protection Agency (EPA) estimates there

J.-L. Morel et al. (eds.), Phytoremediation of Metal-Contaminated Soils, 109–190.

are 30,000 candidate sites for hazardous waste treatment services in the USA [1]. Whilst a mere 15% of the US superfund sites are contaminated solely with trace elements, a further 64% of these sites contain metals mixed with organic wastes. Metal-contaminated soils may have serious consequences for adjacent ecosystems, agricultural productivity, and human health [2]. The impact of soil trace elements, either present as contaminants or naturally occurring, on living organisms depends on their concentrations and chemical speciation in solid, liquid and gaseous phases. This varies between soils, as a function of both soil properties and trace element source. At many sites, no action has been taken to assess, and, when necessary, reduce risks to human health and the environment. When sites are left barren, risk can result from the re-entrainment of contaminated soil particles, which bypass many of the constraints to mobility that are present in the soil-plant-consumer exposure pathway. Remediation of such sites is both economically and socially desirable.

Several practices based on civil engineering are used to remediate contaminated soils: excavation, stabilization of the soil with cement or similar material, landfilling, soil washing, capping [3]. These are environmentally invasive, expensive, and may generate wastes that may prohibit their use at large areas and small sites within residential areas. Electroreclamation is based on an electrokinetic process, which occurs when a direct current is applied between a cathode and anode placed in the soil: cations migrate to the cathode and anions to the anode [4]. This interesting technique is however expensive to install, requires substantial amounts of electricity to run and requires regular maintenance. There is a need for cost-effective remediation technologies, especially those which are non-invasive and employ simple technology [5].

Mild remediation options for reducing trace element exposure are: deep ploughing, phytoremediation, and chemical immobilisation [6]. Deep ploughing can give a positive effect if only the upper soil layer is contaminated and the underlying soil has a sufficient fertility and a buffer capacity. This also does not address the problems of leaching of metals to groundwater. Phytoremediation holds great promise for unobtrusive, cost-effectively treatment of metal-contaminated soils [7]. The duration of treatment and lack of commercially-available plant strains that can act in the absence of chemical modification of the contaminated soil currently limit the implementation of phytoextraction. Rigorous evaluation of excluder, tolerant and metallophyte plant species is in progress, and there is increasing interest in exploiting biodiversity to phytoremediate a contaminated-soil. In soils that are contaminated with a number of metals, one or more elements may limit phytoextraction potential: *e.g.* high soil Cu content can reduce the growth of *Thlaspi caerulescens*, and thus Zn and Cd phytoextraction. Increasing metal concentration in soil pore water using organic ligands such as EDTA has been considered to promote phytoextraction [8], but some metal complexes persist for several weeks, posing an environmental risk in the form of groundwater contamination [9].

A developing risk-based phytoremediation concept involves phytostabilisation. It is also called inplace inactivation, *in situ* immobilisation, and phytorestoration. This technique uses soil amendments combined with revegetation and/or microorganism manipulations [3, 10]. Through the use of soil amendments, metals undergo transformations or sorption reactions in the soil, which induce change in their speciation and concentration in the soil compartments. Formation of insoluble contaminant species reduces leaching through the soil profile and biological interactions with living organisms. Plants are used to cover the soil surface to prevent erosion, to reduce water percolation, to prevent direct contact with the soil, to increase biodiversity, and some may help to extract the remaining soluble forms. Phytostabilisation can be adaptabled for a range of situations from barren industrial sites to urban areas. Metals remain in place, and the challenge is the sustainability or enhancement of processes leading to the lowest exposure for non-essential elements, without inducing deficiencies of major nutrients [10, 11].

2. THE PREMISES FOR PHYTOSTABILISATION

Soil chemistry, crystallography, plant biology, agricultural practices, and ecotoxicology give a broad background for developing phytostabilisation strategies. Valuable information on promoting plant growth in harsh or suboptimum conditions was gained by management of trace element deficiency and toxicity in agricultural fields, land application of sludges, and mine spoil revegetation [2]. Use of lime, gypsum, basic slags, fertilizers and organic matter are traditional agricultural practices to optimise soil properties and plant growth. Liming was used in combination with organic matter to attenuate Cu exposure in the topsoil of French vineyards [12]. The initial attenuation decreased with elapsed time, but its sustainability was sufficient to allow the establishment of grapevine root systems in less contaminated deeper soil layers. At numerous sites with low or high trace element contamination, soil acidification results in the mobilisation of soil metals and plant exposure. Soil pH under grass in plots at the Park Grass Experiment has been allowed to decrease to < 4. Consequently, Cd, Pb, Cu, and Zn have been mobilised into available forms in unlimed plots and were taken up by the grass cut for hay [13]. While liming can rapidly reverse topsoil acidity, treating subsoil acidity is a more protracted process. In some cases, an undecomposed mat of organic matter (thatch) may restrict incorporation of applied lime into the upper soil horizons.

Many tailings and waste materials contain high concentrations of trace elements, low levels of plant macronutrients, and acid-generating sulphides.

To improve soil fertility or to reduce metal toxicity to plants, sewage sludge, papermill sludge, powerplant fly ash, fertilizers and lime were used, in particular for the remediation of abandoned toxic mine spoil in southeastern Ohio [14]. Attenuation experience using alkaline amendments, in combination with biosolids, has been gained at two large tailings sites in Canada: Copper Cliff (2225 ha) and Elliot Lake [15, 16]. The release of metals by airborne particles due to the "freeze-dry" effect in early spring and late fall required the establishment of a stable vegetation cover to complete the attenuation. Several principles and procedures have arisen from work at these sites [15]:

1. Sufficient agricultural limestone should be added to the acid-generating material to raise the pH of the seedbed strata to approximately 4.5 - 5.5. Half the limestone should be applied before seeding and the balance at the seeding time.

2. Seeding should begin in the area closest to the source of the prevailing winds to minimize the covering or damaging of young plants.

3. Nitrogen, P, and K fertilizers should be added, to account for the loss of nutrients due to reactions in the material, in order to ensure plant establishment at the seeding time.

4. Grass seed germination should have the advantage of cooler temperatures and available moisture, *e.g.* late summer in most Northern countries.

5. Legumes should be seeded in the spring, 1 or 2 years following the initial establishment of grasses.

A companion crop (*i.e. Secale cereale*) should rapidly provide a protective canopy, reducing surface wind velocity and providing shade, for the slower-growing grass. Additionally, it will act as a seed trap for volunteer plant species from adjacent vegetated areas.

6. Frequent small doses of a nitrogenous fertilizer should be applied.

7. Mulching should be applied especially when seeding areas exposed wind or slopes.

**In case a layer of natural soil is used to cover the tailings and provide a seedbed, the change in surface porosity must be considered, and adequate drainage should be installed to eliminate water erosion.*

Prior to seeding, 25 t ha^{-1} of lime was incorporated at Copper Cliff. A straw mulch and a chemical binder were combined as a surface stabilizer. The straw acts as a source of organic matter, a trap for seeds, and carries natural soil particles inoculating the tailings with various soil organisms, which will accelerate the soil-building process. At seeding time, additional limestone was spread and mixed into the surface. Fertilizer (8-24-24 N-P-K) was applied at 740 kg ha^{-1}. Fall rye was seeded at 60 kg ha^{-1}, along with a mixture of other grass seeds (68 kg ha^{-1}: 25% *Poa compressa*, 25% *Agrostis gigantea*, 15% *Poa pratensis*, 15% *Phleum pratense*, 10% *Festuca arundinacea*, 10% *Festuca rubra*). Additional fertilizer (350 kg ha^{-1}) was placed in bands along the seed row at the same time. Slow release nitrogen fertilizer was spread over the seedling area after the initial germination. *Lotus corniculatus* was seeded the following spring. Based on soil tests, the long-term maintenance program included additional limestone and fertilizer applications. Ten years after seeding, an organic horizon existed (2-3 cm) and a podzolic profile was in the process of formation. Volunteer *Betula papyrifera* and *Populus tremuloides* were the first tree species to colonise the grassed area. In addition, *Pinus banksiana*, *Pinus resinosa*, and *Robinia pseudoacacia* were included in a tree planting program. The vegetative cover evolved from a prairie like grassland to a scattered tree savanna, then to a forest of indigenous species. The abundance of wildlife had considerably increased. The metal content of *P compressa* and *A. gigantea* fell within the normal range for grasses, and metals did not accumulate in the food chain. The vegetation resolved some water quality problems, such as acid drainage, by intercepting precipitation, evapotranspiration and formation of an oxygen-consuming barrier of decomposing plant residues. Unfortunately, data for metal and As leaching at this site were not available.

Sludge-amended soils are often maintained at near neutral to slightly alkaline pH using liming agents because metals will become less soluble and less mobile, due to precipitation or binding by organic matter [17].

3. GENERAL CONSIDERATIONS FOR PHYTOSTABILISATION AND LIMITATIONS

Natural attenuation generally means dilution, dispersion, (bio)de-gradation, irreversible sorption, volatilisation, and/or radioactive decay of contaminants in soils and groundwaters (http://www.sandia.gov), and causes a net reduction of contaminant toxicity and both human and ecological risk. Dilution and dispersion are seldom acceptable methods in compliance with regulations. Sorption is rarely irreversible, and may take too much time. Sorption (bio)geochemical processes in contaminated-soils can be enhanced by the application of amendments. Inactivation, *in situ* immobilisation and chemical immobilisation, cover several methods that aim to reduce potential exposure *via* the soil or wastes. Definitions have been previously provided [5, 18, 19]. Here "*in situ* inactivation of trace elements in contaminated soils" is defined as decreasing concentrations of trace elements or rendering their chemical forms less bioavailable or toxic in the exposure pathways, *i.e.* the soil solution, the gaseous phase or ingested particles, by adding an amendment to the soil or the waste. At least two options based on the use of amendments can be adopted, using changes in abiotic and/or biotic conditions.

The first approach is to enhance naturally-occurring processes involved in transformation or sorption reactions that can alter both speciation and concentration in solid phases and the soil solution. This might not always be possible, or can be too time consuming though. The second alternative approach is to introduce one or more compounds *via* an amendment, leading to either new solid reactive phases or the presence of an essential element for the transformation process. Several mechanisms can be involved: *e.g.* sorption, acid-base reaction, precipitation, oxydo-reduction and demethylation. Solidification is not considered. If inactivation is opted, it should be done only after action has been taken to remove, treat, or contain the contaminant source. At polluted sites, soils are usually contaminated with several trace elements. In this case the use of additives can immobilise a range of elements, including some essential nutrients. For example, MnO_2 can bind metals such as Cd, Zn, and Pb, but appears less effective in Cu contaminated soils [20]. Calplus (clay-aluminium hydroxides) was found effective for Zn and Cu, but not for Cd [6]. Synthetic zeolites can have a high affinity for Ca, decreasing metal binding and increasing leaching of organic matter [6]. Beringite combined with compost reduces Cd and Zn exposure, whereas it can increase As exposure and leaching [21]. A detailed knowledge of the binding mechanisms for additives and their derived products is desirable to treat a specific site. Changes in the ionic composition of the solution may influence solid-solution partitioning.

Table 1. Plant species established at field trials with phytostabilisation of metal-contaminated soils.

Plant species	Sites
Agrostis castellana, Agrostis delicatula *Holcus lanatus, Cytisus striatus*	Jales, Portugal
Lotus corniculatus, Poa compressa, Agrostis gigantea *Poa pratensis, Phleum pratense, Festuca arundinacea* *Festuca rubra*	Copper Cliff, Canada
Agrostis capillaris, F. rubra	Maatheide-Lommel, Belgium

Phytostabilisation is an adaptable technique, which only requires standard agricultural equipment. Wherever possible voluminous solid wastes are removed and consequently the soil is ploughed to incorporate soil amendments and to prepare a seedbed. Plants are either seeded or transplanted. Irrigation is utilized if necessary. Plant roots physically stabilize the soil, thus preventing erosion, and reduce water percolation through the soil. Sorption or precipitation of metals can occur in the rhizosphere. Berti and Cunningham [3] suggested exploring whether some plants could contribute to metal immobilisation by inducing the formation of insoluble compounds inside plant tissues or on the root surface. A plant cover minimizes human contact from re-entrained particulates and reduces soil erosion induced by rainfall. Plants used for this purpose should exhibit several characteristics (Table 1): quick and easy establishment, fast growth, tolerance to the soil metal levels as well as the other limiting factors (low soil pH, salinity, coarse soil texture, drought), excellent plant cover, and low translocation of metals in edible aboveground plant parts. A relatively long life, ability to self-propagate, and a readily commercial availability are advantages.

Any phytoremediation process requires contaminants be in contact with the root zone of plants. In the majority of sites where the use of immobilisation has been adopted, treatments have been confined to the uppermost regions of the soil. When oxygen input is not a problem, a partial soil excavation can allow treatment of depths up to 70-100 cm; which is generally sufficient for the viability of the majority of plant roots. A geo-textile barrier can be included to improve the separation. Chemical impact of contaminated, not processed deep layers on the upper part is important, especially when trees are used for revegetation. Experience from the Lommel-Maatheide Experiment (see below), showed that trees died as their roots came into contact with soil solution from untreated contaminated lower soil layers. Conversely, revegetation may enhance organic deposition, and nutrient cycling might have a beneficial impact on the subsurface soil. More information on physical, chemical and biological interactions between the phytostabilised surface and the subsurface is required. Woody and herbaceous plant materials that fall on the surface can be removed to accelerate the remediation. Knox et al. [22] propose to immobilise contaminants released from decomposing plant materials in a mineral-containing mat (geomat) deployed at the ground surface. Among several materials, metallic iron and a Fe oxide waste were the most efficient for lowering the aqueous Cr, Hg, Co, Ba, Eu, Pb, and U concentrations. Unterköfler et al. [23] have investigated the immobilisation of metals leached from fallen leaves and OM (surface layer) from *Salix caprea*, *Populus tremula* and *Betula pendula* on site using a 2 cm-layer of vermiculite. This material was able to adsorb and immobilize more than 99% of the leached metals.

Remediation duration can be defined in different ways, depending on major targets, the exposure route, land use, and acceptable risks. It can be time required to reduce trace element concentrations *e.g.* in aqueous phase to threshold values in drinking water, in edible plant parts to standards in foodstuffs, or in mammal and human fluids such as blood and urine. In an ecotoxicological viewpoint, it can be time consumed to eliminate all unacceptable risks, *e.g.* (geno)toxicity on living organisms, especially regarding long-term biogeochemical cycles and life. At least, remediation can be considered achieved using the ALARA principle ("As Low As Reasonably Achievable"), but based on the risk assessment process and future land use, not on total metal concentrations in ecosystem compartments. The time required for metal immobilisation may be a limiting factor depending on proposed future land use. Most field experiments have demonstrated a rapid effect of soil amendments, which enables restoration of grassland in year 1. Isotopic dilution can be one tool to discriminate between chemically-reactive (labile) metal and fixed (no-labile) metal pools in soils, and predict the reaction time to reach the minimal solubility [24]. In metal-spiked soils incubated for 900 days, under aerobic moistened conditions, metal fixation trends were described by reversible first-order kinetic rate equations, which suggested Zn fixation was reversible and soil metal lability did not decline to zero, even under neutral to high pH conditions [25].

It is difficult to define the success of an ecosystem recreation by productivity and biodiversity, because metal-polluted soils are well known to support diverse and highly adapted plant communities. In addition, when soil remediation policy is only based on soil quality standards such as total concentrations, site stabilisation will never be considered successful. A shift in emphasis to Soil Quality Standards based on human exposure and ecological risks is a prerequisite for the possible adoption of remediation based on site stabilisation. The definition of 'success' in these circumstances is a complex issue, based upon human health and ecological and commercial considerations. For example, in the Dutch part of the Kempen region where the soil contamination was classified as "urgent", a successful soil remediation had to address the following points: leaching of metals to the groundwater, transport of metals by the groundwater, transport of contaminated sediment and sedimentation of contaminated materials, genetic adaptation of plants and invertebrates, accumulation of litter, kidney dysfunction in humans and plant contamination [6].

The long-term effectiveness of phytostabilisation can be addressed using the following three key issues [26]:

(1) What are the risks posed by the contaminant concentration and speciation over time?

(2) What are the processes that may affect the long-term fate of contaminants?

(3) What is the long-term fate of contaminants following the use of soil amendment?

Some points must be examined prior to the design of a treatment: Are all contaminants, in particular trace elements, amenable to inactivation? Are contaminants limited to the root zone? Is bioavailability sufficient to allow bioaccumulation and adverse effects? Do the amendments alter the exposure route or transport to groundwater?

Due to the lack of biogeochemical models for the use of soil amendments, the sustainability of phytostabilisation cannot currently be subject to valid assessment. Practices and plant covers vary from site to site, and depend on specific primary concerns, *i.e.* human exposure to dust, plant toxicity, food-chain contamination, environmental quality or leaching. Soil amendments are very diverse. The most effective amendments and application rates vary for individual metals and soils.

4. NATURAL AND SYNTHETIC SOIL AMENDMENTS

How are soil amendments selected? A potential amendment should rapidly attenuate exposure to trace elements, prevent leaching, plant uptake, toxicity to animals, etc. A long lasting, if not permanent, effect is required. Amendments should be: inexpensive, readily available, easy to apply with low associated risks, devoid of disruptive or adverse effects, especially on the soil structure and fertility, compatible with plants used for revegetation, suitable for different contaminants, and in compliance with regulations [3]. Amendments supplying plant nutrients, such as phosphate fertilizers, or increasing soil water holding capacity, such as organic matter (OM), are useful to optimise plant growth. Phytostabilisation should be suitable for combination with other techniques, adaptable to agricultural management and suitable for different soil types. Osté [6] postulated that the material must have a high metal binding capacity at common pH (*i.e.* 4-8) and needs to be durable under the environmental conditions occurring in the soil. The latter is not a general prerequisite though. Some materials such as steelshots and zerovalent iron can have no binding capacity, but their alteration in soil can lead to sorption processes or chemical transformations. Aging of materials such as iron oxides, allophans, and silica gels, together with crystallisation may increase the inactivation of trace elements.

Numerous organic and mineral materials, selected for their lack of negative influences on key aspects of soil health, have been used to reduce solubility and bioavailability of trace elements in contaminated soils (Table 2) [6, 10, 20, 27]. Physical and chemical properties of these amendments have been discussed in reviews [27, 28]. These materials potentially reduce exposure by one or more of the following processes: sorption, redox reaction, precipitation, ion exchange, complexation, excesses of competing elements, humification [28]. Some mechanisms are not-well known or unpredictable, and may result in undesirable side effects.

Some of the solid phases and processes used for the remediation of metal-contaminated soils exist in natural ecosystems. Materials such as lime, phosphates, and organic matter have a lengthy history of agricultural use. $CaCO_3$, $(Ca, Mg)CO_3$, $Ca(OH)_2$, and CaO are used as liming agents. An increase in soil pH ionises pH-dependent exchange sites, raising cation exchange capacity (CEC) and metal sorption to soil particles. Carbonate formation can also be favoured. If soil pH is increased, the mobility of anions such as MoO_4^{3-} and AsO_4^{3-} and dissolved organic matter (DOM) can increase, resulting in increased leaching. Liming is mostly effective in the short-term with repeated applications required to maintain metal immobilisation. A range of P-containing materials, (*i.e.* hydroxyapatite, phosphate rock, phosphoric acid, tri super phosphate, Thomas basic slags) may substantially reduce metal bioavailability [4], based on the formation of

Table 2. Natural and anthropic amendments input into contaminated soils to immobilise trace elements.

Alkaline materials	**Fe, Mn, Al (Hydr)Oxides**
Calcium, magnesium carbonates	Ferrihydrite
Calcium, magnesium oxides	Hematite
Limed sludges, alkaline biosolids	Lepidocrocite
Fluidised bed coal fly ashes	Magnetite
	Maghemite
Phosphate minerals	Mud of water-treatment
Apatites	Smelter wastes
Hydroxyapatite	Wastes of TiO_2 treatment
Basic slags	Red muds (bauxite, Al)
Phosphoric acid	Hydrous Mn oxides
Phosphates salts (K, Na, Ca, NH_4, etc)	Birnessite
Natural phosphates	Chalcophanite
	Iron grit (steelshots)
Alumino-silicates	**Zeolites**
Clays (smectites, bentonite, etc...)	Natural zeolites
Red muds	Synthetic zeolites
Coal fly ashes, *e.g.* beringite	
Gravel sludge	
Organic matter (compost, farmyard manure, etc)	**Salts** (Cu, Zn, iron sulfate, etc)

(Hydro)pyromorphite as one important reaction. Competition between phosphates and anions such as arsenates cannot be ignored. Amorphous Al, Fe, and Mn oxide minerals are ubiquitous as both discrete particles and coatings. Porosity, micropores and surface areas, together with adsorption of metal ions are significant for these phases. When metal ions diffuse in hydrated micropores, intraparticle surface diffusion is a rate limiting mechanism in the sorption process. This has been found to range between 10^{-16} and 10^{-9} cm^{-2} s^{-1} [29]. Surface diffusivity is a function of the mean distance between sites and the energy required to jump to the neighbouring site. In the case of a divalent metal ion, this is related to the square of the inverse of the hydrated radius multiplied by the number of water molecules in the primary sheath. Where the hydrated sheath is retained, Coulombic forces appear to dominate metal adsorption to amorphous Al, Fe and Mn oxides. Synthetic birnessite and hydrous Mn oxides are very reactive forms of manganese oxides with metals such as Pb, Zn, and Cd. A range of materials containing Fe oxides can affect metal speciation in soils, *e.g.* non-crystalline and crystalline Fe oxides, Fe-rich (a TiO_2 by-product), steelshots (iron grit), red muds and sludges from water treatment [4, 28]. Organic materials such as composts, biosolids, and peat, which form complexes with metals such as Pb and Cu, have been investigated. Aluminosilicates with potential immobilizing effects on trace elements have been used: clays with high CEC such as montmorillonites and bentonite, natural and synthetic zeolites with fixed charged sites throughout interconnecting channels and cages and coal fly ashes including by-products from fluidised bed burning of coal refuse such as beringite.

Another way to immobilise metals like Cd involves the inoculation of soils with specific microorganisms. Some of these organisms can accumulate a considerable proportion of the soil pool of particular metals, despite the fact that they constitute a minor fraction of the total soil biomass. Cadmium biosorption by free- and immobilised cells in alginate beads are investigated using soil extracts [30]. Also under consideration is increased biosorption capacity using genomic tools. The mouse metallothionein I (MT) protein has been targeted to the cell surface of the metal-tolerant bacterial strain *Ralstonia eutropha* CH34, isolated from metal-contaminated soils [31]. The resulting strain, named *R. eutropha* MTB, exhibits an enhanced ability to immobilise Cd ions from external media. Inoculation of Cd-polluted soil with this bacterial strain decreased the toxic effects of the metal on the growth of *Nicotiana bentamiana*. However, there is no evidence for the long-term efficacy of such a treatment for a field soil. Microorganisms may assist metal immobilisation and/or plant development, and their activity therefore deserves further consideration.

Efficacy of materials to attenuate the exposure of plant species and other organisms has been evaluated using many short-term batch studies, dynamic leaching studies and microcosms [4, 10, 20, 32, 33, 34]. There is a clear need for long-lasting demonstrations of phytostabilisation to convince both the general public and legislators of the value of this technique. Such examples are required both to stimulate private initiatives and cofinancing and to amend legislation. New soil remediation policy, based on risk-assessment, exposure attenuation and risk reduction, and site-specific circumstances needs to be encouraged.

Below, special attention is paid to materials tested in long-term field experiments and outdoor mesocosms: lime, zeolites, phosphate compounds, metal oxides and other minerals, biosolids, and industrial by-products. Abbreviations used in the text are: Unt (Untreated contaminated material), C (compost), B (beringite), SS (steelshots), Z4A (zeolite 4A), CB (compost and beringite), CSS (compost and steelshots), CBSS (compost combined with beringite and steelshots). Application rate is expressed as % air-dried weight of soil.

5. TESTS FOR DETERMINING THE NEED OF SOIL AMENDMENTS AND MONITORING THE EFFICACY

There are three main routes of exposure to soil-borne trace metals. Soluble contaminants migrate with soil water, are taken up by plants and aquatic organisms and, in some cases can be lost through volatilisation. The other pathways are from direct ingestion, and dermal contact with contaminated particulates. In view of this, any thorough evaluation of the overall effect of additives and the sustainability of trace element inactivation should combine physico-chemical and biological methods. No consensus is currently available that recommends a standard battery of tests to assess the extent of hazard reduction. Current practice to determine the need for amendments and demonstrate their efficacy involves the use of chemical, biological, and spectroscopic tests.

Several chemical tests are frequently used (Table 3). Leachability is estimated under landfill conditions by TCLP (toxicity characteristic leaching procedure) and under acid rain conditions by SPLP (simulated precipitation leaching procedure). Single and sequential chemical extractions estimate the extractable fraction of one element at a given moment with a well-defined substrate/solution ratio. Even though it is difficult to predict the mobile pool of metals in a natural situation on this basis alone, it can be used as a rapid means of evaluating changes in extractable fractions after addition of amendments. One advantage of neutral salt solutions such as $Ca(NO_3)_2$ is that these have a limited effect on both the operative pH at the exchange sites and on complex formation [35]. Sequential extractions are intended to dissolve various chemically defined pools, usually identified as the following: soluble, exchangeable, carbonate bound, oxide bound, organic matter bound, and residual [4]. A mild extractant, such as pure water or a dilute salt, is generally the first step. Extraction continues with progressively harsher extractants until the total digestion of solid phases is complete. These methods can more or less quantify changes in exposure, but do not necessary measure the risk attenuation while organism response is not taken into account. Soil pore water can be directly sampled using the Rhizon sampling system, and diffusive thin gel (DTG) is a promising method to kinetically estimate the root exposure. *In vivo* animal feeding studies are used to determine to what degree ingested soil is solubilised in the gastro-intestinal (GI) tract and taken in the blood. This type of study is both expensive and time consuming. The PBET (physiologically based extraction test) test was designed to predict metal availability following ingestion. It incorporates GI tract parameters representative of a human (including stomach and small intestine pH and chemistry, soil-to-solution ratio, stomach mixing and stomach emptying rates). Swine and rat dosing studies correlate well with PBET data for Pb, but the correlation for As is less good [3].

Table 3. Chemicals tests to determine hazard reduction in metal-contaminated soils.

TCLP (toxicity characteristic leaching procedure)	US EPA (1990)
SPLP (simulated precipitation leaching procedure)	US EPA (1995)
Sequential extractions	Tessier et al., 1979
PBET (physiologically based extraction test)	Ruby et al., (1993)

References cited in Berti and Cunningham (2000), Geebelen (2002).

Chemical forms of soil metals can be investigated by X-ray diffraction (XRD), X-ray absorption spectroscopy (XAS), scanning electron microscopy with energy dispersive X rays (SEM/EDX) and extended X-ray absorption fine structure spectroscopy (EXAFS) [36]. These techniques provide valuable information on both chemical forms and oxidation states. Limitations are metal concentrations in the soil owing to detection limits and the determination of metal fate in the soil if the amount of metals immobilized after soil amendment is small compared to other fractions. Techniques such as XRD can only be used in case the precipitation has crystalline structure, and not when it is amorphous. Spectroscopic techniques are unsuitable to determine small changes in soluble soil metal pool fractions.

Plant tests are useful to determine the need for amendments. These may also detect other limitations for plant growth, such as high sodium levels. Other bioassays are used with target specific organisms such as earthworm, swine and rat (used to estimate bioavailability to humans), and bacteria [3, 4, 7].

For ecotoxicological evaluations, it is important to survey several endpoints, *e.g.* biodiversity, bioaccumulation in living organisms, metabolomics and proteomics as well as genotoxicity, at different trophic levels, with a progression from a well-defined battery of tests to more complex conditions representative of real ecosystems. A combined use of chemical extraction, microbial biosensors, phytotoxicity and zootoxicity tests was illustrated by Vangronsveld et al. [37]. An important criterion to judge the efficacy of phytostabilisation is to assess the acceptable residual risk. For example, is exposure of organisms at a subtoxic level acceptable? This question is not solely a scientific one, and particularly depends on the future land use and public acceptance.

Once added to the soil, the amendment generally cannot be removed. Iron grit is an exception as the oxidised steelshots and some soil particles such as magnetic iron oxides can be physically separated using a magnetic field [38]. Physical separation is also possible with some aluminosilicates.

Below, we discuss beneficial and adverse effects of soil amendments from the following aspects: extractability and chemical speciation of trace elements, leaching of trace elements to groundwater, toxicity and uptake in the soil-plant system and toxicity and uptake by animals.

6. CHEMICAL TESTS

Several investigations have provided strong evidence that soil amendments can bring about reductions in soluble and extractable pools of metals in long-term field experiments. One of the oldest field trials on barren soils contaminated by industrial dust fallout is located at Lommel-Maatheide, Limburg province, Belgium, a region with a long history of Zinc smelting. The sandy soil of this region is characterized by an acid pH (4.5-6.0), a low organic matter content (< 2%) and a low cation exchange capacity (< 2 cmol kg^{-1}) In 1990, 3 ha of a highly metal-contaminated soil were treated with a combination of cyclonic ashes (beringite, B, 5%) and compost (C, 5%) [34, 37, 39]. Soils were sampled 5 years after the use of soil amendments. Physico-chemical parameters of the untreated and treated soils were compared. Analysis of the treated soil showed a change in pH to slightly alkaline (7.3-7.9) together with increases in organic matter content and cation exchange capacity. Water-extractable and total Zn concentrations (HNO_3:HCl, 3:1, v/v) in soil were determined in untreated and CB-treated plots, and in an uncontaminated garden soil with a similar sandy texture as a reference [40]. The ratio of water extractable *vs.* total Zn was highest in untreated soils (Table 4). This ratio decreased significantly (up to 70 times) in the CB-treated soils, and its value was below that in the uncontaminated garden soil. Similar results were obtained for Cd. The mechanisms behind these changes are considered to be increase in soil pH, precipitation, sorption on Ca-phosphates, ferrihydrite, allophans, and other minerals added in the beringite amendment. However, EXAFS and X-ray diffraction indicated that silicates such as hemimorphite ($Zn_4Si_2O_7(OH)_2.H_2O$) and willemite (Zn_2SiO_4), and clays are also involved in Zn speciation [41].

The zinc smelter at Overpelt, located in Limburg province, Belgium as well, produced 250 tons of Cd waste in 1950, with approximately 50% of this total as atmospheric emissions (340 kg of Cd day^{-1}).

The smelter converted to an electrolytic process between 1969 and 1974, but even though emissions were significantly reduced, there were still elevated concentrations of Cd, Zn and Pb in the local environment. A field trial and a small-scale semi-field experiment have been established at this site as part of the EC PHYTOREHAB project using beringite (5%), steelshots (1%) and compost (5%) as soil amendments [42]. Soluble and exchangeable Cd and Zn fractions in the small-scale Overpelt Experiment were assessed 3.5 years after initial treatment, using a 0.01 *M* $Ca(NO_3)_2$ extraction [43] (Fig. 1). Soil pH values were: 6.5 (control soil), 6.4 (C), 6.9 (CB), 6.9 (BSS), 7 (CBSS), and 6.1 (Unt). Compared to the Unt soil, Cd concentration was up to 4 times lower in CBSS treatment. Despite a similar pH, CBSS more efficiently reduced extractable Cd than CB, demonstrating an additional mechanism due to steelshots oxidation. The CBSS and BSS treatments showed the greatest reduction in extractable Zn (4 times

Table 4. Ratio of water-extractable Zn on total Zn concentration in soil measured 5 years after the soil treatment at the Lommel-Maatheide experiment (SD, n = 3).

	Total Zn (aqua regia, mg kg^{-1})	Water-extracted Zn (mg kg^{-1})	ratio (%)
Treatment			
Uncontaminated garden soil	106 (7)	0.7(0.06)	0.660
Untreated	1.234	11425 (506)	14(11)
CB	0.029	7639 (455)	2.2(0.22)

Source: Vangronsveld et al. (1996), CB: Compost (5%) + beringite (5%)

to the Unt treatment). Zinc inactivation was higher in BSS-treated soil than in CB-treated soil, although soil pH was similar in both treatments. Metals have been effectively immobilised at Lommel-Maatheide for more than 10 years after CB application, but if treatment efficacy is measured on the basis of the initial decrease in trace element exposure, then Cd and Zn immobilisation is superior using CBSS. A column extraction procedure, to simulate 60 years leaching using a 0.01 M $Ca(NO_3)_2$ solution, has been conducted on untreated and treated Overpelt soils. Extractable metal pools decreased in the following order: Unt > C > CB > CBSS for Cd, Cu, Pb, and Zn.

Metal immobilisation can be demonstrated by metal redistribution between solid phases using sequential extraction. An example of this is found in data from the small-scale Reppel Experiment. Reppel, a village in Limburg province Belgium is the site of a former As(III) refinery, which operated from 1910 to 1965. Up to 8 ha of this site were highly contaminated by the storage of Zn and As products. The soil has been sampled in the adjacent agricultural area contaminated by fallout from this refinery [24]. Post-4 years soil was amended with B, SS, and BSS, Zn redistribution among organic and mineral phases was associated with a decrease in Zn exposure.

Comparison of B-treated and Unt soils showed that the percentage of Zn associated with amorphous minerals and other compounds increased whereas Zn associated with organic compounds decreased (Fig. 2) [44].

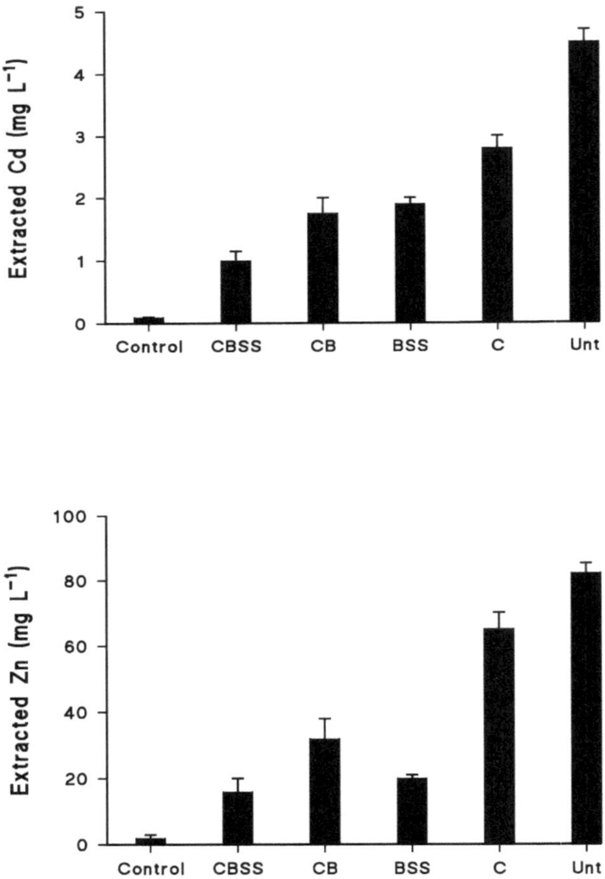

Fig. 1. Cadmium and zinc extracted by 0.1 M calcium nitrate in soils sampled at small-scale Overpelt Experiment, 3 years after soil amendment addition (Unt: untreated, Control: control soil, C: compost (5%), CB: C + beringite (5%), BSS: beringite (5%) + steelshots (1%), CBSS: C + beringite (5%) + steelshots (1%). Source: [43]

Much less data is available for metalloids. It is accepted that the use of soil amendments may affect their speciation in soils. Several elements such as arsenic may undergo a change in oxidation state, which can influence exposure and toxicity. In soil solutions sampled from the small-scale Reppel Experiment, As(V) was mainly present in the (1%) steelshots-treated soil before and after the oxidation of this amendment in the contaminated soil [24]. Concentration of As(III) represented only 0.17% of total As in solution, and was 1.6 $\mu g\ L^{-1}$ in the Unt soil solution. Monomethyl-arsenic and dimethyl-arsenic were identified in the soil pore water but their concentrations were not relevant compared to As(V). Although the relative As(III) concentration increased to 1.5% in the SS-soil solution, As(III) concentrations even decreased (1.35 $\mu g\ L^{-1}$). This may reflect As(V) oxidation by newly formed Mn oxides. But if this mechanism occurred, it is not a significant one regarding As exposure and risk assessment.

Sorption studies showed that a range of iron oxide-producing systems including iron grit, iron II sulphate + lime, and iron III sulphate + lime and goethite+lime, exhibited a high affinity for arsenic in solution [45]. Iron III sulphate+lime provided the greatest reduction in leachable As in 3 contaminated soils originated from North West England. In treated soils, As was mostly bound to the iron oxide fraction. Iron oxides produced *in situ* are more effective at immobilizing soluble As compared to goethite.

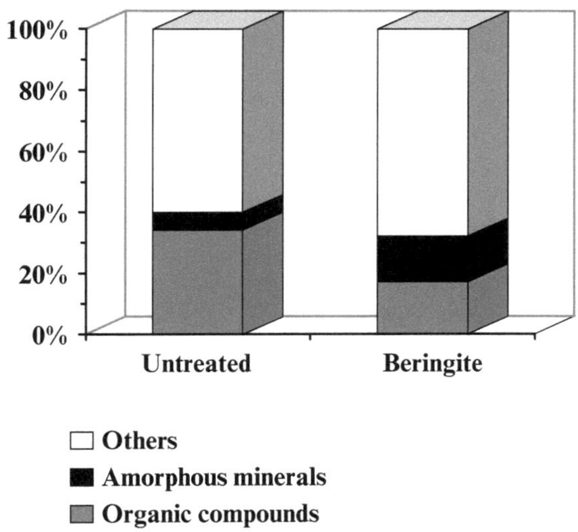

Fig. 2. Distribution of Zn in the compartments of the beringite-treated and untreated small-scale Reppel Experiment soils. Source: [44]

The efficacy of steelshots (1%), beringite (5%), steelshots (1%) + beringite (5%) and hydroxyapatite (HA, 0.5, 1 and 5%) to immobilise metals and As was investigated in 6 contaminated soils (*i.e.* Overpelt 1 & 2 Reppel, Louis Fargue, Jales, Pyhäsalmi) following a 3-week reaction time [46]. The most effective amendments in reducing 0.1 *M* calcium nitrate-extractable Cd, Cu, Ni, Pb, and Zn in all soils were B, BSS, and 5% HA. The lowest water-extractable As fraction was found in soils amended with B, SS or BSS. The B effect was, however, soil dependent. HA addition led to higher extractable As fractions and phosphate-arsenate competition for the sorption complex of the soil solid phase was suggested. The use of HA at combined metal-As contaminated sites needs to be approached with care.

Incidental soil ingestion due to hand-to-mouth activity represents a significant direct exposure pathway for non-dietary sources of trace elements in contaminated areas [47]. Metal-contaminated soils can pose a human risk through the direct ingestion of soil by children in residential areas. Human dosing to establish the efficacy of amendments is not feasible but batch experiments deliver some information. Prediction of metal availability can be gained using an *in vitro* physiologically based extraction test (PBET). Ten metal-contaminated soils were collected from various locations in the USA and Europe, reflecting eight different sources of Pb enrichment [4]. Cyclonic ash (5%) + steelshots (1%) (CASS), North Carolina phosphate rock (PR, 1%, except for Joplin and Leadville soils where 2% was incorporated), and lime (1%) were separately added as soil amendments. Metal bioaccessibility (MBA: fraction of total ingested metal concentration solubilised in the GI tract and absorbed into systematic circulation) was measured using a modified PBET procedure (0.4 *M* glycine pH 2.2) [48]. Lead MBA ranged from 1% (Hoboken) up to 76% (Belin-Beliet) in untreated soils (Table 5). Phosphate rock was the only additive able to reduce PBET Pb. Its value decreased in three soils, *i.e.* Joplin (–21%), Pribram (–26%) and Hoboken (–50%). Phosphate rock addition to soils generally results in the formation of Pb pyromorphite, which is fairly insoluble even at low pH. Sorption processes and formation of Pb carbonates may be also involved in the treated test soils. However, PR efficacy to reduce PBET Pb depends on soil parameters such as soil pH and Pb speciation, raising questions about its application in general. Due to the low solubility of PR, much higher amounts (5%-10%) may be applied to immobilize Pb through pyromorphite formation. Cyclonic ash + steelshots and lime did not reduce Pb MBA. Conversely, lime addition increased PBET Pb in Joplin (+26%) and Leadville (+43%). Amendment efficacy to decrease Cd and As MBA was different. Cadmium MBA varied in a range from 4%-93%. CASS decreased PBET Cd in three soils (–29% Joplin, –57% Hoboken, –18% Leadville) whereas $CaCl_2$ extractable Cd was elevated. Binding to Fe and Mn oxides is suggested while the alkaline effect of cyclonic ashes in soils would be neutralized at low pH in the PBET test. Phosphate rock decreased PBET Cd only in Leadville soil (–25%) whereas slight increases were found for Paimboeuf and Belin-Beliet soils. Lime addition increased PBET Cd in Leadville soil (15%). Arsenic MBA was low for all soils (2%-9%). A decrease in PBET As was only obtained with one amendment, *i.e.* CASS, in four soils (–44% Pribram, –41% Hoboken, –39% Bazoches, > 15% Aiken). Arsenic sorption on Fe oxides is a possible mechanism. Conversely, PBET As was increased by PR addition to Pribram (31%), Hoboken (155%) and Belin-Beliet (65%), and by lime addition to Hoboken (98%). These amendments did not contain any significant As content. Release of phosphates and competition for sorption sites would occur following PR incorporation. Change in soil pH is the suggested mechanism for lime addition.

Table 5. PBET Pb (mg kg^{-1}) on untreated soils and on soils amended with cyclonic ash and steelshots (CASS), phosphate rock (PR), and lime (detection limit = 0.02 mg Pb kg^{-1}) and Pb bioaccessibility (MBA = PBET extracted/ total Pb, in %) in untreated soils. Mean values (n = 3) followed by the same letter are not different (Kruskal-Wallis test, $p < 0.05$). Source: Geebelen (2002).

	Total Pb	MBA	Unt	CASS	PR	Lime
Paimboeuf	188	21	39^{ab}	39^{ab}	36^{a}	48^{b}
New Orleans	6065	22	1323	1267	1185	1230
Joplin	5444	61	3307^{b}	3345^{b}	2596^{a}	4158^{c}
Pribram	2663	32	848^{b}	800^{ab}	626^{a}	801^{ab}
Hoboken	656	1	4.6^{b}	3.9^{ab}	2.3^{a}	6.5^{b}
Moreno Valley	236	23	54	64	62	60
Leadville	3857	13	464^{a}	536^{ab}	449^{a}	662^{b}
Aiken	678	56	381	420	388	421
Bazoches	598	27	164	164	135	152
Belin-Beliet	1859	76	1415	1427	1571	1556

Exposure of children to Pb accumulated in residential yards is an acute problem in distressed urban areas in the USA. Biosolids application has been chosen to remediate 25 residential yards in Baltimore, Maryland [49]. An 8-10 cm layer of a sterilized and composted Baltimore municipal sewage sludge, approved by US EPA for unlimited use in gardens, was used to treat soils. Sites were seeded with grass. Initial data suggests that application of biosolids has not generally changed bioaccessible Pb fractions, assessed by the simulated *in vitro* GI method, except at two sites where it decreased by 12% and 25%. Follow up will assess the effectiveness of biosolids in treating soil contaminated with lead-based paint. However, the organic fraction of the biosolids will decrease with time, so treatment efficacy may not be sustainable.

The influence of P treatments, *i.e.* triple superphosphate (TSP), phosphate rock, acetic acid followed by TSP and phosphoric acid, on *in vitro* GI available Pb was studied in five metal-contaminated soils and mine wastes [50]. Reduction in bioavailable Pb was found for all treatments. Phosphate rock was equally or more effective than TSP or phosphoric acid in four out of five soils tested. In one material, preacidification prior to P application resulted in a lower Pb bioavailability. Lime increased GI-available As in a contaminated soil [47]. Only the Fe(II) sulphate plus lime treatment decreased GI available As compared with the limed control. Both Al sulphate + lime and Mn sulphate + lime reduced water soluble As, but these forms of immobilised As were dissolved by the acidic conditions of the *in vitro* GI extraction.

These selected data demonstrate that metal and arsenic exposure *via* the soil solution (soil-plant-animal pathway) and *via* soil ingestion, can be decreased in several different types of amended contaminated soils. Currently, the effect of a particular amendment is often unpredictable. For example, when there is a decrease in trace element concentration in the soil solution, it is uncertain whether this is reflected in reduced mobility to groundwater.

7. LEACHING

Some long-term experiments have reported time-dependant changes in metal and As concentrations in water percolating through amended soil layers. This has produced information on transport to deeper untreated soil layers and groundwater. Changes in chemical parameters such as proton flux, SO_4^{2-}, Cl^- and ionic strength may also affect trace element speciation in deeper soil layers, but this point is rarely considered. The small scale Overpelt Experiment showed a 12 month decrease in Cd and Zn leaching in the following order: Unt > C > CB > CBSS (Fig. 3). The lowest amounts of these metals leached in the CBSS-treated soils (–99% Zn, –98% Cd) were mirrored by the lowest amounts of Cd and Zn extracted by a 0.01 *M* calcium nitrate solution. Compost addition, used to reverse the low organic matter content, increased Cu and Pb leaching. Increases in soil pH and organic matter may change the solid-solution partitioning of the organic matter and increase dissolved organic matter (DOM) concentration and/or present colloids in the soil solution able to transport metals. Copper and lead bound to DOM would leach to deeper soil layers or groundwater. The combination of organic amendment with alkaline materials such as beringite and other cyclonic ashes increased Ca concentration in solution, leading to competition and a higher ionic strength, though this was not sufficient in the CBSS treatment to reduce Cu and Pb leaching compared with Unt. When these metals are of primary concern at a contaminated site, an additional approach, *e.g.* phosphates added in the amendment combination, may solve this side effect.

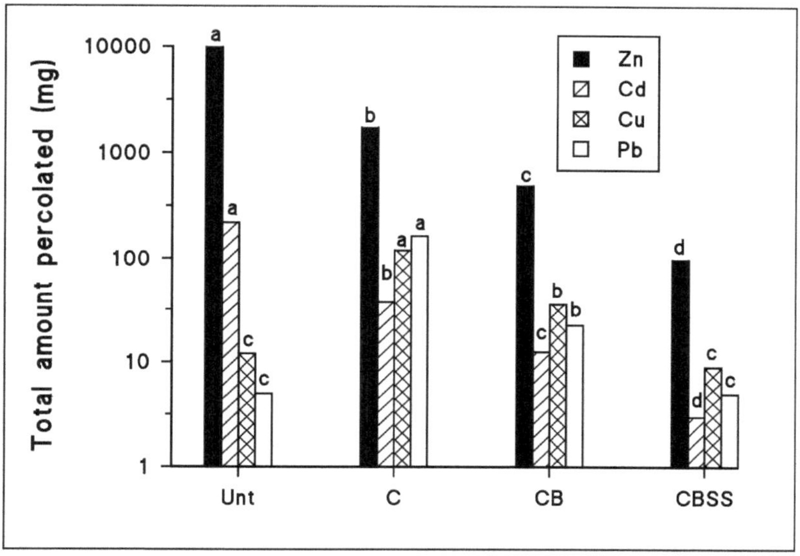

Fig. 3. Total percolated metal content (mg) at the small-scale Overpelt Experiment, during the 8 month period after soil amendment. For each metal, mean values with the same letter are not statistically different at the 5% level. Source: [111]

The effects of natural rainfall (600 mm yr^{-1}) were simulated and accelerated over a 30-year period to investigate the sustainability of metal immobilisation by beringite in a kitchen garden soil from the Lommel-Maatheide site [37, 51]. Column tests were undertaken with a sandy soil containing Zn (730 mg kg^{-1}), Cd (8 mg kg^{-1}) and Pb (300 mg kg^{-1}), untreated or amended with 5% beringite. Slightly acid rain water was used as the percolating fluid. The simulated annual rates of Zn and Cd percolation sharply decreased, *i.e.* –80% Zn, –90% Cd, when the soil was treated with beringite [51]. Between 0 and 8 years, the amount of Cd and Zn percolated from the untreated soil stabilized at about 70 μg and 10 mg per year, respectively. After a simulation period of about 20 years, a slight but significant increase in the yearly amount of percolated Cd and Zn was observed. This tendency was not observed in the B-treated soil, and percolated Cd and Zn amounts were slightly reduced from the preceding period. At the end of the 30-year observation period the mean total amounts of Cd and Zn percolated from the untreated soil were 2.37 mg Cd and 296 mg Zn. For the B-treated soil, the amounts were 0.38 mg Cd (16% compared with the Unt) and 27 mg Zn (9% of the Unt). Extraction using a $Ca(NO_3)_2$

solution was carried out on the untreated and treated soils during the 30-year simulation period (Fig. 4). Immediately after the B amendment, there was a 10-fold decrease in extractable Zn and the immobilisation process continued to occur with time. Extractable Cd from the B-treated soil decreased more than 2-fold after 1 year, and extractable Cd was 3 times lower in B-treated soil than in Unt soil after the 30-year simulation.

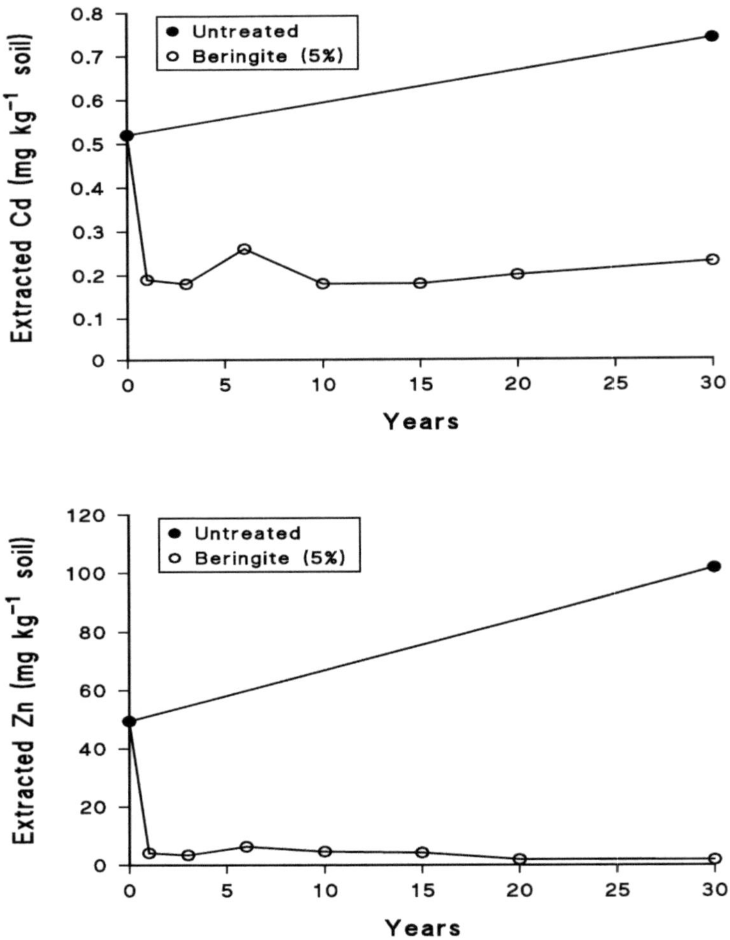

Fig. 4. Changes in Zn and Cd extracted by a calcium nitrate solution during a simulated 30 year period in treated (beringite 5%) and untreated sandy metal-contaminated soil originating from a kitchen garden at the Lommel Maatheide site. Source: [51]

Table 6. Changes in percolate pH with elapsed time at the small-scale Jales Experiment.

	Control soil	Unt	C	CSS	CB	CBSS
07/1998	6.4	4.6	7.4	7.5	8.0	8.0
01/1999	6.4	4.0	6.7	7.0	7.2	7.4
10/2000	6.3	4.1	7.0	7.0	7.3	7.4
09/2001	5.8	3.5	5.9	6.5	7.4	7.7

At Jales, a village in NE Portugal are located a former goldmine and a large volume of spoil characterized by an excess of trace elements (As 1325, Zn 165, Pb 170, Cd 3.8 $\mu g\ g^{-1}$), an acid pH (4.1) and a lack of organic matter. Results for As and metal leaching through the amended soil layer are available from the small-scale Jales Experiment [21]. Initially, trace element composition and volume of percolates were determined at daily intervals over a 443-day-period following initial spoil treatment. Starting in 2000, percolates were collected during each rainfall period. The pH of percolates from Unt spoil was acid and decreased by 1.1 unit (Table 6). The occurrence of sulphide oxidation is supported by increasing sulphate concentration in percolated water (Table 7).

Increases in salt concentrations can have adverse effect for revegetation and aquatic organisms. In year 4, sulphate concentration was lowest in CB and CBSS treatments compared with C and CSS (Table 7). Highest tree development in CSS- and CBSS-spoils may enhance oxygen diffusion along root systems and sulphide oxidation. All amendments increased the percolate pH (Table 6). In year 4, pH was still alkaline in the CB- and CBSS-treated spoil, whilst it decreased in CSS and especially in C. The increase in pH reflected the alkaline properties of B added in the soil amendment. Metal leaching, especially Zn and Cu, was a matter of concern for the Jales spoil (Table 7). In year 3, Cu concentration was 43 times higher in Unt compared with the control soil. Copper leached was greatly reduced to a value in the range of the control soil in all amended spoils. The combination of mineral additives with compost did not offer any additional effect, as compost initially increased the spoil pH to 7.4. Among the amendments, CBSS decreased Cd, Zn, and Pb concentrations in the percolate and only resulted in a small increase in leached As (Table 7). In year 3, Pb leaching was lowest in CB, CSS and CBSS. Compost addition alone initially increased Pb leaching. In contrast to data for Cu, this effect was counter-balanced by incorporation of mineral compounds in the amendment. In fall 2001, Cd leaching was lowest in CB, but was also decreased in other treated spoils. Zinc leaching was largely attenuated in CB and CBSS, both treatments having the highest pH values. Carbonates added with B can react with Zn. The pH values of Jales spoil and percolates were correlated. Consequently, As leaching increased in the CSS and C amended spoils as water-soluble As in spoil reached its maximum value in the pH range 5.5–6.5 for the Jales spoil [52]. Arsenic concentration in Unt was in the range 18-122 $\mu g\ L^{-1}$ and

Table 7. Concentrations of metals, arsenic, and sulphates in the percolates, at small-scale Jales Experiment.

		Control soil	Unt	C	CSS	CB	CBSS
As[a]	min[c]	0	18	218	59	117	26
	Max[c]	30	122	54 500	1 730	10 587	441
	10/2000	21	36	5 475	4 255	4 435	1 305
	09/2001	21	44	2 272	224	3 760	893
	04/2002	12	67	937		1852	168
Cd[a]	min	0	143	0.2	0	0	0
	Max	8	1732	26	21	5	3
	10/2000	< 0.2	781	4	2	3	1
	09/2001	< 0.2	192	14	53	1	38
	04/2002-06-02	0.75	86	8		0.75	1
Pb[a]	min	0	2	0	0	0	0
	Max	25	120	1669	119	119	54
	10/2000	4	26	17	9	4	2
	09/2001	3	18	10	< 2	< 2	< 2
	04/2002	3.6	7.1	6		2	2
Zn[a]	min	6	7 071	30	4	2	4
	Max	1534	59 421	2 150	579	782	564
	10/2000	22	36 600	364	55	177	33
	09/2001	35	20 700	4 330	6 380	63	39
	04/2002	1071	5478	1264		127	496
Cu[a]	10/2000	6	3 475	15	27	34	16
	09/2001	24	1 051	36	90	53	59
	04/2002	61	204	20		6	17
Ca[b]	04/2002	9.4	7.2	41.1		85	83
Fe[b]	04/2002	0.53	0.31	0.38		0.21	0.11
Mn[a]	04/2002	49	1144	599		63	193
Ni[a]	04/2002	12	2.6	2		3	8
SO_4^{2-}[b]	10/2000	2.0	24.7	5 .5	8.3	5.8	7.7
	09/2001	14.6	183	140	568	89	74

a: $\mu g\ L^{-1}$; b: $mg\ L^{-1}$; c: from 06/04/1998 to 19/5/1999; means values from duplicates levelled off on 36-67 μg L^{-1} after year 2 (Table 7). These values are 2-5 times higher than those for the control soil whereas spoil As concentration was 1204 fold higher. Arsenic concentration increased in C and CB spoils compared with Unt, and to a lesser extent in CBSS and CSS (Table 7). Quantification of As percolated on the first 13-month-period led to the following range (base 100 for Unt): C (31 509) > CB (10 801) > CSS (1 382) > CBSS (661) > Unt (100). Spoil amendment resulted in As concentration in percolates higher than maximum permissible concentration (MPC) value (50 μg L^{-1}) for drinking water in all treatments. Soluble organic C was enhanced in percolates from C-spoil. The increase in As percolation could be triggered by several possible mechanisms following compost addition. Under the mechanisms suggested were: an increase in spoil pH and As desorption from solid phases such as Al oxides, As binding to DOM, competition with DOM and with inorganic anions such as phosphates for binding sites, *e.g.* iron oxides. Adsorption of As(V) on amorphous Al oxides as a function of solution pH was greatest at pH values of 4-5 and decreased as solution pH continued to increase [53]. Beringite contains CaO, so one would expect phosphate and arsenate to form 'apatite' or some calcium phosphates/ arsenates in the CB-treated spoil, all of which are sparingly soluble. However, if this mechanism occurred it was insufficient to prevent As leaching on B amended spoils.

Boisson et al. [54] monitored leaching over a 9-month-period at the small-scale Reppel Experiment. In the Reppel soil, As concentration in percolated water ranged from 500 to 700 μg L^{-1} [24], being 15 times the average As concentration in percolates from the Jales spoil, despite the latter having an 11.7 fold higher As content. This demonstrates that As transport to groundwater is more related to mineral speciation and physico-chemical parameters than to total content in the topsoil. Arsenic concentration in percolated water was higher for the B-treated soil than for BSS- and SS-treated soils. This observation was in agreement with the Jales data. Total percolated As was 13.5 mg in B-treated Reppel soil and 9 mg in Unt. Thus, B amendment dispersed As contamination. Similar effects may occur with other alkaline additives. Reductions in As concentration in leachates from the Reppel soil were 8.6 times and 12.0 times greater in the SS- and BSS-amended soils compared with Unt. Taking into account the volume of percolated water, these additives led to a reduction of total As leached by 3.4 times and 4.8 times. Phases added with beringite and/or changes in geochemical processes induced by this material largely increased As leaching as in the C-treated Jales spoil.

A large difference occurred in As leaching between B- and BSS-treatments at each sample time in Jales spoil (Table 7) and Reppel soil. This highlighted that amendment combinations can avoid such side effects. Mechanisms involved in the case of SS have been discussed, but are still uncertain [10, 24, 46]. Oxidation of SS may draw down the extractable metals and/or As by co-precipitation or sorption with newly formed Fe and Mn oxides. Arsenate mainly forms inner-sphere bidentate binuclear complexes on both goethite and ferrihydrite, and the fraction of monodentate bonds decreases as the As(V) coverage increases for the amorphous Fe oxides [55]. One question regards the effect of ageing or residence time on the kinetics of As(V) sorption on the newly formed hydr(oxides) after SS oxidation. Sorption studies over a 12 month period showed that As(V) sorption on goethite was initially rapid, with over 93% As(V) being sorbed in a 24-h period at pH 6. In time, sorption increased while the molecular environment did not change [55]. When ageing progressed, there was no significant change in the amount of As(V) desorbed from goethite by PO_4^{2-}. As the As(V) ion is larger in size and interacts more strongly with some of the surface OH groups, a significant amount of As(V) remained bound to the goethite after 5 months of desorption, even though the PO_4^{2-} desorptive solution was three times stronger than the initial As(V) sorptive solution. Oxidation of sulphide minerals over time would produce sulphates, but sulphate was much less effective than PO_4^{2-} at promoting As(V) desorption [55]. Acidification will promote As(V) sorption on amorphous Al oxides [53].

From the monitoring of small-scale Overpelt, Jales and Reppel experiments, it is concluded that amendments such as CB, CBSS, and BSS have beneficial time-dependant effects in reducing the leaching of Cd and Zn. Leaching of other cationic metals such as Ni, which show a pH dependent solubility, would also probably be attenuated. Transient Pb dispersion can occur with C addition. Attention must be paid to As leaching, and to other trace elements that form anionic species in solution. Compost and beringite separately added to soils can significantly increase As mobility. Trace elements such as Mo, Cr(VI), Se and Boron present at a site might be considered.

Long-term effects of synthetic zeolite addition to contaminated soils on metal and As transport to the groundwater is not well known. Osté [6] investigated leaching from zeolite-treated soils in a batch experiment. Among six synthetic zeolites, (mordenite-type, faujasite-type, zeolite X, zeolite P, and zeolite A) zeolite A had the highest binding capacity for Cd and Zn between pH 5 and 6.5 and was stable above pH 5.5. In the leachate from soil columns, DOM strongly increased particularly at the start of the experiment. This resulted in a higher metal leaching caused by formation of metal-DOM complexes, while free ionic concentrations of Cd and Zn were decreased in soil pore water. The increase in DOM concentration was explained by pH increasing from 6 to 8 and Ca binding after zeolite A application. Zeolite-type materials can be synthesized from fly ash. Using this product, Lin et al [56] reported a strong reduction of Cd concentration in the leachate of zeolite-treated soil columns.

Many short-term batch and column experiments have been done with other additives [10, 20], but fall outside the scope of this review as fate and behaviour can largely change over time following initial soil amendment.

8. FIELD PRACTICES AND CROPPING ASSAYS

8.1. Liming

Applying chalk or limestone ($CaCO_3$), quicklime (CaO) or hydrated lime ($Ca(OH)_2$) to raise soil pH (liming) has been common practice for centuries [13]. Materials containing some Mg such as dolomitic limestone are often used to ensure an adequate supply of Mg to crops. Rothamsted Experimental Station has a set of long-term experiments (*e.g.* Park Grass Experiment) established in the mid 19th century that are particularly useful in understanding the relationship between land use, acidification, lime use and metal mobilisation [13]. Liming effects in agricultural soils with low Cd and Zn concentrations was illustrated by data from two long-term liming experiments established in 1963 [57]. Soil pH in a heavy, silty clay loam ranged from 4.5 to 7.0 in plots at Rothamsted Experiment Farm. It varied from 4.3 to 7.7 in a sandy loam soil at Woburn Experimental Farm. At both sites, total concentrations of Cd and Zn in the soil solution decreased as solution pH increased from 4 to 8. This could be explained by an increase in the adsorptive capacity of the soils due to deprotonation of surface-bound H^+. Immobilisation of low Cd concentrations by liming is sometimes ineffective, as exemplified by increased Cd concentration in potato tubers at 3 sites in South Australia [58]. Seasonal climatic conditions and their consequences on roots distribution between limed surface soil and the deep layers, as well as the occurrence of chlorocomplexes go some way to providing an explanation. The efficacy of dolomitic limestone was compared in a field experiment with that of other amendments, *i.e.* synthetic zeolites prepared from fly ashes, organic loamy shales (top layer of coal beds), muck (material of sapric Histosols), and acid peat in a Fluvisol (pH 5.3) contaminated by flood water (15 mg Cd, 1900 mg Zn, and 1200 mg Pb kg^{-1} soil) [59]. Amendments were mixed in the 0–20 cm depth soil layer, and mustard, rye and oats were cultivated. With the exception of acid peat, amendments improved plant yield. Best results were found with muck and dolomitic limestone. Both amendments decreased Cd and Zn concentrations in plant tissues, whereas their influence was marginal for Pb. Synthetic zeolites and organic loamy shales were less effective.

Some field trials deliver information on the use of lime and similar materials for phytostabilisation of Hg-contaminated soils. A decrease in Hg plant uptake was demonstrated following the addition of sugar cane foam [60]. This waste lime was applied at 3 levels, *i.e.* 0, 0.075 and 0.15 kg m^{-2} on a Hg-contaminated acid paddy soil (0.595 mg Hg kg^{-1}, pH 4.62) located in the Baiyun District of Guangzhou, China. Soil pH ranged from 4.7 to 6.5. Extractable Hg in soil was reduced from 3.5 (untreated) to 1.6 μg kg^{-1} (0.15 kg waste lime m^{-2}). Chinese cabbage, *Brassica parachinensis*, *cv.* Sanjiu-19 was grown in plots. Plant yield was similar between treatments, but Hg content of plants decreased from 0.011 (untreated) to 0.005 mg Hg kg FW^{-1} (0.15 kg waste lime m^{-2}). Some amendments can induce biotic demethylation, reduction, and volatilisation of Hg. In a greenhouse study, the interactions of lime, N, P, and OM additions with respect to plant growth and Hg uptake in the annual grass Zorro fescue (*Vulpia myuros* L.) was examined [61]. In a mine soil at the Sulphur Bank Mercury Mine, Clear Lake, California, the addition of lime and OM showed a negative correlation with soluble- and tissue Hg concentration, due to either Hg adsorption to OM or to inorganic surfaces. Liming increased dry matter yield and consequently uptake of Hg by plants. An efficient alternative method that could be used at Hg-contaminated sites is Hg phytovolatilisation if plants are able to establish themselves in the harsh environment [62].

The practice of liming to remediate contaminated soils and mine tailings has the potential to mobilize arsenic due to the pH dependence of As sorption reactions on oxide minerals and layer silicates [63]. This point is extended in the side effect section below. It may as well be of concern for other trace elements such as Mo, U, V, and Se.

8.2. Biosolids and Liming

Use of biosolids and similar organic wastes such as papermill sludge, alone and in combination with other materials, is a long-standing practice, by which many sites have been restored for 25 years or more [64]. This application provides organic matter necessary to improve soil physical properties, water infiltration and water holding capacity. It delivers micro- and macronutrients necessary for plant growth, and decreases bulk density. It is possible to combine biosolids with other materials that display a high calcium carbonate equivalent, in order to restore metal affected ecosystems, despite the fact that biosolids contain metals [32]. Biosolids can be combined with alkaline materials such as limestone, and cyclonic ashes. Application of biosolids plus lime should reduce metal availability, and increase soil pH. Decrease in direct exposure *via* soil ingestion would be less evident as metal immobilisation could be reversed in the acidic conditions of the stomach. Experiments at Pronto Mine, Ontario, Leadville, CO, Bunker Hill, ID, Palmerton, PA, and Baltimore, MD investigated the efficacy of biosolids and high calcium carbonate to restore a vegetative cover and/or to decrease human exposure at metal contaminated sites.

8.2.1. Pronto Mine Experiment

Pronto mine near Elliot Lake, Ontario was a former uranium mine, operating till 1960, where copper ore was processed until 1970 [16]. Some factors contributed to the lack of vegetation establishment on the fine-grained copper tailings (23 ha) at Rio Algom's Pronto Waste Management Area, *i.e.* low existing pH and continued acid generation, high levels of soluble metals, compaction, erosion, poor nutritional status, lack of organic matter and a high water table which limits the tailing oxidation but causes salts to accumulate near the surface. Semi-operational test plots were constructed on these tailings using papermill sludge as a cover material in combination with lime, with and without a capillary barrier (gravel and blast furnace slag), or through direct incorporation. All plots received 6-24-24 NPK slow-release fertiliser at an equivalent rate of 400 kg/ha, and were seeded with a grass-legume mixture. Over a 3-year period, acceptable establishment of vegetation only occurred in the case that cover materials were incorporated directly. The effectiveness of papermill sludge as a vegetative medium covering oxidized Cu tailings is being determined. Elemental concentrations in the aboveground parts of grass samples were similar to values listed for vegetation growing on background soils in Ontario. Covers of papermill sludge show the following benefits: consumption of atmospheric oxygen through carbon oxidation, reduction of a rise of the water table due to their large water-retentive capacity, reduction of erosion and encouragement of a more diverse and stable vegetation.

8.2.2. Leadville Experiment, CO

Pyrite rich wastes from historic Pb and Zn mine tailing piles entered the Arkansas River and contaminated areas along an 18 km stretch. Oxidation of the reduced sulphur has resulted in acidic soil pH (1.5 – 4.5) [65]. Several contaminated areas were amended with a mixture of 224 Mt ha^{-1} municipal biosolids and 224 Mt ha^{-1} agricultural limestone, and then seeded with a mixture of native grasses. A field experiment was established with 90 – 180 Mt ha^{-1} biosolids alone, or in combination with 224 Mt ha^{-1} lime, and cultivated with annual rye grass. In year 1, treatments of biosolids + lime, increased pH from 3.9 up to 6.4. Calcium nitrate-extractable Cd, Pb, and Zn (–99%) decreased in the surface horizon. Metal concentrations in annual rye grass decreased: Cd (–54%), Zn (–40%), and Pb (–51%). Initial results suggest that biosolids, in combination with lime, were effective in restoring a plant cover on the amended area. But metal concentrations in plant tissue remained too high to prevent animal exposure. Soil organisms, *i.e.* flagellates, ciliates, amoebae, and nematodes were present in higher concentrations in the amended soils than in the uncontaminated control soil, collected upstream of the contaminated area. Earthworm assays indicated a similar survival and biomass in the uncontaminated control soil and in 3 of the 4 amended soils. Using soil packed in columns, lime kiln dust plus biosolid demonstrated the highest pH neutralization capacity and the lowest cereal rye Zn concentration [66].

8.2.3. Bunker Hill Experiment, ID

Mining and smelting of Zn and Pb ores has resulted in extensive metal contamination of the surrounding hillsides and waterways, with more than 500 ha being barren of vegetation. Mining waste material with little or no organic matter contain 5500-14700 mg Zn kg^{-1}, 1500-4900 mg Pb kg^{-1}, and 7-28 mg Cd kg^{-1}. Soil pH ranged from 4.6 to 7.0. Surface application of biosolids (112 Mt ha^{-1}) in 1997, in combination with wood ash (220 Mt ha^{-1}) and log yard debris (20% by volume) was able to restore a vegetative cover on the metal contaminated materials for three years [32, 67]. High N (4.4-5.3%) and low N (2.8%) biosolids were also investigated as applied at 55 and 110 Mg DW ha^{-1} plus 140 Mg wood ash ha^{-1} (Table 8). Plots were seeded with a mixture of ryegrass and vetch. The high N biosolids amendments produced higher biomass than the low N amendment. In the biosolids amended plots, increased soil pH and decreased $Ca(NO_3)_2$ extractable metals permitted plant roots to enter the untreated subsoil. Metal concentrations in plant tissues were within 'normal' concentrations ranges for all treatments during the 3-year period. Calcium nitrate-extractable Zn was reduced in the subsoil from 159 in the untreated plot to 10 mg kg^{-1} in the amended plots. In the wetland site, surface applications of a biosolid compost (60% DW) and wood ash (40% DW) to a depth of 15 cm was sufficient to enable a volunteer plant community to re-establish in the treated area. In this case, metal concentrations of reeds (*Typha latifoli*) were within the normal range. Based on nutrient concentrations in water from the outflow portion of the site, amendment application has not impacted the water quality. In a pot experiment, a mixture of biosolids and alum appeared to be the most successful of five treatments, in reducing soluble Zn species and increasing plant yield *via* an improvement in soil structure, nutrient status and organic matter content [68].

Table 8. Soil pH and 0.01 *M* $Ca(NO_3)_2$ extractable Zn (mg kg^{-1}) of amendments and subsoil, plant biomass (Mg ha^{-1}) and Zn concentrations (mg kg^{-1}) in Western wheatgrass (*Agropyron smithii*) at Bunker Hill Experiment, 3 years following soil amendment.

Treatment	pH	Zn extracted	Plant Biomass	Plant Zn
Untreated	5.8 ± 0.9	150 ± 60		169 ± 10
Low N 55				
Amendment	7.8 ± 0.1	0.8 ± 0.1	1.6 ± 1.0	63 ± 3
Soil	7.3 ± 0.4	11 ± 1		
High N 55				
Amendment	7.8 ± 0.2	1.7 ± 0.6	4.0 ± 1.3	47 ±10
Soil	7.5 ± 0.5	9 ± 2		

Source: Brown et al. (2000)

8.2.4. Palmerton Experiment, PA

A zinc smelter that operated from 1890 until 1980 before modern environmental regulations were implemented has contaminated this site. In the surrounding area, zinc contamination causes serious problems with growing lawns and crops. At Blue Mountain, ARS agronomist R. Chaney has designed test plots that compare the growth of various turf grasses on high-zinc soils treated with different soil additives (http://www.soils.wisc.edu/~barak/soilscience326/agres.htm). Mixing high-iron and high-lime sewage sludge compost into high-zinc soil reduces zinc toxicity in grasses while soil zinc binds to iron in the compost. This decreases zinc uptake by plants. The grass will obtain adequate iron, which further reduces zinc uptake. Tall fescue (*Festuca arundinacea*) varieties grew well on soil treated with the sludge compost. However, planting turf types, which have high endophytes, can cause health problems for foraging animals.

8.3. Cyclonic Ashes

8.3.1. Lommel-Maatheide and Overpelt Experiments

Cyclonic ashes proved to reduce plant exposure to metals and to restore a vegetation cover at several different sites. In the case of the Lommel-Maatheide Experiment, Belgium, 12 years after the CB treatment, currently the vegetation (mainly *Agrostis capillaris* and *Festuca rubra*) is healthy and regenerating by both vegetative means and seeds (Photograph 1) [40, 51]. Applying fertilisers can further enhance plant development (Photograph 2). Growth of test plants (*e.g. Phaseolus vulgaris*) on untreated soils was strongly inhibited. Induction of "stress" enzymes was very pronounced and metal-specific isoperoxidases were induced in both primary leaves and roots. Zinc and Cd accumulated in the primary leaves [51]. The typical features of phytotoxicity as cited above, were absent in test plants grown on the CB-treated soils. Based on these results, contaminated soil in the playground area of the Lommel school was treated with CB. The vegetation immediately recovered and is still developing after 3 years [34, 37] (Photograph 3). Following evidence of metal attenuation in the small-scale Overpelt Experiment at LUC, a field experiment was established at Overpelt. After 3 years, vegetation is well established in the B-, SS-, CSS-, and CBSS-treated plots. Plant biomass production was highest in the CBSS plots.

The efficacy of cyclonic ashes for reducing trace element exposure was investigated in cropped soils. Decreases in Cd and Zn uptake in vegetables was found in test plots established in kitchen gardens. Ten kitchen gardens with sandy soils contaminated by aerial deposition from the former Lommel-Maatheide pyrometallurgical zinc smelter, were treated with cyclonic ashes (100 ton/ha). Zinc contents were between 360 and 983 mg kg^{-1} dry soil; cadmium content varied between 3.1-9.4 mg kg^{-1} dry soil. In year 1, gardens were partly treated with cyclonic ashes (5%, incorporated by rototilling to a depth of about 25 cm). The same vegetables were grown on both untreated and cyclonic ashes-treated plots. Comparison of Cd contents in edible parts of these plants at harvest time (after normal washing) showed marked (2 – 4 times) reductions in Cd content (Table 9). In year 2, the Cd guideline value (0.2 mg Cd kg^{-1} FW) for these high Cd accumulating species (spinach, celery, and lettuce) was reached in almost all cases (except garden 6; results not shown), indicating a progressive amelioration of the Cd immobilization.

Zinc and cadmium contents have been determined in lettuce grown on the Lommel-Maatheide kitchen gardens soils, originating from the percolation experiment, which simulated a 30-year period [51]. Initially, Zn and Cd uptake in the lettuce leaves was 12.0 and 0.31 mg kg^{-1} FW. Gradually, Zn and Cd uptake decreased to 2.48 and 0.09 mg kg^{-1} FW in the B-treated soil, when plants were grown during a 30-year period. These values were very similar to those for lettuce grown on a reference uncontaminated soil, *i.e.* 4.82 mg Zn kg^{-1} FW and 0.08 mg Cd kg^{-1} FW. As metal leaching was lower in B-treated soil, an increased immobilization of both Zn and Cd during the time period under investigation was suggested. In the untreated soil, Zn and Cd uptake was 5.82 mg Zn kg^{-1} FW and 0.55 mg Cd kg^{-1} FW after the simulation period. Zinc leaching and Zn/Cd competition at the root surface may explain decrease in Zn uptake and increase in Cd uptake.

Table 9. Cadmium contents (mg kg^{-1} fresh weight) in Spinach, Lettuce and Celery cultivated in situ on garden soils 2 years after 5% cyclonic ash treatment. * = no plant material available. Given are means of 3 measurements.

Garden	Spinach		Lettuce		Celery	
	original	treated	original	treated	original	treated
control	0.11	*	0.06	*	0.08	*
1	0.96	0.32	0.18	0.06	0.39	0.18
2	0.46	0.23	0.14	0.05	0.26	0.13
3	0.77	0.47	0.25	0.09	1.28	0.42
5	1.48	0.57	0.78	0.18	1.37	0.81
6	0.68	0.40	0.13	0.07	0.55	0.30
7	1.85	0.71	0.15	0.08	0.43	0.24
8	*	0.53	0.63	0.15	0.68	0.32
9	0.87	0.42	0.33	0.12	0.46	0.25
10	1.23	0.42	*	*	*	*

8.4. Iron Oxides

Iron, Al and Mn oxides commonly occur in soils and react with metals and As [22]. The OH-OH distance in Fe, Mn, and Al oxides matches well with the coordination polyhedra of many trace metals. Such hydroxyl groups form an ideal template for bridging trace metals [36, 41]. Reactions with metals can be promoted when these (hydr)oxides are combined with alkaline materials [28]. Three field trials in UK were established to investigate the efficacy of iron oxides as immobilising agents for As [17]. These were an agricultural field, adjacent to a derelict As smelter in Cornwall, long-term sludge spreading plots at a sewage works in Northampton, and a domestic garden in St Helens, Merseyside. Sources and soil concentrations of As differed between the sites. Table 10 lists the soil characteristics of each site, plot dimensions, details of treatments applied at each site, together with the test crops cultivated in successive seasons. Amendments were applied to supply a specific percentage of Fe oxides, calculated as Fe_2O_3. Calculations were based on the 0-10 cm soil horizon and assumed a bulk density of 1.3 g cm^{-3}. On this basis, a 1% treatment amounted to 1.3 kg m^{-2} (13 t ha^{-1}). A single treatment was applied before the first crops were planted, with no further amendment for the duration of the trial. Ferrous sulphate (commercial grade) was applied in solution at St Helens (both treatments), Cornwall and Northampton (0.2% treatment) and as a solid at Cornwall (0.5 and 1% treatments). Because Ferrous sulphate acidifies the soil, it was applied with 0.6% lime at Northampton. Lime was applied at 0.2, 0.5 and 1.0% to the 0.2, 0.5 and 1.0% Fe oxide treatments respectively and at 1% and 4% to the 1% and 2% Fe oxide treatments at St Helens. Three replicate plots were set up at each site for each treatment and amendments were fully incorporated into the soil by tillage. Experimental plants, with the exception of potato, were raised in compost modules and transplanted into the field plots. As far as possible, plants were grown to a saleable size before harvest. Plant As concentrations were measured and data was expressed in terms of plant transfer coefficients [AsTC = Concentration of As in crop (mg kg^{-1}) vs. Total Soil As concentration (mg kg^{-1})].

Plant As concentrations from the unamended soils are given in Table 11, together with the AsTC values for the amended plots (Table 12). Treatments with 0.2% Fe oxides caused a general reduction in AsTC for most crop components at the Cornish site, compared with the control plots, but effects were only significant ($P < 0.05$) for calabrese leaf (1999), cauliflower, radish skin and radish (peeled). AsTC was very variable for potato, but the values were very low for this crop and subject to high variability around the treatment means. There were no significant responses for the Northampton soil. Iron grit treatments had no effect on AsTC values at both these sites. The AsTC values for both Fe oxide treatments at St Helens showed no significant differences in either 1999 or 2000, despite the fact that As was apparently more bioavailable at this site than at either of the other test sites. The test plot at St Helens was a linear strip and there were indications that one end of the plots contained more bioavailable As than the other. Fe oxide treatments stimulated uptake of other metals. At the Cornish site, 1% Fe oxides (2000 data) gave significant uptake (* $P < 0.05$, ** $P < 0.01$) in calabrese leaf (Cu*, K**, Mg**, Ni** and Zn**), lettuce (K*, Mg**, Mn*), spinach (Mg**, Ni**, Zn*), beetroot leaf (K*, Mg**, Mn*) and beetroot tuber (Cu**, K**, Mg**, Zn**). Data from another site in Somerset, which had an alkaline soil, contaminated with As, Cd and Zn (As experimental data not reported here) showed significant increases in Cd uptake following the application of 1% Fe oxides and increased Cd transfer coefficients in lettuce and spinach.

Table 10. Characteristics of the UK field trials with iron oxides.

a) Soil characteristics			
	Cornwall	St Helens	Northampton
Total As (mg kg^{-1})	748	65	72
Total C (%)	2.6	7.3	16.5
LOI (%)	6.3	11.1	31.6
pH (water)	7.24	7.37	5.36
Sand (%)	40.8	71.0	15.1
Silt (%)	37.5	19.0	48.3
Clay (%)	21.7	10.0	35.5
b) Plot dimensions (m^2)			
	Cornwall	St Helens	Northampton
	50	2	50
c) Soil amendments and year of application			
Amendments	Cornwall	St Helens	Northampton
No treatment	1998	1999	1998
Iron Grit (2%)	1998	-	1998
Fe oxides as $FeSO_4$ (0.2%)	1998	-	1998
Fe oxides as $FeSO_4$ (0.5%)	2000	-	-
Fe oxides as $FeSO_4$ (1.0%)	2000	1999	-
Fe oxides as $FeSO_4$ (2.0%)	-	1999	
d) Vegetable crops grown (- indicates no crops grown that year)			
Year	Cornwall	St Helens	Northampton
1999	Lettuce, Calabrese, Cauliflower, Radish	Lettuce, Calabrese	Lettuce, Calabrese
2000	Lettuce, Calabrese, Spinach, Beetroot	Lettuce, Calabrese	Lettuce, Calabrese, Spinach, Beetroot
2001	Potato	–	–
Cultivars	Beetroot (Libra), Calabrese (Regilio (1999), Marathon (2000), Lettuce (Cindy (1999), Kermit (2000), Potato (Lady Rosetta), Radish (China Belle), Spinach (Perpetual)		

Table 11. Crop As concentrations (mg kg^{-1} dry matter) in unamended soils.

Crop	Site Cornwall	St Helens	Northampton
Calabrese (leaf) 1999	8.58	1.62	0.18
Calabrese (leaf) 2000	3.91	1.42	0.08
Lettuce 1999	6.77	17.81	0.41
Lettuce 2000	0.97	6.85	0.08
Beetroot (root)	1.13	–	0.05
Beetroot (leaf)	1.41	–	0.39
Spinach	0.49	–	0.03
Cauliflower	0.64		
Potato tuber (peeled)	0.10		
Potato tuber (skin)	0.35		
Radish leaf	14.4		
Radish root (peeled)	8.39		
Radish root (skin)	21.3		

Direct comparisons of plant AsTC values as related to treatments and soil types are also compounded by differences in crops grown, climate at each of the field sites and the local environment. However, a comparison can be made for calabrese leaf and lettuce (2000 data). AsTC did not differ between Cornwall and Northampton, but was 30 times higher at the St Helens site. The order for As contamination was St Helens<Northampton<<Cornwall, whilst As bioavailability increased in the order Northampton<Cornwall<<St Helens. This may be a function of soil texture, as the order of increasing AsTC mirrored the increase in soil sand content (Northampton<Cornwall<< St Helens) and decreasing clay content. Plant As uptake is relatively low on clays and silts when compared to sands and sandy loams [69]. An additional source of variability may be found in the origins of As in each soil. The St Helens site was contaminated by long distance (> 1 km) aerial deposition of particulate material from historical copper refineries, glass works, tanneries and alkali works. The Leblanc waste from the latter is found in many parts of St Helens and may be the source of much of the re-entrained material. The site in Cornwall was directly adjacent to a historic smelter; As would have originated in particles or dusts emitted from this process. Arsenic at the Northampton site was introduced in sewage sludge and may be affected by the elevated soil organic matter content.

Table 12. Selected AsTC values (x 10^{-3}) for Fe oxide and Fe grit treatments at 3 UK field experiments.

Crop	Site								
	Cornwall			Northampton			St Helens		
	Control	Fe oxide (0.2%)	Fe-grit	Control	Fe oxide (0.2%)	Fe-grit	Control	Fe oxides (1.0%)	Fe oxides (2%)
Calabrese leaf 1999[a]	12.9[a]	6.6[a] *	9.4[a]	2.33[b]	2.90[b]	2.64[b]	2.59	1.58	1.95
Calabrese leaf 2000	5.51	4.93	4.98	1.11	0.58*	0.84	2.34	4.6	3.31
Lettuce 1999	1.01[a]	0.86[a]	0.73[a]	0.54	0.41	0.47	27.3	20.5	17.7
Lettuce 2000	0.14	0.12	0.14	0.10	0.06	0.11	9.67	11.2	8.9
Beetroot (tuber)	1.5	1.2	1.5	0.7	0.42	0.83			
Potato tuber (peeled)	0.15	0.55	0.11						
Radish tuber (peeled)	1.27[a]	1.00[a] *	1.04[a]						
Radish skin	3.16[a]	1.45[a] *	2.62[a]						
Cauliflower	0.94	0.5*	0.75						

Values followed by * are significantly different from unamended control at $P < 0.05$ (n =10, except for [a] n = 6, [b]n = 9).

Results from these trials illustrate the problems of using prescriptive amendments to reduce As mobility across widely differing sites. Soil properties would appear to have a major influence on the As bioavailability, which does not appear to be counteracted by Fe-based soil amendments. The chemical form in which As contaminates the soil may also be important, but in some cases (*e.g.* Northampton) this is inseparable from the pollution source. A more worrying occurrence is the increase in metal bioavailability that results from soil acidification following Fe oxide applications. Increases in plant uptake of phytotoxic (Cu, Ni, Zn) and hazardous (Cd) metals followed incorporation of Fe oxides at 3 different sites. This is unacceptable at sites where As is present with other metal contaminants. Reductions in lettuce yield accompanied by increases in Al and Mn concentrations could be accounted to the acidifying effect of the $FeSO_4$ treatment in the case of the 0.5% Fe oxides without lime, and 1% Fe oxides with less than 1.25% lime. The data for As uptake by lettuce shows that a judicious choice of cultivar can result in a much more significant reduction in plant As uptake and transfer coefficient, than the incorporation of any of the tested amendments. This response is consistent across the three test sites.

8.5. Cyclonic Ashes and Fe and Mn Bearing Materials

8.5.1. Louis Fargue Experiment, Domaine INRA de Couhins, France

Sewage sludges were applied to a coarse sandy soil from 1976 to 1980. In 1995 one block was amended with beringite (5%) and another one with steelshots (1%) [10]. Based on corn grain yield, there were no differences between SS-treated and untreated plots, but a continuous decrease was observed for the B-treated plot (Fig. 5). Corn ears were better filled in the SS-plot, whilst those from the B-treated plots showed poor filling due to Mn deficiency, which resulted from Mn inactivation by beringite in the sandy soil. The addition of SS appeared to result in a sustainable decrease in Cd bioavailability to maize and reduced grain Cd content, at moderate rates of soil Cd contamination (Fig. 6). In 2000, Cd concentration in corn grain from the SS plot, reached a roughly 40% lower, minimum value, compared to the 1980-1994 period before soil treatment. This data indicated amelioration over time, as Cd may become occluded in the newly-formed Fe- and Mn-oxides, and Mn increased in the deeper soil layers due to agricultural practices. The short-term reduction in Cd exposure observed in the B-treated plots (–50%) was not sustainable. This may be related to a Cd/Mn antagonism, as Mn-deficient maize plants showed elevated Cd contents in the leaves and especially in the ears, with a correspondingly reduced biomass. Additionally, this site is contaminated with sludge-borne Ni. Both SS and B amendments were able to cause a sustainable decrease Ni concentration in corn grain, probably due to an increase in soil pH. Nickel behaves differently from Cd in the B-treated plot. In 2001 (year 6 post soil treatment), metal toxicity to, and metal uptake by lettuce was investigated. This plant was selected while it is more sensitive to metals than corn and is recognised as a foliar metal accumulator. Dry matter yield showed a beneficial effect of SS amendment (+32% compared with Unt), whereas a detrimental effect of B (–27%) was again observed (Table 13). Lettuce shoots revealed elevated Cd and Ni, and low Mn when the untreated plots were compared with the uncontaminated plots. Decreases in Cd (–40%), Ni (–57%), Zn (–21%) and P (–30%) concentrations, and elevated Mn concentration (3 times) were found in plants grown on the SS-plots. Both dilution due to increased biomass and changes in the soil solution concentrations were suggested to be responsible for these findings. Lead and copper concentrations were not modified by SS amendment. Based on SS composition, Fe, Mn, Ni and Cu were added to the soil, but only Mn exposure and uptake was increased. The beneficial effects of SS may be explained by the adequate Mn supply together with decreasing exposure to Cd and Ni. Lettuce biomass reduction in the B treated plots could be the result of low P, Cu and possibly Mn concentrations. This may indicate decreasing concentrations of these elements in the soil solution and/or a dysfunction of root uptake and root-shoot transport.

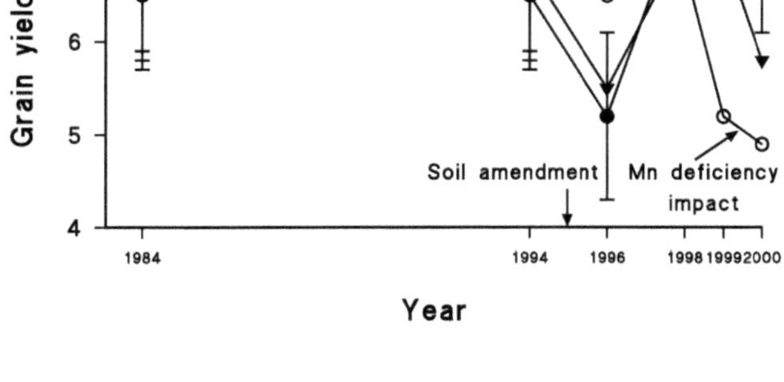

Fig. 5. Changes in corn grain yield at the Louis Fargue Experiment, in the untreated-, beringite- and steelshots-treated plots originally contaminated by a cumulative 50 Mg sewage-sludge ha^{-1} application

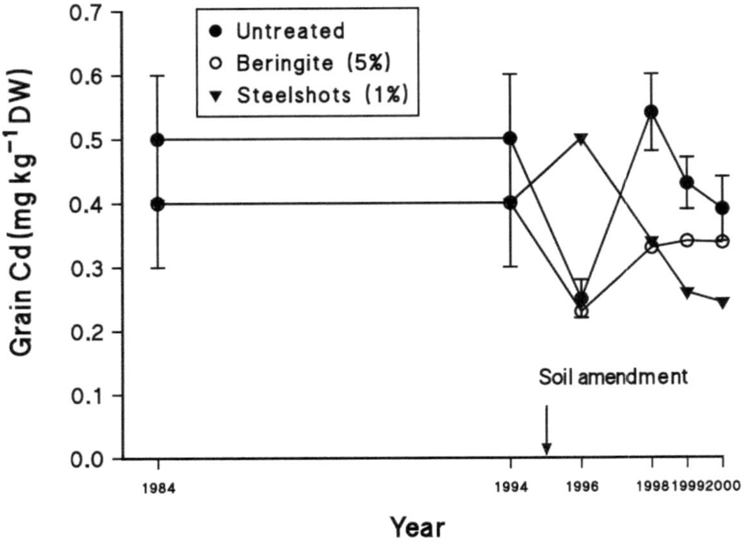

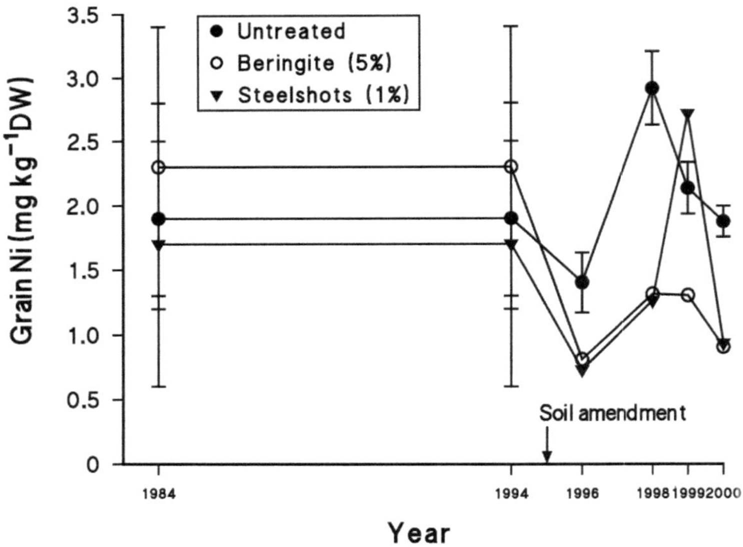

Fig. 6. Cd and Ni concentrations in corn grain at Louis Fargue Experiment, before and after soil amendment using either beringite or steelshots

Table 13. Yield DM and shoot metal concentrations in lettuce cultivated at Louis Fargue Experiment, 6 years after the amendment.

		Untreated	Beringite	Steelshots	Uncontaminated
Yield	Mg ha^{-1}	37.2 ± 7.9	27.2	49.4	63.9
Cd	mg kg^{-1}	56.9 ± 10.4	68.8	34.7	1.6
Cu	mg kg^{-1}	8.0 ± 0.4	5.4	8.3	9.3
Mn	mg kg^{-1}	8.7 ± 1.6	9.5	26.2	40.1
Ni	mg kg^{-1}	15.5 ± 3.4	7.6	6.7	1.6
Pb	mg kg^{-1}	0.33 ± 0.11	0.37	0.43	0.88
Zn	mg kg^{-1}	70.5 ± 2.4	71.8	55.8	82.1
P	g kg^{-1}	8.0 ± 0.2	5.6	5.6	8.6

8.5.2. Jales Experiments

Previous attempts to establish a vegetation cover on the fine-grained spoil at the former Jales Gold mine, Portugal, were unsuccessful. Colonisation by vegetation was limited to a few isolated spots. The grasses *Holcus lanatus, Agrostis castellana,* and *A. delicatula* were the sole colonisers, growing in small isolated tufts [70]. Consequently, erosion of the spoil heaps by wind and water results in a permanent pollution of surrounding terrestrial and aquatic ecosystems. A small-scale semi field experiment was established at Bordeaux, France, and a field experiment was set up *in situ* on the spoil pile at Jales.

Small-Scale Semi Field Trial. Holcus lanatus, originating from the mine site was seeded following implementation of the soil amendments. Grasses germinated but did not grow on Unt (Table 14). *Holcus lanatus* germinated and grew on all treated spoils. In 1998 (year 1), *H. lanatus* shoot yield was similar in all treated spoils, but low compared to the control soil. In year 2, higher biomass production was found in *H. lanatus* from all treated spoils. Greatest biomass was produced in CB and CSS, but differences between amendments were not significant. In year 3, from the middle of the growing season onward, *H. lanatus* declined within the herbaceous vegetation cover and formed isolated tufts. Regular destructive harvesting was halted. An index assessing the presence and development of *H. lanatus* was adopted. *Holcus lanatus* disappeared mainly from the control soil and the CBSS-treated spoil (Table 14). *H. lanatus* was most frequent in the C- and CB-spoils, where pine trees declined. Spoil pH values in these two treatments are different (Table 6), but they display highest As concentrations in percolates (Table 7), and by inference, in soil pore water. Arsenic exposure as well as competition between plant species for mineral nutrients, water, and light, may explain the unexpected response of *H. lanatus*. Seeds of *H. lanatus* and volunteer plant species were dispersed across the plots by natural agencies, such as wind. In spring 2001, some *H. lanatus* seeds germinated on the Unt spoil. All rapidly died in both Unt plots, except two seedlings that developed a small tuft in one plot. In year 2, volunteer plant species established themselves in treated spoils. These included trees (*Salix caprea* L.), vascular plants (*Erigeron Canadensis* L.) and bryophytes (*Funaria hygrometrica* Hedw), increasing plant diversity. *Salix caprea* was the pioneer tree coloniser developing on CBSS, CSS, and C, but not on Unt, control soil, and CB. Birches (*Betula* sp.) are present on CBSS, and maples (*Acer negundon* L.) on CB, C, and CBSS. *Gnaphalium spp.* was identified on B-treated spoil and *E. canadensis* grew well in the CBSS, CSS, and C-spoils.

Table 14. Plant growth parameters (in % compared with reference soil) at small-scale Jales Experiment.

Plant species	Plant parts	Year	Reference	Unt	C	CSS	CB	CBSS
Holcus lanatus	shoot yield	1	100 a	0 c	14 b	14 b	13 b	9 b
Holcus lanatus	shoot yield	2	100 a	0 c	55 b	76 ab	79 ab	61 b
Holcus lanatus	vegetation cover[b]	3	100 d	32 e	432 ab	399 b	499 a	249 c
Pinus pinaster[a]	shoot length	2	100 b	36 c	58 c	120 ab	106 b	125 a
Pinus pinaster	trunk diameter	2	100 a	34 b	41 b	99 a	90 a	107 a
Pinus pinaster	shoot length	3	100 ab	29 d	60 cd	123 a	89 bc	119 a
Pinus pinaster	shoot length	5	100 ab	19 d	63 c	108 a	75 b	108 a

a: 3 year-old, shoot length for reference 487 ± 23 mm; mean values followed by the same letter are not different at the 5% level. b: vegetation cover index: calculated as mean value for a grid of 9 squares each noted as following 0 absent, 1: present - 10%, 2: 11-50%, 3: 51-75%, 4: 76-100%.

Two *Pinus pinaster* Ait. *subsp. atlantica* Villar (6 month-old) were transplanted into each plot in year 2. Differences in shoot length and trunk diameter at the end of the growing season demonstrated a beneficial effect of CB, CSS, and CBSS on tree growth compared with Unt and C-treated spoils (Table 14). Needle necrosis was marked in pines grown from Unt and C plots, and one of the four trees died in both treatments. In year 3, all pines died in Unt. In year 5, pine growth was similar in CSS, CBSS, and control plots. From year 3 onwards, needle necrosis developed on all pines grown in CB plots. Shoot length in year 5 became lower in CB plots than in CSS and CBSS plots. This adverse effect was not apparent in year 2 demonstrating the need for long-term experiments. Spoil pH was similar in CB and CBSS plots (Table 6) and did not explain the pine response. High As concentrations were found in CB percolates (Table 7). Similar results were found in C plots where the pines died more rapidly. Arsenic involvement in the decline of CB-pines is supported by As accumulation in needles (Table 16). The presence of SS in the amendment may explain differences in pine growth. Several explanations can be put forward: *e.g.* low water-soluble As due to sorption on newly formed Fe oxides, nutrient inactivation due to B addition and prevention of this by SS-borne solid phase, or Mn release by SS. Due to competition for uptake between PO_4^{2-} and AsO_4^{2-}, reduced P availability induced by ageing of the CB-spoil may enhance As phytotoxicity expressed in the CB pines (Tables 14 and 16). Non-crystalline iron hydroxide lowered DOM and soluble P and S in a contaminated Cu soil, suggesting that the oxide surfaces effectively removed both inorganic anions (sulphate, phosphate, and likely As) and organics from solution [20].

Table 15. Trace element concentrations in the shoots of *Holcus lanatus* at small-scale Jales Experiment (mg kg^{-1} DW)

		Year		Ref. soil	C	CSS	CB	CBSS
HL	shoot	2	As	0.8 a	9.0 d	4.3 b	6.7 c	3.2 b
HL	shoot	2	Zn	19.8 b	26.4 a	19.7 b	21.1 ab	19.4 b
HL	shoot	2	Cd	0.04 a	0.59 c	0.28 b	0.20 b	0.4 bc

Mean values followed by the same letters are not different at the 5% level. HL*: Holcus lanatus;* bt: biological test in the laboratory.

From year 3 onwards, carpophores of ectomycorrhizal fungi, *e.g. Rhizopogon roseolus* (Corda) Th. M. Fries associated with *P. pinaster,* were observed in the CSS and CBSS spoils [71]. Combined with the effects of iron oxides, this biotic process may partly explain the better P nutrition in CBSS- and CSS-pines (Fig. 7). *Hebeloma leucosarx* Orton and *Hebeloma mesophaeum* (Fries) Quélet developed in association with *S. caprea* in the CBSS spoils.

In year 2, metal and As contents of *H. lanatus* shoots were lower for CB-, CSS-, and CBSS-treated spoils compared to the C-spoil (Table 14). Lowest shoot As contents were found for the CSS- and CBSS-spoils, which agreed with changes in water-soluble and percolate As concentrations (Table 7). In the control soil, CSS- and CBSS-spoils, As content of *H. lanatus* shoot was below the French MPC value for fodder (4 mg As kg^{-1}, 88% DM) (Table 15). In year 3, the As MPC value was exceeded in old, aerial plant parts of all treatments, but in young plant parts only in the CSS plots. (Table 16). Typical concentrations in grasses sampled in uncontaminated areas are 0.1 - 0.9 mg As/kg DW [72]. Concentration of As in the shoot of *H. lanatus* grown on Jales spoils exceeded these values and As concentrations in the roots were highest in the CB-spoil (Table 16). Low values were found in CBSS and CSS root samples. This agreed with the pattern of As concentrations in pine needles. Compared to *H. lanatus*, As accumulated

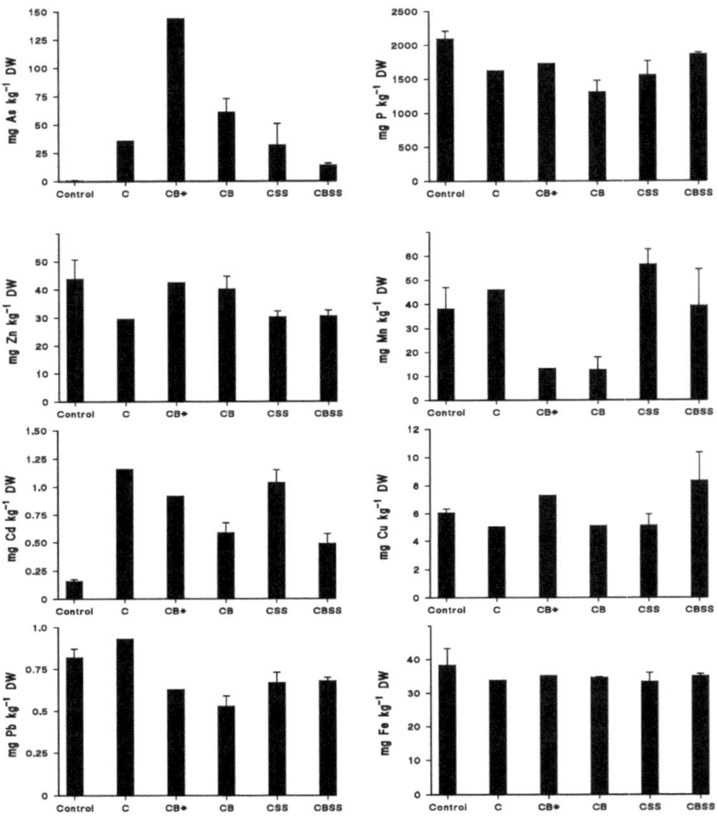

Fig. 7. Mineral composition of the needles of Pinus pinaster Aits grown in untreated and treated spoils at small-scale Jales Experiment, 3.5 years after spoil amendment. (Unt: untreated, Control: control soil, C: compost (5%), CB: C + beringite (5%), BSS: beringite (5%) + steelshots (1%), CBSS: C + beringite (5%) + steelshots (1%)).

more in shoots of *E. canadensis* and foliage of *S. caprea* (Table 16). These two species confirmed decreases in As content of aerial plant parts in the CBSS treatment. In the CSS and CBSS plots, shoot Zn content of *H. lanatus* was decreased to the level of the control (Tables 15 and 16). Zinc accumulation in both *E. canadensis* shoots and *S. caprea* leaves was highest for plants grown on C-spoil. It was reduced in the CSS and CBSS spoils. Changes in Cd content in *H. lanatus* shoots were related to FW shoot yield. Cd content was below the French MPC value (1 mg kg^{-1} DW) for all treatments.

Table 16. Arsenic and Zn concentration (mg kg^{-1} DW) in the plant parts at the small-scale semi-field Jales Experiment, 3 years after spoil amendment

As	*P. pinaster*	*H. lanatus*			*E. canadensis*	*Salix caprea*
	needles	*roots*	*shoots*§	*shoots*§§	*shoots*	*leaves*
Control	0.97	n.g.	n.g.	n.g.	0.5a	n.g.
C	36	73b	3.0	14.1c	25.5c	13.6a
CSS	32	59ab	4.9	7.9a	29.3c	n.g.
CB	61* - 144**	186c	4.5	8.1a	n.g.	n.g.
CBSS	14	57a	3.7	12.5b	15.4b	9.0b

Zn	*P. pinaster*	*H. lanatus*	*E. canadensis*	*Salix caprea*
	needles	*shoots*§§	*shoots*	*leaves*
Control	43.5	n.g.	48 - 89	n.g.
C	29.5	36	438	1126
CSS	30.2	17	112	n.g.
CB	40* - 43**	13	n.g.	n.g.
CBSS	30.5	32	108	388-807

* low visible phytotoxicity; ** high visible phytotoxicity, §§: old, n.g. no growth, nd not determined

Pine needles were collected in year 3. Lowest As and highest P concentrations were found in the control (Fig. 7 and Table 12). High As concentrations were displayed by pines grown in the CB-treated spoil, especially in those exhibiting necrosis of needles. The lowest As and highest P concentrations in pine needles were found for the CBSS plots. The presence of SS in amendments, leading to newly-formed Fe- and Mn-oxides in the spoil, resulted in reduced As accumulation in pine needles. Arsenic accumulation above 62 mg As/kg DW in shoot is toxic for *Pinus sylvestris* [72]. A critical phytotoxic level of 5-10 mg As/kg DW is reported for sensitive crops, and 20 mg As/kg is believed to cause severe growth effects [61]. Arsenic accumulation was suggested to explain the phytotoxicity in *P. pinaster* grown on CB-spoil. Pine trees grown in C-spoil exhibited visible phytotoxicity, and some cases resulted in tree death, whereas As content of needles was lower than that in CB. However, root alteration may affect As uptake and transport to the needles. Highest Cd concentrations in needles were measured in trees grown on C- and CSS-spoils (Fig. 7). Amendments containing B led to low Cd concentrations in trees grown on CB and CBSS spoils. This mirrored low Cd concentrations in percolates (Table 7). The quantification of Mn revealed very low concentration in needles from pine grown on CB spoils, compared with all other treatments (Fig. 7). Arsenic and Mn concentrations varied considerably. The supply of Mn to pines from SS application in CBSS and CSS was remarkable. Manganese–deficient plants suffered impaired growth, as Mn is an essential element for water splitting and oxygen evolution in photosynthesis, and for the activity of certain peroxidase, catalase, and superoxide dismutase enzymes that detoxify oxidative stress. The latter probably occurred here as a result of As accumulation in needles. Copper content of pine needles was highest in both CB and CBSS spoils, but remained below the Cu toxicity range (Fig. 7). These treatments displayed alkaline spoil pH and changes in Cu speciation would probably be a function of organometallic complex formation [20]. Zinc leaching and Zn exposure were high in this spoil, but unexpectedly, low Zn concentrations were determined in pine needles as in *H. lanatus* (Fig. 7, Tables 11 and 12). Zinc concentrations for pine grown in treated spoils were similar or even lower (*i.e.* C, CBSS, CSS) than in the control soil, falling within the 39-65 mg kg^{-1} DW range for Zn concentration in young (7 years) *P. pinaster* needles [73]. There was no evidence of Zn phytotoxicity. Lead concentration in needles was low (0.6-0.9 mg kg^{-1} DW) with no difference between the spoils and the control soil suggesting a significant contribution from atmospheric depositions. Decreases in Mn and P availability and As accumulation in needles are suggested to be responsible for the decline of pine growth in the CB-spoil.

Jales Field Experiment. Based on the laboratory and small-scale experiments, 8 plots were set up on one of the Jales spoil terraces: Unt, B (5%), SS (1%), C (commercial garden compost, 5%), BSS, CSS, CB, and CBSS, with and without P fertilisation [74]. In the 4th year after amendments were incorporated into the spoil to a depth of approximately 30 cm, the maximum increase in spoil pH occurred with the full combination CBSS (Table 17), but this value was low compared with the small-scale semi-field experiment (Table 6).

Table 17. Water soluble As (mg kg^{-1}) and water spoil pH at Jales Experiment, four years after spoil amendment (n = 10).

	Unt	SS	BSS	CSS	CBSS	CB	C	B
pH	4.1± 0.10	4.3± 0.05	4.8± 0.11	4.5± 0.09	5.3± 0.06	4.7± 0.07	4.0± 0.09	5.1± 0.08
As	0.32± 0.02	0.36± 0.01	0.54± 0.02	0.30± 0.02	1.01± 0.02	0.55± 0.01	0.25± 0.02	0.36± 0.02

Source: Bleeker et al. (2002)

Ageing, a higher oxidation rate of sulphide materials and proton flux from adjacent areas and deeper layers may be involved. An increase in spoil pH is important for a reduction in Zn and Cd exposure. The CBSS treatment also resulted in a 3-fold higher water-soluble As concentration (Table 17). Despite a small pH increase in SS and CSS treatments, water-soluble As was similar to Unt. The use of tolerant grasses in combination with spoil treatments resulted in a rapid and effective revegetation of the contaminated spoils. Colonisation and reproduction of both *H. lanatus* and *Agrostis castellana* were most successful (470 days after sowing) when additives were combined, and spoil supplemented with P fertiliser (Table 18). Best results in surface cover were found for both grass species in BSS, CSS, CBSS, and CB (Photograph 4). Phosphate addition resulted in a more profound improvement of surface cover for *A. castellana* than for *H. lanatus* [74]. The addition of P-fertiliser increased tillering in both grass species. The tiller number was highest in CBSS (*i.e.* 110 tillers *H. lanatus* m^{-2}), BSS, CB and CSS. Low As concentration in grass was only obtained when the additives were combined (Table 18). Lowest As concentrations were found in CBSS, CSS, CB and BSS for both *H. lanatus* and *A. castellana*. This mirrored data from the small-scale semi-field (Tables 15 and 16). Volunteer plant species colonised some plots (Photograph 4).

A. delicatula started to colonise the area especially around BSS and CB. Other species have made an attempt to survive the harsh environment, especially in plots with added B: *Chamaespartium tridentatum* (L.) P. Gibbs, *Adenocarpus complicates* (L.) Gay, *Plantago lanceolata* L., *Lolium multiflorum* L. (seed spread with the compost addition), *Cytisus striatus* (Hill) Rothm, and *P. pinaster* (never surviving past seedling stage). In year 3 almost no plant had established itself in the Unt and SS plots. In the CB, BSS, and CBSS plots, *Agrostis* and *Holcus* developed and improved their environment, which in turn accelerated the survival of other plants like *C. striatus*. The As content of *C. striatus*, was reduced by roughly 33% in the CBSS and CB spoils compared with the other treated spoils. In contrast, no effect of the amendments was found on the As content of *Betula alba* L.. It is questionable whether or not side-effects of B-addition, as they were found in the small-scale experiments, will occur in time. Due to a lower initial pH than at the small-scale experiment, this might not occur, but needs to be investigated.

8.5.3. Small-Scale Reppel Experiment

This experiment is located at the INRA Bordeaux-Aquitaine Centre, France. In year 5, soil pH in plots amended with B reflected the alkaline properties of this material (Table 19). Oxidation of SS, combined with B addition had the greatest effect on soil pH. The FW yield of all plants increased in B and BSS-treated soils. Aerial plant parts were examined, but

Table 18. Arsenate (µmol g^{-1} DW) in aboveground biomass of *Holcus lanatus*, *Agrostis castellana*, *Cytisus striatus* and *Betula alba* grown in year 2 on untreated spoil (Unt) and spoil treated with Steelshots (SS), Organic matter (C) and/or Beringite (B) in presence and absence of P fertiliser. n = 8 (± SE), n.g; no plant growth.

	H. lanatus			*A. castellana*		*C. striatus*	*B. alba*
	green	green fertilised	senescent fertilised	green	green fertilised	green	green
Arsenate							
Unt	n.g.	1.39 ±0.12	2.11 ±0.23	n.g.	0.63 ±0.10	n.g.	0.90 ±0.15
SS	n.g.	0.89 ±0.15	1.73 ±0.12	n.g.	0.36 ±0.04	0.38 ±0.04	0.91 ±0.28
SSB	0.47 ±0.05	0.34 ±0.09	0.81 ±0.08	0.27 ±0.05	0.43 ±0.05	0.27 ±0.02	1.34 ±0.16
CSS	0.39 ±0.03	0.23 ±0.02	0.76 ±0.06	0.25 ±0.01	0.28 ±0.01	0.32 ±0.03	0.94 ±0.15
CSSB	0.34 ±0.03	0.25 ±0.01	0.72 ±0.06	0.17 ±0.04	0.28 ±0.05	0.19 ±0.02	1.18 ±0.10
CB	0.38 ±0.05	0.33 ±0.01	0.89 ±0.07	0.29 ±0.03	0.30 ±0.05	0.21 ±0.01	1.35 ±0.43
C	1.06 ±0.04	0.54 ±0.03	1.27 ±0.09	0.50 ±0.12	0.50 ±0.16	0.32 ±0.01	1.44 ±0.14
B	0.61 ±0.04	0.44 ±0.06	1.20 ±0.09	0.59 ±0.03	0.40 ±0.07	0.34 ±0.03	0.86 ±0.06

Source: Bleeker et al. (2002)

no visible toxicity was noticed. In this agricultural contaminated soil, beneficial effects of incorporation of soil amendments on plant biomass were sustainable. Compared to the control soil, the mineral composition of the corn 3rd leaf from the Unt soil, showed contamination by As, Cd, and Cu (Table 20). The tissue As Table 18. Arsenate (µmol g^{-1} DW) in aboveground biomass of *Holcus lanatus, Agrostis castellana, Cytisus striatus* and *Betula alba* grown in year 2 on untreated spoil (Unt) and spoil treated with Steelshots (SS), Organic matter (C) and/or Beringite (B) in presence and absence of P fertiliser. n = 8 (± SE), n.g; no plant growth concentration was below the critical phytotoxic level of 5 mg As kg^{-1} for sensitive crops. Cadmium and Cu tissue concentrations were not phytotoxic for corn. In year 4, a decrease in As concentration was found only with BSS, despite substantial changes in soil pH. Effectiveness of BSS was sustainable, but since year 1 this decreased from 70% to 50% (Fig. 8). This decrease could partly be explained by a reduction in As concentration of the corn 3rd leaf from the Unt plots over time. In years 1 and 2, As contents in the corn 3rd leaf fell within the range 8-9 mg kg^{-1}. Decrease in B-treatment effectiveness was more pronounced (Fig. 8) and in year 4, this amendment did not affect As taken by maize. Initial soil pH was 7.5 after B amendment and soil pH decreased to 6.35 (Table 19). Dissolution of calcium-arsenates formed and As desorption from Fe oxides are suggested. The SS oxidation combined with B, leading possibly to a mixture of newly formed Fe and Mn oxides, calcium phosphates and arsenates, and allophans, may explain the decrease in As exposure *via* the soil solution in the BSS treatment. Steelshots contain 0.2% chromium that might enhance the formation of chromate-arsenate ettringite in the presence of beringite. In previous years, As uptake by corn was reduced in the SS treatment (Fig. 8). In year 4, there was no difference between the Unt and SS treatments. Without a combination with B, SS addition did not sustainably attenuate As exposure to maize.

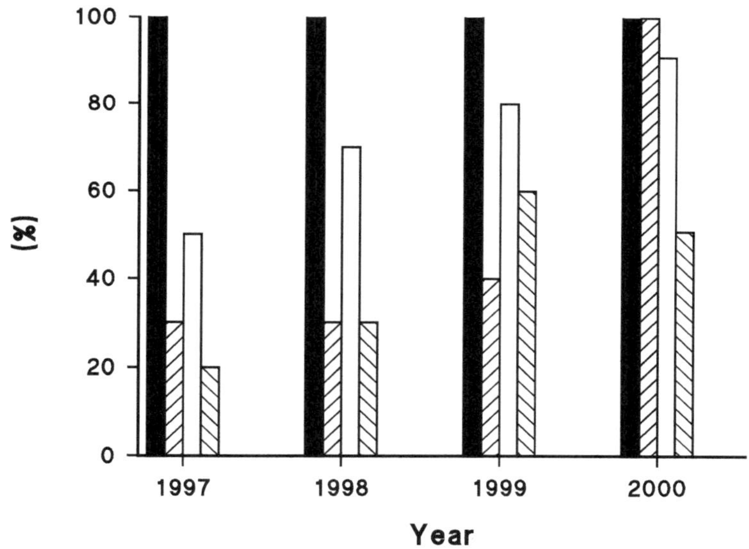

Fig. 8. Changes in As concentration in corn 3rd leaf at small-scale Reppel Experiment (expressed in % compared with the untreated Reppel soil)

Table 19. Plant biomass and soil pH at small-scale Reppel Experiment, Bordeaux, F.

	Corn Ear§	Soil pH	Lettuce Shoot*	Radish Whole plant*
Year	4	5	5	5
Control	204 a	4.9	113a	6.1ab
Unt	157a	4.9	168a	5.7a
SS	230ab	5.15	246ab	8.5bc
B	264b	6.35	283c	7.9b
BSS	249b	6.9	293bc	10.5c

* g FW/plant, § g DW/ear

Zinc and Cd concentrations were reduced in the 3rd corn leaf in plants grown on all B-amended soils, reflecting their higher soil pH (Table 20). Soil CEC increased from 3.8 to 5.3 cmol kg^{-1} in B-treated soils, which was higher than expected, based on the 22 cmol kg^{-1} CEC of beringite. This may be due to the formation of minerals. Chemical extractions inferred that B addition induced *de novo* formation of ferrihydrite and allophans in Reppel soil [44]. The proportions of Mn (3 times), Zn and Cu associated with these amorphous minerals, showed significant increases. Sorption of Cu by amorphous minerals might explain why Cu uptake was lower than expected in corn grown in the B-and BSS treated spoil, compared to SS, despite higher soil pH found in the former. A high pH may enhance formation of organometallic Cu complexes in the soil solution. Non-crystalline Fe oxides were claimed to decrease DOM [20] and one would expect a reduction in Cu and Pb uptake in the SS plots. However, in plants grown in the SS treatment, both Cu and Pb concentrations increased in the corn 3rd leaf. This was not found in the BSS treatment. Steelshots contain Cu (1010 mg kg^{-1}) as an impurity. The source of the additional Cu and Pb taken up by corn has not yet been identified.

Table 20. Trace element concentrations (mg kg^{-1}) in the corn 3rd-leaf grown in the soils at small-scale Reppel Experiment, 4 years following soil amendment.

	As	Zn	Cd	Cu	Pb
Control soil	< 0.5a	69b	0.065a	< 3a	0.41c
Unt	2.16c	69bc	0.668d	4.6b	0.28a
SS	2.16c	86c	0.705d	8.2d	0.51d
B	1.97c	43a	0.315c	5.2c	0.38b
BSS	1.09b	48a	0.233b	5.5c	0.28a

8.5.4. Small-Scale Overpelt Experiment

Laboratory simulations have demonstrated the effectiveness of B and BSS treatments [46]. An increase in vegetation cover, in terms of both biomass and diversity, could be quantified in the CB and CBSS treated lysimeters (Table 21).

Both small-scale semi-field experiments (*e.g.* Reppel, Overpelt) and field experiments (Jales, Overpelt) confirm the sustainable beneficial effect of beringite combined with steelshots to remediate sandy metal-contaminated soils at low and high metal contamination. Beringite application appears to have a long-term effect in suppressing soil acidification. However, the use of amendments such as Beringite that will lead to alkaline soil conditions must be carefully considered in relation to their potential impact on soil As partitioning.

Table 21. Natural colonization of plant species on lysimeters filled with the Overpelt soil.

Soil treatment	Number of species	Plant species
Unt	1	-*Agrostis capillaris* (1 plant in outer part of the lysimeter)
C	2	-*Agrostis capillaris* -*Cerastium fontanum*
CB	5	-*Agrostis capillaris* -*Cerastium fontanum* -*Sonchus asper* -*Poa annua* -*Chenopodum album*
CBSS	6	-*Agrostis capillaris* -*Cerastium fontanum* -*Sonchus asper* -*Claytonia perfoliata* -*Tanacetum vulgare* -*Barbarea sp*

8.6. Zeolites and Other Amendments

The Northampton and Staffordshire field trials were established in UK to evaluate the efficacy of soil amendments in reducing Cd transfer from soil to vegetable crops [33]. Elevated soil Cd at the Northampton site (mean total Cd content 47 mg kg^{-1}) was the result of the application of sewage sludge, whilst at the Staffordshire site (mean total Cd content 16 mg kg^{-1}), deposition of Cd oxide particles from an adjacent pigment manufacturer had contaminated a domestic garden soil. The Staffordshire site was only Cd contaminated, whilst the Northampton soil contained elevated contents of other metals and As. (Table 22). Test plots at Northampton were 10 x 5 m, but at the Staffordshire site restrictions on space restricted plot size to 2 x 2.2 m. Plots were treated in 1998 with the following amendments: $FeSO_4$ 1% (equal to 1% Fe_2O_3, plus 5.7% lime), Iron Grit 1%, Lime, 3% and Zeolite 4A, 1%. The treatments were added at rates calculated as a % of the weight of the top 10 cm of soil, assuming a bulk density of 1.3. Thus, a 1% treatment amounted to 1.3 kg m^{-2} (13t ha^{-1}). Amendments were manually incorporated into the upper soil horizon. Test plants were raised from seed in compost modules, and then transplanted at each site. This mirrors current agricultural practise for field-grown vegetable crops. Crops were grown over 3 seasons at Staffordshire and 2 seasons at Northampton.

Table 22. Crop Cd content (mg kg^{-1} dry weight) in relation to soil amendment in two UK field trials.

1) Staffordshire						
Crop	Treatment					
	Control	Zeolite 4A	$FeSO_4$	Lime	Iron grit	sem
Calabrese (1998)	5.12	5.41	6.67	5.17	7.17	0.8
Calabrese (2000)	3.2	2.5	2.8	2.9	2.8	0.16
Cabbage (1998)	3.58	3.84	4.04	5.59	4.76	1.1
Lettuce (1998)	18.58	12.54	17.78	13.3	19.36	2.9
Lettuce (crop2 1998)	15.7	8.9	9.2	15.8	9.95	-
Lettuce (1999)	28.9	21.7	37.8	25.1	36.2	6.33
Lettuce (2000)	24.7	23.7	23.8	21.6	4.15	3.27
Spinach (1998)	16.7	12.6	21.1	16.4	15.1	3.47
Radish leaf (1999)	16.3	15.4	16.1	16.5	18.9	2.55
Radish leaf (2000)	21.5	20.6	18.2	15.4	18.1	3.13
Radish root (2000)	12.4	13.2	10.6	11.9	13.8	1.67
2) Northampton						
Crop	Treatment					
Calabrese (1999)	2.65	1.8	2.66	2.28	2.29	0.39
Calabrese (2000)	1.58	1.77	1.53	1.42	1.46	0.09
Lettuce (1999)	28.8	15.7	18.04	15.7	18.1	2.95
Lettuce (2000)	12.5	9.93	10.1	9.1	11.0	1.43
Spinach (2000)	13.3	9.26	10.4	9.6	9.4	1.49
Beetroot (2000)	7.91	5.88	6.00	5.6	5.9	0.67

Results were less than conclusive (Table 22). There were no significant reductions in plant Cd at either site following the incorporation of *in situ* soil amendments. Both lime and zeolite treatments achieved some reductions in soil-plant Cd transfer, but samples were too variable within each site to be considered as significant at the replication level employed. Cadmium uptake by the same crop showed well-marked cultivar differences in successive growing seasons. The data for lettuce illustrates this very clearly. Cultivars were changed each season (1998 Cindy, 1999 Sunny, 2000 Kermit), as the seed merchants could not re-supply those used in previous years. Each had the same growth form (Butterhead) but there are clear differences in Cd uptake by these apparently similar varieties.

Some comments can be summarized from UK field trials. The use of *in situ* amendments appears more effective in reducing As mobility in soils than is the case for Cd. Some significant plant responses have been noted with Fe-oxides at the site in Cornwall, but not at other contaminated sites. Whilst the Fe-oxide treatment reduces soil-crop As transfer, it can increase transfers of other heavy metals, due to soil acidification. None of the *in situ* amendments acted to consistently or significantly reduce Cd transfer from soil to crop. These trials used major UK field-grown vegetable crops (Calabrese, Potato, Cabbage, Cauliflower, Lettuce) as well as some potential indicator species (Spinach, Radish, Red Beet), which were known to be efficient Cd and As accumulators. The results demonstrate that whilst transfer of both As and Cd from soils to crop is generally low, incorporation of a range of amendments has little further influence on reducing risk associated with this transfer. A more effective method of achieving the objective might be *via* the use of metal-inefficient cultivars; results show that there is a wide range of As and Cd uptake efficiency in the genome of several vegetable species.

8.7. Red Muds

Red mud is a by-product of the alumina industry that is alkaline and rich in Al/Fe-oxides. Several pot experiments have demonstrated its efficacy for remediation of metal contaminated soils over relatively long time periods [10]. Red mud performed well in a 15-month pot study carried out by Friesl et al [75]. Red mud reduced Zn, Cd, and Ni extractability by 63%, 42%, and 50% as compared to the control. Zinc, Cd and Ni uptake in *Amaranthus hybridus* were reduced by 53%, 40% and 59%. Friesl et al [75] and Lombi et al [76] have designed long-term field experiments to assess the effectiveness of these amendments, but results are still in infancy.

8.8. Phosphates

Phosphates react with many metals, metalloids and radionuclides. Precipitates formed can be stable over a wide range of geochemical conditions. A range of compounds has been evaluated including mineral apatite, synthetic hydroxyapatite, and diammonium phosphate materials. In pot trials, these materials were effective at attenuating metal exposure *via* the soil solution and incidental ingestion, especially for Pb, but also for Zn, Cd, U and Mn through formation of metal-phosphate minerals [77, 78]. Conversion of soil Pb to pyromorphite, a lead phosphate [$Pb_5(PO_4)_3$(OH, Cl, F,...)], could immobilize soil Pb and reduce its bioavailability, especially in the gastrointestinal tract. This would be enhanced by initial acidification of the soil to enhance dissolution of Pb solids thus intensifying the subsequent reaction.

A field trial was installed in a vacant city lot at Joplin, MO in 1997 to evaluate different techniques for Pb inactivation in contaminated soils. Lead was mainly present as Pb-carbonates. The efficacy of TSP (1%, 3.2%), rock phosphate (1%), phosphoric acid (5 000 and 10 000 mg P kg^{-1}), compost (10%, 10% compost + 0.32% P, 10% compost + 1% P) and Fe-rich, a by-product from titanium processing (1% Fe + 0.32% P, 2.5% Fe + 0.32% P, 2.5% Fe + 1% P) was compared [79, 80]. The compost was a Fe-rich product using lime-stabilized biosolids as a feedstock. Three months after amendment application to a depth of 12.5 cm in the upper soil layers, $CaCO_3$ (71% purity) was added to each plot to bring the pH to 7. Soil pH ranged from 6.45 in the control treatment to 7.6 in the 10% compost treatment. The amount of lime required ranged from 156 kg per plot (3.2% P as TSP) to 38 kg per plot (10% Compost + 0.32% P as TSP). Seed (K31 Tall Fescue) was hand scattered over the plot surface. Relative plant uptake of Pb in relation to total soil Pb was calculated. This ratio ranged from 0.0013 (3.2% P) to 0.0085 (2.5% Fe + 1% P). Pb availability measured by both relative plant Pb uptake and *in vitro* accessible Pb indicated 3.2% P as the most effective amendment. Reductions in Pb availability were also evident with 10% compost + 0.32% P and 2.5% Fe + 0.32% P.

A field demonstration was carried out at a Pb-contaminated battery-recycling site in Jacksonville, Fl [81]. Three phosphate combinations were used: H_3PO_4 only, H_3PO_4 + CaH_2PO_4, and H_3PO_4 + phosphate rock. TCLP (Toxicity Characteristic Leaching Procedure)-Pb was determined in soil samples over a 300-day period. All P amendments were effective in reducing soil TCLP-Pb, Cu and Zn to below the regulatory limit. However, Pb levels in groundwater showed a transient increase up to 0.25 mg L^{-1} in the first 4 weeks after application of amendments.

Environmental models are needed both to increase the acceptance of *in situ* attenuation and to help identify the limiting factors. One model has been developed for Pb-contaminated soils amended with hydroxyapatite and simulated to describe the effects of initial contaminant concentration, diffusion coefficient, and aggregate diameter on the time of remediation defined as the time required to reduce the aqueous phase lead concentration to < 1 μg L^{-1} [82]. The model showed that aggregate size plays a significant role in determining the effectiveness of remediation.

Arsenic movement with additions of phosphates has been reported [46, 83]. Phosphate input can increase the size of the mobile As fraction and As uptake by higher plants [84].

8.9. Clays

Addition of clay to soil can produce physical and chemical changes that could affect contaminant fate and transport. These include increased cation exchange capacity, increased mineral surface areas, and sorption within the clay interlayer. Illitic materials may be effective stabilizing agents for Cs^+ due to their ability to fix this cation under a range of moisture conditions. This was investigated using batch and dynamic leaching to evaluate the effectiveness of illite alone, and in combination with hydroxyapatite, and zeolite, as amendments to reduce mobile pools of ^{137}Cs and U in contaminated sediments [85]. Addition of illitic clays reduced Cs^+ extractability by 1 *M* NH_4Cl, 0.5 M $CaCl_2$ and TCLP. Zeolite was ineffective in reducing Cs^+ mobility when subjected to aggressive extractants. Hydroxyapatite was less effective than illite at reducing NH_4^+-extractable Cs^+. Hydroxyapatite and mixtures of hydroxyapatite with illite or zeolite were highly effective in reducing U extractability in both batch and leaching tests. This immobilization was extremely rapid.

Montmorillonites modified by aluminium display a preferential sorption of trace metals compared to other bivalent cations and have been proposed as binding agents [28, 86, 87]. Metal binding to clay-aluminium complexes is higher than binding to clay or aluminium alone. Reduction in CEC due to the positive charge of the aluminium hydroxide is compensated by an increase in high affinity sites for metals and increased surface area. Al- and Al_{13}-montmorillonites immobilised Ni, Cu, Zn and Cd whereas their effect on dissolved Pb was rather small [86]. The sorption of Ni and Zn increased with time, probably due to structural rearrangement in the Al interlayer. This was not found for larger cations such as Cd. Cadmium was hypothesized to sorb on surface sites of the Al hydroxide lattice, and consequently little or no incorporation may occur. Al-montmorillonite was effective in the following pH ranges: 6 – 8 for Ni and Zn, 4 – 6 for Cu, and 7 – 9 for Cd. It did not induce nutrient deficiency in red clover. Addition of HCl led to a partial remobilisation of the sorbed metals. Under neutral and acidic conditions, reactions with metals are due to the permanent negative charges of the montmorillonites, while edge site complexation is not important in this pH range. Sorption and desorption were hysteretic with pH which may retard remobilisation. Zinc desorption from aged Al_{13}-montmorillonites after acid addition showed a positive linear correlation with Al in solution. Within 6 months, the interlayer Al_{13} converted to solid $Al(OH)_3$, where Zn is probably incorporated.

Gravel sludge, a waste product of the gravel industry, contains illite (29%), calcite (30%), and quartz (18%). Its efficacy as an *in situ* immobilizing additive was investigated at two application rates in three field trials with sandy loam soils at Dottikon, Rafz and Giornico, Switzerland, contaminated by Zn, Cu, and Cd [88]. Gravel sludge application increased pH in all three topsoils by up to 0.6 units, and reduced $NaNO_3$-extractable Zn concentration by more than 65%. No effect was found on the $NaNO_3$-extractable Cu concentration at Rafz, and an increase in this Cu fraction was evident at Giornico. In the Dottikon soil, $NaNO_3$-extractable Cu concentration decreased from 0.97 to 0.7 mg kg^{-1}, while Zn and Cu concentrations in ryegrass were reduced by more than 35%. Lettuce Zn and Cd tissue concentrations were decreased by 22 to 48% at Giornico and Dottikon, whereas no effect was found at Rafz. Gravel sludge efficacy was highest in soils with high $NaNO_3$-extractable metals, and higher for ryegrass than for lettuce.

8.10. Competitive Uptake at the Root Surface or Competitive Transfer into Plant Parts

Attenuation of trace element exposure and bioaccumulation may operate through an alteration of the uptake of an element from the soil, due to competition for uptake at the root surface or competition during transfer to organs [58]. There are several examples of this type of interaction, *e.g.* Cd/Zn, Cu/Fe, As/P [24, 89], and these may be of use in treating contaminated agricultural soils. Use of P fertilizers with Cd impurities has resulted in a slow accumulation of Cd in agricultural soils in Australia, producing crop Cd uptake that exceeds regulatory limits. Several amendments were evaluated under field conditions at 3 sites. These were Alfisols (clay mineralogy dominated by kaolinite) and pH values between 5-7: *i.e.* natural clay, alum-based water treatment residuals (Alum WTR, 15 t ha^{-1}), lime (15 t ha^{-1}), magnesite ($MgCO_3$, 15 t ha^{-1}), clay, Fe-rich fines from a ferrous smelter (FRF), zinc sulphate (25 kg ha^{-1}) and copper sulphate (10 to 50 kg ha^{-1}) [58]. Large differences in tuber yield occurred between sites, but amendments had no significant effect on "within-site" potato tuber yield or leaf Cd concentrations. Cadmium concentrations in tubers grown on soils amended with a mixture of Cu/Zn salts and with natural clay were lower (−18%) than those amended with lime. All other treatments at all sites were ineffective in reducing Cd concentrations in tubers. From all treatments, especially liming performed poorly under field conditions, compared with the good results obtained in the laboratory at similar application rates and under similar conditions of irrigation water salinity. Results were not sufficient to warrant recommendation the use of Cu/Zn salt application to reduce crop Cd concentrations.

9. OTHER ASSAYS

9.1. Soil Microorganisms

Presence and behaviour of microbial communities are one bottleneck in the process of restoring soil functions that enhance phytostabilisation. Studies have been made with biosensors and other options, but relevant data is very limited. Metal exposure to microorganisms was assessed in year 4 at the Louis Fargue Experiment using a biosensor kit (Biomet™) [90]. In this assay, bacterial luminescence intensity is related to metal exposure in the soil. The Cd/Zn specific indicator strain of *Ralstonia metallidurans* CH34 demonstrated a decrease in Cd exposure in B- and SS-treated soils in both the farmyard manure and 300 Mg ha^{-1} sewage sludge plots (Fig. 9). This compared well with decreased Cd concentrations in corn grain in the B- and SS-treated farmyard manure plots. However, data were not consistent with corn grain and lettuce shoots from the 300 Mg sewage sludge ha^{-1} plots as Cd concentrations were similar in plants grown on untreated and treated soils. For the 50 Mg ha^{-1} plots, Cd exposure to microorganisms showed a slight decrease in the B-treated soil. This was reflected by corn grain Cd in year 4 (Fig. 6), but not by Cd in lettuce shoots cultivated in year 6 (Table 13). Similar Cd exposures in the Unt and SS-treated 50 Mg ha^{-1} plots did not relate to decreases in Cd concentration in corn grain (Fig. 6) and lettuce shoots (Table 13). The Ni strain showed a general decrease in Ni exposure for both B-and SS-amended soils compared with the Unt soil. Lowest Ni exposures were found in SS-treated soils in the farmyard manure and 50 Mg sewage sludge ha^{-1} plots (Fig. 9). This agreed with lettuce Ni concentration at 50 Mg sewage sludge ha^{-1} (Table 13). Beringite treatment of the farmyard manure plots led to a lower than expected Ni content in lettuce. Lowest Ni exposure in the B-treated soil at 300 Mg sewage sludge ha^{-1} exactly mirrored the responses of corn grain and lettuce shoots.

The Cd/Zn Biomet strain bacterial biosensor demonstrated a decrease in soil Cd and Zn availability in year 1 following the use of amendments in soils from the Lommel-Maatheide site [34]. The soil bacterial ecology was evaluated. In untreated soils, the metal resistant subpopulation was high. In contrast, a $>90\%$ decrease in Zn-resistant bacteria was observed for soils treated with the most efficient additives such as cyclonic ashes combined with iron grit. These results were confirmed in the long-term field experiment started up in 1990 [39, 40]. Extractable (0.01 M $CaCl_2$) Zn decreased from 525 mg kg^{-1} to 16 mg kg^{-1} in the treated plots. Diversity of higher plant species and saprophytic fungi was extremely low in the untreated area due to the high soil toxicity and the absence of metal-tolerant ecotypes of plants

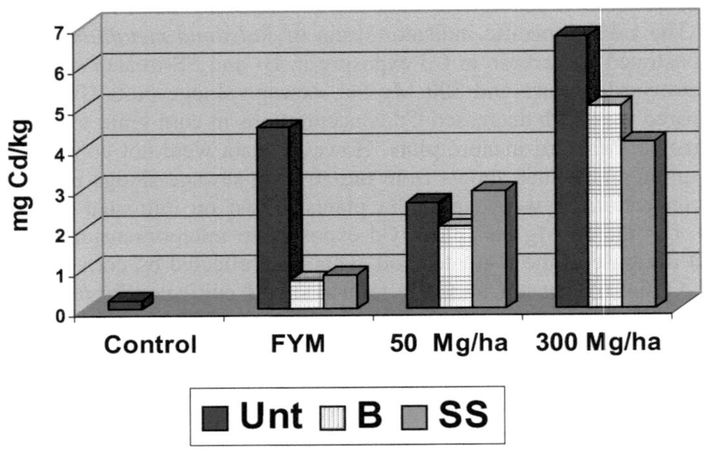

(a)

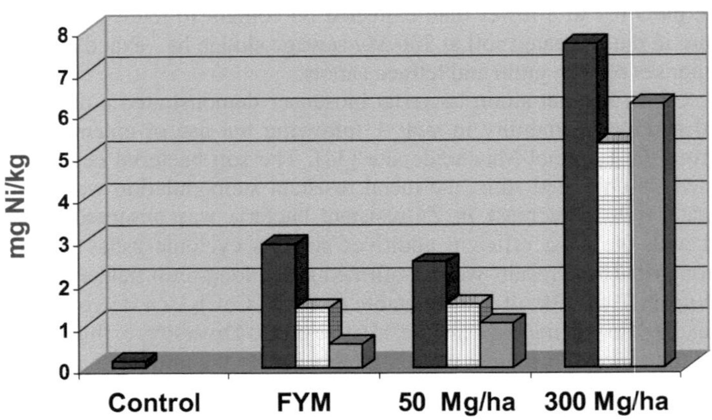

(b)

Fig. 9. Response of the Biomet™ biosensor (Ralstonia metallidurans CH34) exposed to soils sampled at the Louis Fargue Experiment; (a) strain Cd/Zn, (b) strain Ni

Table 23. Increase in plant species diversity at the Lommel-Maatheide Experiment and at the small-scale Jales Experiment.

Lommel-Maatheide						
	Treated soil			Untreated soil		
Frequence in the vegetated quadrats (%)	Number of plant species			Number of plant species		
90	2			1		
1 – 10	3					
< 1	8			2		
presence	13					
Jales	CBSS	CSS	CB	C	Unt	Reference
Number plant species	8	5	5	5	1	4
Soil cover by *H. lanatus**	249	399	499	432	32	100

Sources: Vangronsveld et al. (1996), Mench (unpublished data).
* expressed in % compared with the reference soil

and fungi (Table 23) [51]. Only one plant species was frequent (Agrostis capillaris), while the frequency of two other ones (Betula pendula, Stelleria media) was less than 1%. In contrast, two plant species (A. capillaris, Festuca rubra) were mainly present in the CB-treated plots, and several non-metal tolerant perennial forbs colonised the revegetated area (e.g. Cerastium fontanum, Centaurium erythraea, Plantago media) (Table 23). Most of these species belong to mycotrophic families so that the presence of a mycorrhizal network in the soil promotes their establishment. In Unt plots, the arbuscular mycorrhiza (AM) infection percentage of grass roots ranged from 0 to 42% (Table 24).

Table 24. AM infection percentages in the roots of the grasses along the transect lines in the beringite-treated and untreated site at the Lommel-Maatheide Experiment; pH and total concentrations of Zn, Cd and Cu in the corresponding soil cores.

Distance (m)	AM infection (%)	pH	Zn	Cd	Cu
				(mg kg^{-1})	
CB-treated					
50	81	7.3	12750	18	475
100	69	7.9	3600	12	895
150	76	7.4	1168	8	206
200	38	7.6	9875	53	495
250	37	7.9	13250	55	765
300	65	7.6	4750	15	1393
Untreated					
25	0	4.5	2400	16	160
108	0	5.1	2080	8	80
150	3	5.5	2720	14	400
200	42	5.7	4960	61	395
250	24	5.9	4160	56	405
300	14	5.9	800	12	80

This percentage was greatly increased in treated soils, ranging from 37 to 81%, while total metal concentrations in soils were similar or even higher. The soil food web, expressed by the numbers and diversity of organisms and by metabolic function such as bacterial growth and soil respiration, clearly normalized [91]. Under grass, approximately 10^9 bacteria, 6 x 10^4 protozoa, 5 m length of fungal hyphae and 27 nematodes were found per gram of dry soil, each of which was 10 to 100 times higher than in the non-treated plots (Table 25). Also the functional diversity of soil bacterial populations, measured as the capacity to metabolise a number of different substrates, had almost doubled after soil treatment [91]. Chemical (*i.e.* addition of cyclonic ashes) and biological (plant growth) manipulation of soil, therefore, resulted in a marked decrease in the toxicity of soil metals and an increase in resources for soil organisms. Consequently, the soil food webs were restored.

Table 25. Organisms, metabolic process-rates and zinc-availability in incubated (30 days) soil samples from the Maatheide site: bare soil, bare soil plus organic substrates and "grass" soil.

Organisms etc.	*Bare Soil*	B.S.+S	*"Grass" soil*
Bacteria $x10^9.g^{-1}$ [a,b]	0.12 ± 0.02	0.42 ± 0.09	0.99 ± 0.32
Fungi $m.g^{-1}$ [b]	0.0	0.4 ± 0.6	5.4 ± 2.6
Protozoa $x10^3.g^{-1}$ [a,b]	2.5 ± 0.75	104 ± 47	63 ± 30
Nematodes $.100\ g^{-1}$ [b]	13 ± 12	13 ± 4	3416 ± 233
^{14}C Leucine $ngC.g^{-1}.h^{-1}$ [a,b]	8 ± 3.5	49 ± 8	73 ± 15
Respiration $ppmCO_2\ .30d^{-1}$	50	6000	4200
Biolog % (95)	46	-	83
Bacteria $^{10}log\ CFU.g^{-1}$ [a,b]	5.6 ± 0.2	7.8 ± 0.2	7.5 ± 0.1
Streptomycete $^{10}log\ CFU.g^{-1}$ [a,b]	4.8 ± 0.2	6.5 ± 0.3	7.4 ± 0.1
Fungi $^{10}log\ CFU.g^{-1}$ [a,b]	5.0 ± 0.1	6.4 ± 0.4	6.0 ± 0.1
$Zn_{CaCl2}\ mg.kg^{-1}$	525	-	16

[a,b] Significantly different Fig.s between bare soil and bare soil plus substrate (a) and between bare soil and grass soil (b), ($p < 0.001$). Source: Bouwman et al. 2001

The immobilisation efficacy of 3 amendments, lime, cyclonic ashes + steelshots (CA+SS), and a North Carolina mineral apatite was evaluated in 10 Pb-contaminated soils using the Pb Biomet strain [92]. An adverse effect was evident in seven of these soils. The combination of cyclonic ashes and steelshots induced a decrease in the bioluminescence response for five treated soils indicating a lower metal stress. Cyclonic ashes combined with steelshots appear more effective in reducing bacterial exposure to Pb than lime and apatite.

In a pot experiment, the bioluminescence response of two biosensors, *i.e. Escherichia coli* HB101 (pUCD607) and *Pseudomonas fluorescens* 10586s/r (pUCD607) increased after soil amendment addition [76]. Toxicity decreased and the ability of the soil microbial community to utilise various carbon substrates increased in metal-contaminated soils amended with either lime, coal fly ash or red muds. The microbial community may have increased to a greater extent if the soil had been reinocculated with a suspension from an uncontaminated soil.

In a series of pot trials, soils from the Louis Fargue, Reppel, and Jales Experiments were inoculated with solution from a leguminous rhizosphere, and then cultivated with dwarf bean. The presence of rhizobium nodules on roots was enumerated. The number of Rhizobium nodules was enhanced from 15% to 50% (100% based on control soil) by SS amendment in the 50 Mg ha^{-1} Louis Fargue sludged-plot. Rhizobium symbiosis was inhibited in all Jales and Reppel soils except for a 33% restoration in the BSS-Reppel soil.

9.2. Earthworms and Mites

Earthworms are thought to be highly exposed to soil metals as their diet consists of organic material in soil. Toxicity testing using earthworms is a well-developed method for studying bioavailability and toxicity of soil contaminants. Cadmium uptake by earthworms in a sandy metal contaminated soil treated either with lime or cyclonic ashes (CA) was investigated under laboratory conditions [6, 93]. Four weeks after initial exposure, earthworm Cd contents in lime- and CA-treated soils did not differ from untreated soils, although soil pH had increased. In an uncontaminated, Cd-spiked soil, Cd accumulation in earthworms was reduced by increasing the number of binding sites (δ-MnO_2, 1%) rather than by increasing proton competition (lime, 0.135%). Lowest Cd exposure was found with δ-MnO_2 combined with lime. This demonstrated the value of combining materials and including amendment materials able to form inner sphere complexes with metals. Three chemical immobilisation amendments were examined in a Cd, Zn, and Pb contaminated soil, using the earthworm (*Eisenia fetida*) toxicity test: a lime-stabilised municipal biosolid, a North-Carolina rock phosphate, and an anaerobically digested municipal biosolid [94]. Biosolids combined with lime, reduced the toxicity of the smelter soil to earthworms (*i.e.* cumulative mortality amounted to only 10%), whereas other soil amendment-combinations were lethal. Earthworms appeared to be more sensitive than lettuce. Biosolids plus limestone reduced earthworm mortality in alluvial mine tailings deposits at Leadville [95]. Amendments (B, SS, BSS) reduced the toxicity of Reppel soil to earthworms (*Lumbricus terrestris*), but As, Cd, and Zn contents were not decreased. Consequently, earthworm consumers may be exposed.

Soil ecosystem development on four vegetated tailings sites at Copper Cliff, Ontario, 0, 8, 20 and 40 years post rehabilitation, was assessed in terms of mite (Acari) population and compared with those from four control sites [96]. Mite density on older and more botanically diverse tailings sites was similar to that on control sites, but species richness of oribatids and mesostigmatics was lower. Species richness and diversity on tailings were lower at less botanically diverse sites regardless of age. Similarity of tailings-mite communities to control-site communities increased with age, but it was always less than 60%. Mite assemblages on tailings were dominated by a few colonising species, whereas control sites had a diverse assemblage of species.

9.3. Mammals

Soil to Plant to Animal (herbivore exposure) and Remediated soil to Animal are potential pathways for metal accumulation. Soil amendment may solve the exposure problem *via* soil ingestion, but vegetated remediated sites may increase the potential harm to wildlife. The extent of Pb bioaccumulation was determined into the blood, bone, liver and kidney of young rats fed with Pb contaminated soil treated with P plus Mn oxide (cryptomelane, $K_2Mn_2O_{16}$) [50]. Mean blood Pb concentrations ranged from 8.3 to 13.4 μg dL^{-1} for the unamended soil mixed diet, and from 5.2 to 12 μg dL^{-1} for the P and/or Mn oxide treated soil mixed diets. Phosphate rocks were more effective in reducing Pb bioaccumulation when combined with cryptomelane.

If earthworms can survive in vegetated, amended soils, then insectivorous mammals can be exposed to risk [95]. Unfortunately, there has been very little full scale ecosystem monitoring at phytostabilised sites. Making use of existing information on unremediated sites, Brown et al [95] have suggested 6,700 mg Pb kg^{-1} as an average permissible soil Pb level for shrews with no symptoms of Pb stress. Based on kidney cortex Cd of shrews living in contaminated environments, estimated safe soil Cd for earthworm-consuming wildlife ranged from 11 to 316 mg Cd kg^{-1}. Zinc risk to livestock and wildlife is limited where amendments, such as biosolids and alkaline products, have been used to restore Zn-Pb contaminated soils [97]. Firstly, Zn concentrations in vegetation have been reduced below the diagnostic threshold for phytotoxic concentrations, and fall within the normal range for healthy crops. Secondly, plants show a phytotoxic response to Zn at lower concentrations than those that may induce toxic effects in herbivorous mammals. Nonruminants such as horses may suffer from Zn-induced Cu deficiency, and dietary Cu must be taken into account.

9.4. Birds

Excess Cd (10 mg kg^{-1}) in willow buds on trees growing in alluvial mine tailings rich in Zn, Cd, and Pb was associated with reproductive problems in ptarmigan, as the birds consumed willow buds in early spring [98]. But in general Zn content in forages limits Cd absorption, increases soil pH and controls plant Cd uptake. Again, if earthworms and/or insects can survive in vegetated, amended soils, then insectivorous birds might be exposed to risk.

10. FAILURES AND SIDE EFFECTS

One hazard associated with the development of vegetation cover is that metal uptake by plants may expose primary consumers to contaminants. In addition, if the soil cover is insufficient, soil particles can be re-entrained to plant foliage, reducing the efficacy of soil amendments when judged on metal transfer *via* fodder consumption. Restoration of dewatered metalliferous fluorspar tailings in Derbyshire, UK is an example [99]. Air-dried, digested sewage sludge with additional phosphate fertiliser was added in 1974 to help establish vegetation cover directly on the tailings surface. Many volunteer plant and animal species have established, and an ecosystem developed. Vegetation and plant litter had high Zn, Pb, Cd and F levels compared to an uncontaminated reference site. Levels for Pb, Cd, and F were at least ten times greater than normal in earthworms, while Zn was no more than three times normal. As a result, Zn was regulated in the field vole (*Microtus agrestis*) and the shrew (*Sorex araneus*) but both mammals bioaccumulated Cd, Pb and F. Shrews and field voles accumulated Cd in the kidney and the liver, but the voles were less efficient. Lead and F were mainly accumulated in bones in both mammal species. Are these trace element bioaccumulations acceptable in terms of ecotoxicological risk? A part of the answer may not derive from bioaccumulation, but from studies of biomarkers and functions, and the use of post-genomic tools, *e.g.* changes in transcriptomic (up- or down-regulation of genes), and proteomic (functional analysis) to detect sublethal and/or genotoxic effects. Most phytostabilisation experiments currently involve crops that could directly enter food chains. To complete the panel of strategies, and to avoid consumers' exposure, it would be necessary to manage demonstrations using alternative crops that represent a sustainable, low-hazard economic use for remediated sites. These may include plants used for fuel, fibre, oil and construction materials.

An additional hazard relates to changes in trace element speciation due to geochemical process(es) and/or the development of a biotic activity. Trace element removal from the soil solution (*e.g.* chemisorption, plant uptake) may initiate release from residual insoluble forms. Cadmium in particular can return to the soil labile pool after lengthy periods of immobilisation/ insolubility. The failure to reduce activity or phytotoxicity of Cu^{2+} ions using both mineral and organic amendments in a severely Cu-contaminated soil was attributable to the dynamic interrelationship between labile and nonlabile pools [58]. It was possible to reduce the size of the total soluble and labile pool of Cu using additions of high-surface area oxides of Al or Fe, but Cu^{2+} activity was not modified without changing the pH. A low Cu^{2+} activity at high pH however may be insufficient to prevent soil phytotoxicity if labile soluble forms of Cu remain high and buffer Cu^{2+} activity. This example poses the question as to whether free metal activity or soluble concentration is a better predictor of phytotoxicity or bioaccumulation.

Data from 5- to 12-year-old experiments demonstrate that phytostabilisation can be sustained, but that one soil amendment can give different responses in different soils. Some failures are due to metal source and soil processes, *e.g.* rapid oxidation of sulphide materials in the Pyhäsalmi spoil, Finland, which leads to soil acidification. In the small-scale Jales Experiment, the sulphate concentration in the water drainage was reduced by amendments such as CB and CBSS, but it was still high (6 times) compared with a control soil. The development of a root mass along with a decaying vegetation blocking oxygen was not able to reduce oxidation of the material. In soil with an elevated pool of soluble organic carbon, this material can react with the sorbant phase, and may account for the lack of responses in test crops [33].

Saturation of the sorptive sites can lead to failure as well. An amendment may reduce solubility of the contaminant, but this sink could become saturated either already in the short-term, or over an extended time course. In the case of soils with a long history of contamination, bioassays can show, that while total soluble and labile concentration of the contaminant can be reduced by amendment, this decrease is not sufficient to avoid uptake and associated adverse effects [20]. A decrease with time in the efficacy of both B and SS amendments for attenuating As exposure in the small-scale Reppel Experiment is a case in point. Despite good results obtained in the laboratory in conditions similar to the field, amendments to reduce Cd bioavailability *i.e.* lime, clay, Fe-rich fines, Zn and Cu sulphate and alum-water treatment residuals, performed poorly under field conditions in 3 Cd-contaminated agricultural soils in South Australia [58]. The application rate of the amendment is also very important. A decrease in 0.01 *M* $Ca(NO_3)_2$ extractable Cd, Ni and Zn was enhanced after increasing the application rate of SS from 0.1% to 10%, and did not peak at a 1% level [38]. However, the 10% application rate was expensive and produced undesirable side effects, *e.g.* sorption of P, Mn exposure.

The methods used to apply and incorporate the amendment can be very important for the immobilisation process. In the Northampton Experiment, the SS material (iron grit) was applied at the soil surface, but not immediately mixed with the soil. The surface of this material oxidizes very rapidly, and newly formed oxides were not properly located to immediately react with trace elements in the soil solution. Consequently, there would be a reduced potential for metal immobilisation in the soil. This may partly explain the lack of clear-cut effects of this material on plant Cd uptake in this experiment. The degree of the amendment mixing with the soil is probably another key for a successful attenuation. In particular, SS application must be fractionated to avoid the formation of iron oxide nuggets.

There is a lack of information on the potential of phytostabilised sites to revert to their pre-stabilisation soil chemistry, particularly under field conditions. Consequences of soil acidification on amended soils have been investigated in laboratory-based batch experiments. Biosolid induced metal immobilization became de-stabilised as acidic soil conditions developed (pH < 6) [47, 100]. Immobilization products of Rock phosphates (*i.e.* metal pyromorphites) were more stable as soils acidified (pH < 5), even though they have less effect on metal extractability and phytoavailability compared with biosolids. Use of carbonate-rich materials (*e.g.* ground limestone) has had some success in reducing solubility and crop uptake of Cd and Zn in acid soils, but due to leaching, these responses are often not sustainable. In some parts of the Pronto Experiment, applied limestone has been more or less consumed due to oxidation of the Cu tailings and associated soil acidification. It would be interesting to follow the consequences for established vegetation and microbial communities if additional lime is not administered.

Several oxyanions such as AsO_4^{3-} may substitute for SO_4^{2-}in sulphate minerals in alkaline environments [101], however, initial results from an old industrial soil do not demonstrate direct SO_4 substitution by AsO_4, but may indicate AsO_4 coprecipitation or adsorption onto the sulphates and the formation of pharmacolite ($CaHAsO_4.2H_2O$). Gypsum ($CaSO_4.2H_2O$), and similar sulphate compounds, *i.e.* bassanite and anhydrite [44], and ettringite [28] are present in the mineral composition of beringite. Arsenic trapping in these sulphate minerals would be enhanced just after the beringite application, when soil alkalinity was at its maximum extent. Data showed that As immobilisation reversed with time in the B-treated Jales and Reppel soils. In one hand, soil acidification, desorption, and solubilisation of newly formed calcium arsenate are possible mechanisms. On the other hand, As-trapping by sulphate-minerals would be counterbalanced over time by several additional mechanisms, *e.g.* oxidation of sulphide minerals, SO_4 release, Ca leaching and uptake, etc. These processes would be rather displaced in favour of the formation of crystalline iron arsenate and adsorbed As species when beringite was combined with iron grit.

Well-aerated freely drained soils are considered to harbour oxidising conditions. Waterlogging, compaction, high organic matter content and capping can result in reducing conditions. Oxygen depletion can occur within the rhizosphere and inside aggregates when root and microbial consumption exceed the oxygen supply. Several trace elements, *e.g.* V, Cr, As, Se, Mo, Hg, may change their oxidation state within the range of redox conditions commonly encountered in soils [102]. In addition, solid phases may undergo changes in their mineral composition, structure, and stability. Both can directly influence the concentration and speciation of trace elements, and their partitioning between different physical phases. Reducing conditions may occur with high volume applications of amendments such as sewage sludge (biosolids). Volatile organic acids produced during the anaerobic decomposition of organic matter may form metal complexes, increasing metal concentration in the soil solution. There was a very short-term increase in Mn and Ni exposure following SS application, but there was little conversion of arsenate to arsenite [24]. Clearly there is a need for the ability to predict the duration of these conditions. Additional problems may arise in the reverse situation, where newly formed solid phases such as Fe(III) and Mn(IV) oxy-hydroxides, resulting from SS oxidation, may undergo reductive dissolution and release adsorbed and co-precipitated trace elements in the soil solution. Increases in Fe^{2+} and Mn^{2+} concentrations would displace trace elements from exchange sites in the soil, while competition from these ions may limit bioavailability. The reductive dissolution kinetics of birnessite by oxidisable organic ligands such as catechol was rapid, independent of pH, and essentially complete within seconds under conditions of excess of catechol at pH 4 to 6 [103]. Sorbed contaminants could be released into solution, whereas reactive compounds might have been sorbed or co-precipitated as a consequence of hydrolysis of the Mn(II) that was released. If reducing conditions or increases in oxidisable organic ligands are temporary, it would have few consequences, as on re-oxidation, Fe and Mn are again oxidised and precipitate from solution as oxy-hydroxides, affording trace element adsorption or co-precipitation [102]. However, if the environment becomes permanently reduced, immobilisation could be in question. In such circumstances, it may be appropriate to enhance strong reducing conditions and metal availability may be decreased by precipitation as low solubility sulphide minerals.

Among potential unwanted side effects of soil amendments, the following aspects should be considered [6].

1. Toxicity of the material;
2. Toxicity of contaminants in the amendments;
3. Toxicity and deficiency, imbalance in nutrients as a result of reactions induced by the amendments;
4. Toxicity, deficiency, side effects due to impact on soil structure and soil organic matter;
5. Toxicity, deficiency due to changes in soil conditions and release of trace elements from dissolved phases.

Toxicity of the additive can occur immediately after dissolution or degradation. Usually, it can be detected in laboratory experiments, if it is not masked by artificial conditions. A transient adverse effect was found on bacteria and earthworms following the rapid oxidation of steelshots in soil and the release of Fe and Mn in soil pore water before the newly Fe and Mn oxides formed. The sodicity or radioactivity of red muds can be an immediate problem.

The contaminant level in the amendment can be of concern. Several additives such as incinerator ashes, coal fly ashes, biosolids, and Fe-rich are waste products, and can contain either inorganic contaminants such as trace elements or organic contaminants. Contaminants in the additive do not necessary lead to higher exposure *via* the soil-plant-animal system. Dermal contact and solid phase ingestion should be considered. Fe-rich contains Cd and Cr. It also contains a lot of chloride, so for this reason it should be washed up to 10 times before application [104]. Many coal fly ashes contain high concentrations of Boron and SO_4 that may rapidly impact on plant growth. Contaminants and their derivatives may be released in the long term. Steelshots contain Ni, which raised Ni content of maize grain in the Louis

Fargue Experiment upon application. A similar effect is possible for Cu, taken up by corn in the SS-treated Reppel soil.

Interactions between organic and mineral constituents can lead to the formation of a variety of sinks, which govern the behaviour of trace elements in soils and waste materials. Numerous changes in soil chemistry, soil ecology and biological activity result from soil amendment. Unwanted side effects can arise from these abiotic and biotic reactions. Amendments may change contaminant speciation in soil, thus affecting either bioaccumulation, or toxicity. Reactions may fix an element and result in unexpected changes in the exposure to another element. Most experiments have only focused on major contaminants, due to a cost. Therefore, the possibility exists that additional mechanisms may be missed. In contaminated soils with low nutrient and water-retentive capacity, increasing fertility is necessary to enhance the establishment of a vegetation cover. This can be achieved using compost, as there is a lack of organic matter. The increase in DOM from this source may result in reaction with solid phases and/or an elevated labile pool of contaminants. In the small-scale Jales Experiment, compost increased As leaching (Table 7) and plant exposure (Tables 15 and 16), while other materials such as beringite were unable to counter-balance this side effect. Higher leaching and organism exposure to Pb and Cu can also occur after organic matter application. Arsenic mobility can increase after hydroxyapatite addition, as was observed in Overpelt soil [46]. Increased soil pH, resulting from alkaline amendments and manganese oxide amendments (K-birnessite), can increase leaching of non-essential metals due to increases in DOM [20]. The practice of liming was investigated with six mine tailing samples from an abandoned copper smelter near Anaconda, Montana [63]. The soluble As concentration did not correlate with total As, but was closely related to soil solution pH. Saturated pastes of all samples were below saturation with respect to the known metal arsenate solid phase. Liming treatment increased As mobility in unsaturated laboratory columns. Adsorption-desorption reactions on oxide minerals were believed to be responsible for the pH dependence of soluble As concentrations rather than precipitation-dissolution reactions of solid metal arsenates. Consequently, the pH of the post-limed tailings and soils is important for determining the amount of As mobilised after liming. The use of alkaline amendments should be evaluated with respect to potential effects on the mobilisation of trace elements such as As into the soil solution, especially at contaminated sites, hydraulically connected to surface- or groundwater. This aspect would also be of concern when high concentrations of phosphate fertilisers are used. Some long-term experiments recommend prudence in the use of combinations of fly ash and biosolids [105]. DTPA-extractable Pb, Cu, Zn, Fe and Mn were higher in acidic mine soils receiving fly ash and biosolid mixtures. Alkaline additives poor in Ca, such as some zeolites can strongly increase the DOM concentration, resulting in an increased metal leaching [6]. Decreased solubility of essential trace elements and/or macronutrients was noticed after amendment application. For example, beringite was effective in reducing plant exposure to Ni in the Louis Fargue Experiment. Over time however, Mn deficiency and low P-availability developed, both affecting plant yield [10]. Attenuation of Mn availability, induced by B addition to the soil, was also monitored at both the small-scale Jales and Reppel Experiments. Three manganate minerals, *i.e.* hexagonal birnessite, chalcophanite, and todorokite, were expected in the B-treated plots at Lommel-Maatheide Experiment based on the EXAFS studies of the clay fraction [41]. This could be a frequently occurring side effect using this material. Other materials may induce either Mn or P deficiencies. Beside salinity, sodicity, and Cr content, red muds have the capacity to decrease Mn availability.

Lack of soil invertebrates and microorganisms involved in decomposition processes and lack of residual nitrogen mineralisation may be a problem. In several experiments, during the early stages of re-vegetation, there were requirements for repeated N applications, preferably in a non-acidifying form. Maintaining sufficient phosphorus is sometime difficult because of phosphate fixation by iron oxides present in the soil following amendment, or in the amendment itself. Phosphorus concentration in pine needles is decreased in trees grown on CSS-treated Jales spoils compared to C-treated ones (Fig. 7). This could be due to phosphate sorption on newly formed iron oxides. However, trees also grew better in the CSS treatment, and therefore a dilution effect may also be part of the response. Phosphate fixation can be counteracted by adjusting the application rate and by repeated application, but competition with As and induction of As leaching must also be considered.

High application rates of some additives can affect soil properties. Synthetic zeolite in Na-form can damage the soil structure [6]. Steelshots application, over 10% soil weight, reduced the soil porosity [48].

Phytostabilisation can fail due to the interaction between the physical properties of the waste material, and local environmental conditions. Many wastes and contaminated soils behave as sandy loam, and are therefore prone to moisture deficiency. Consequently, water supply, fluctuating soil moisture and temperature are additional constraints to plant establishment and survival. Several procedures must be kept in mind in order to avoid surface desiccation, and freeze dry effects [15], which may interact with geo-chemical processes and the sustainability of the revegetated area. Drainage at different slope levels affects soil profile development, resulting in layers that may limit root penetration. When material such as papermill waste was incorporated into tailings, a good ground cover did not establish, probably due to salt accumulation at the surface, which created an osmotic environment unfavourable for the establishment of glycophyte seedlings, although seedlings of volunteer halophytes did start to colonise [16].

11. REGULATORY CONSIDERATIONS

The lifetime or the target of a successful remediation is something that is often largely ignored. Remediation can be considered successful when risks formerly associated with the site are removed, the residual risk is satisfactory regarding future land uses, protection is maintained over time and untreated wastes are minimized. In addition, spreading of contaminants by natural agencies such as lateral wind and water erosion and by percolation to the groundwater is prohibited or must be strongly reduced. Remediation must have a permanent character with no or minimum maintenance, while many changes may occur and some might challenge the technique used. One could also consider that remediation has been successful when regulatory authorities pass the remediated site as 'fit for purpose'. The administrative and regulatory responsibilities include all work related to filing the remediation program, quality assurance and control, health and safety, documenttation, and compliance with regulatory issues.

Important legislation related to soil contamination is that concerning soil protection or soil remediation, soil and waste management, public health and water. Present environmental regulations on trace element-contaminated soils are frequently based on total contaminant concentrations. However, from ecological, toxicological, and health viewpoints, the bioavailable fraction in exposure pathways should be considered. This leads to regulator evaluation on a case-by-case basis. Most European countries favour the less rigid fitness-for-use approach, but there appear to be large differences in the practice of dealing with land contamination in the European countries [106]. The US EPA uses nine criteria to determine the preferred remedy (Table 26) [107].

Table 26. US EPA criteria to evaluate and determine the remedy preference (Rock and Sayre, 2000).

Overall protection of human health and the environment
Compliance with applicable or relevant and appropriate requirements (ARARs)
Long-term effectiveness and permanence
Reduction of contaminant toxicity, mobility, or volume through treatment
Short-term effectiveness
Implementability
Cost
State acceptance
Community acceptance

The 'Interim Canadian Environmental Quality Criteria for Contaminated Sites' take into consideration future land use and appropriate human exposure conditions as guidance for establishing soil and groundwater clean-up goals [108]. Two remediation standard options can be used. In the numerical standard option, a contaminated site is considered to have been satisfactorily remediated if (a) the soil, surface water or groundwater does not contain any substance with a concentration greater than or equal to the applicable site-specific numerical standard, or (b) the soil, surface water or groundwater at the site does not contain any substance with a concentration greater than or equal to the local background concentration of that substance in the soil, surface water or groundwater respectively. Canadian soil quality guidelines retained cleanup values for trace elements depending on future land uses of the remediated sites, *i.e.* residential/parkland, agricultural, commercial and industrial. The use of *in situ* amendments does not suit this objective and should be combined with other techniques such as the phytoextraction. In the risk-based standard option, a contaminated site is considered to have been satisfactorily remediated if (a) for any non-threshold carcinogenic substance, the calculated human lifetime cancer risk due to exposure to that substance at the site is less than or equal to a risk value recommended by the local medical health officer for the site and (b) for any substance for which a hazard index is calculated, the hazard index due to exposure of a human to that substance at the site is less than a maximum hazard index recommended by the local medical health officer for that site [108].

Research continues in genetic engineering of plants for phytoremediation. These plants should be in compliance with the present regulations [107, 109].

Previous successful applications on a similar site are one persuasive indication. Regulatory acceptance of phytostabilisation is thus coupled to the development of long-term field experiments. It is also related to the evolving scientific and social consensus over how bioavailability should be measured, and which organisms specifically used as models.

One obstacle to the acceptance of the phytostabilisation of metal-contaminated soils by regulators and the public are effective and efficient methods for monitoring progress and judging efficacy. Clearly, a long term monitoring guidance for trace element attenuation using amendments has to be established in the same fashion as the one published by the USEPA for organic contaminants. It would include the following:

1. Proper site characterisation, definition of contaminant source, definition of land reuse;
2. Determination of nutrient source and attenuation mechanisms;
3. Definition of monitoring tools for danger and risk assessments
4. Comparison of attenuation with an engineered alternative;
5. Evaluation of remedial time frame, cost sustainability, and side effects.
6. Characterisation of trace element speciation in the remediated soils, and model for predicting trace element status as a result of changes in systems and climatic conditions.

In protocols for field demonstrations, more frequent investigation of the soil microflora would be required, to evaluate the need for any re-inoculation and to monitor the development of volunteer microbial communities. Small-scale heterogeneity is frequent at contaminated sites. Its potential impact on phytostabilisation deserves more attention but may affect the time course of successful phytostabilisation.

The future land use has to be in compliance with the phytostabilisation option, and eventually land-use restrictions should be considered. If phytostabilisation is attempted and does not succeed, an adequate backup technology with a high chance of success must be ready and available for use at the site [107].

12. ECONOMIC CONSIDERATIONS

In many cases, the party responsible for a contaminated site no longer exists and the costs of remediation are transferred to the public. In other cases, the contaminated areas are not high priorities on remediation agendas. Thus, any remediation technique should be economical to implement. Total projected 5-year cost estimated for the treatment of US sites contaminated by trace elements with or without organics was estimated at $ 42.5 billion [1]. Use of phytoremediation for metal-contaminated soils presented a US market opportunity of just below $ 1 billion per year [110]. Berti and Cunningham [3] provided economic comparison of phytostabilisation with other remediation options (Table 27). Calculations consider a soil contaminated by more than 0.2% Pb to a depth of 30 cm and exceeded the US regulatory TCLP limit of 5 mg Pb L^{-1}. The economic ranking of the decontamination techniques was excavation and landfilling > soil washing > phytoextraction. The costs to implement stabilisation techniques were less expensive and the ranking was asphalt capping > soil capping > phytostabilisation. Differences in chemistry and physical properties at metal-contaminated sites induce variable cost estimates. Assuming similar efficacy levels and accounting for side effects, soil amendments having little economic value such as waste products or by-products are preferred to more expensive materials (Table 28). Usually, the time to observe a risk reduction using soil amendment is relatively short, *i.e.* 2 to 3 years, while currently phytoextraction by natural hyperaccumulators may take 13-16 years to clean a typical site [110].

Table 27. Economic comparison between phytostabilisation and other selected remediation techniques for metal-contaminated soils.

Remediation technique	Cost (Euro ton^{-1})		
	1	2	3
Site decontamination			
Excavation and landfilling	533	150-350	100-500
Chemical treatment			100-500
Soil washing by particle separation	263		75-200
Thermal treatment			170-300
Electrokinetics			20-200
Incineration			200-1500
Phytoextraction	93	20-80	25-100
Site stabilisation			
Asphalt capping	54		
Soil capping	44		
Vitrification (reagent)			75-90
Phytostabilisation	20		

Source: 1 Berti and Cunningham (2000), based on 1 euro = 1 $ and 3000 tons per ha, 2. Ensley (2000), 3. Glass (2000).

Table 28. Economic comparison between soil amendments for metal-contaminated soils.

	€ kg^{-1}
Hydroxyapatite (purified)	1500
Steelshots (iron grit)	1.75
Lime	0.25-0.37
North Carolina phosphate rock	0.2
Compost	0.01
Cyclonic ash	negligible

(Geebelen, 2002)

13. CONCLUSION

Many short-term experiments have demonstrated that the use of soil amendment can attenuate the exposure of plant species and other organisms to metals and metalloids [10, 20, 32, 33; 34]. But for public acceptance of phytostabilisation, and especially to be considered as a permanent remediation technique, long-lasting demonstrations are needed. Several current small-scale semi field trials and field experiments show that successful phytostabilisation can be sustained for over a decade for a range of contaminated soils (Table 29). However, other trials in UK deliver less conclusive results. Experiences gained for major amendments ranged roughly from 5 years to more than 25 years, the oldest usually being with liming. Successful phytostabilisation, equally protective of human health and the environment, is frequently based on a combination of several additives. The oldest field

Table 29. Summary of the beneficial and side effects demonstrated by long term monitoring of trace element attenuation in field experiments.

Sites	Jales	Reppel	Overpelt	Louis Fargue	Maatheide	Northampton	Staffordshire	Cornwall
Monitoring duration	4	5	5	7	12	3	3	3
Beneficial effects								
Decrease in metal exposure	+	+	+	+	+			
Decrease in As exposure	+	+		+				+
Decrease in leaching	+	+	+					
Increase in plant growth	+	+	+	+	+			
Increase in plant diversity	+		+		+			
Decrease in metal content	+	+	+	+	+			
Promote microbial communities	+			+	+			
Side effects								
Induced-deficiency	+			+				
Enhanced metal leaching	+							
Enhanced As leaching	+							
Increase in As exposure	+							
Increase in Ni exposure				+				
Increase in metal exposure								+

experiments confirm that phytostabilisation is an unobtrusively and cost-effective technique for treating metal-contaminated soils. University research teams having relatively few means have mostly established trials, but lack funds to deliver crucial long-term assessments. Soil amendments can initiate a healthy and diverse ecosystem at a contaminated site. Inputs of soil amendment are usually economically reasonable and environmentally benign. Selection of additives in batch experiments is necessary and can avoid increasing contaminant toxicity to living organisms, but may not forecast some long-term effects. Materials found to be promising under batch experiment and/or glasshouse conditions can be less effective in the field. Decline of effectiveness can occur over time. Failures and side effects such as induced deficiencies can arise and tend to be site specific. To avoid failures, a thorough evaluation of materials must be conducted, in particular in field lysimeters and/or in plots for each site, in combination with or without the choice of biological material. A long-term monitoring programme to examine changes in speciation, leaching, and ecotoxicity must be carried out, before any large-scale site treatment is implemented. Consequently, parts of the site may require containment before being considered for *in situ* metal inactivation, when imminent risks are present.

The regulatory acceptance of phytostabilisation in metal-contaminated soils will require more long-term demonstrations. This should be accomplished under a range of environmental conditions that could occur over time. Field experiments should compare several technologies together. While *in situ* immobilisation may be attractive in some locations, it should not be viewed as an exclusive remedial strategy. It may be combined with other options, in particular with the use of metallophytes (phytoextraction, phytovolatilisation), of tolerant species (phytostabilisation), and of plants associated with endobacterias as a secondary sink.

Environmental models are needed to help identify the limiting factors, which can be biotic, and/or abiotic, *e.g.* initial contaminant concentration, diffusion coefficient, aggregate size. Rapid and reliable exposure tests, using bioindicators and biosensors are needed for routine low-cost site monitoring. A range of biosensors, specific for different key contaminants need to be developed. The monitoring programme should take into account different exposure pathways, *e.g.* soil solution, direct ingestion, and the gaseous phase. It should assure human health and environmental protection. Monitoring should be designed to verify that the treatment has not created materials at levels that pose a threat to human health. In most inactivation experiments, human or environmental receptors are not exposed to greater risks during the attenuation process. Currently, there is however an imbalance, while most studies focus on plant contamination *via* the soil solution and the mobility to the groundwater. More long-term studies are needed dealing with soil ingestion, dermal contact and consumption of trace element enriched-products. However, the long-term monitoring programme should also be designed to minimize cost, otherwise the potential economic benefits compared with other techniques would be lost. The level of ecosystem function in a restored as opposed to an uncontaminated site is another question. It may not be possible to restore to full functionality. Facing the challenge to attenuate trace element exposure, to restore a vegetation cover and a microbial community, and/or to re-establish the food compliance, each site exhibits its own unique problems and potential solutions. Public concerns about the long-term effectiveness can be addressed through the assurance that monitoring will provide protection and that alternative remedies will be implemented as needed [26].

REFERENCES

1. Ensley, B.D. (2000) Rational for use of phytoremediation, in I. Raskin and B.D. Ensley (eds.), *Phytoremediation of Toxic Metals: Using Plants to Clean Up the Environment*, John Wiley & Sons, Inc. New York, pp. 3-11.
2. Adriano, D.C. (2001) *Trace Elements in Terrestrial Environments: Biogeochemistry, Bioavailability and Risks of Metals,* Springer-Verlag, New York.
3. Berti, W.R. and Cunningham, S.D. (2000) Phytostabilization of metals, in I. Raskin and B.D. Ensley (eds.), *Phytoremediation of Toxic Metals: Using Plants to Clean Up the Environment*, John Wiley & Sons, Inc. New York, pp. 71-88.
4. Geebelen, W. (2002) *Remediation of Pb Contaminated Soils by Phytoextraction and Amendment Induced Immobilization: Biological Aspect,* Doctoraatsproefschrift, Faculteit Wetenschappen, Limburgs Universitair Centrum, Diepenbeek, Belgium.
5. Vangronsveld, J. and Cunningham, S.D. (1998) *Metal Contaminated Soils: In situ Inactivation and Phytorestoration*, Springer-Verlag, Berlin and R.G. Landes Company, Georgetown, TX.
6. Osté, L.A. (2001) *In situ Immobilization of Cadmium and Zinc in Contaminated Soils : Fiction or Fixation ?* Wageningen Universiteit, The Netherlands.
7. Van der Lelie, D., Schwitzguébel, J.P., Glass, D.J., Vangronsveld, J., and Baker, A. (2001) Assessing phytoremediation's progress in the United States and Europe, *Environ. Sci. Technol.* 35, 447-452.
8. Blaylock, M.J., Salt, D.E., Duschenkov, S., Zakharova, O., Gussman, C., Kapulnik, Y., Ensley, B.D., and Raskin, Y. (1997) Enhanced accumulation of Pb in Indian mustard by soil- applied chelating agents, *Environ. Sci. Technol.* 31, 860-865.
9. Lombi, E., Zhao, F.J., Dunham, S.J., and McGrath, S.P. (2001) Phytoremediation of heavy metals-contaminated soils: natural hyperaccumulation versus chemically enhanced phytoextraction, *J. Environ. Qual.* 30, 1919-1926.
10. Mench M., Vangronsveld, J., Clijsters, H., Lepp, N.W., and Edwards, R. (2000) In situ metal immobilisation and phytostabilisation of contaminated soils, in N. Terry and G. Banuelos (eds.), *Phytoremediation of Contaminated Soil and Water*, Lewis Publishers, Boca Raton, USA, pp. 323-358.
11. Geebelen, W., Adriano, D.C., Mench, M., Clijsters, H., and Vangronsveld, J. (2003) Amendment induced immobilization of Pb in contaminated soils: effect on Pb, Cu, Zn, Cd, Ni, Fe and Mn phytoavailability and phytotoxicity, in G. Gobran et al. (eds.), *7th ICOBTE Conference*, Uppsala, Sweden (submitted).
12. Delas, J. (1963) La toxicité du cuivre accumulé dans les sols, *Agrochimica* 7, 257-288.
13. Goulding, K.W.T. and Blake, L. (1998) Land use, liming and the mobilisation of toxic metals, *Agriculture, Ecosystems, Environment* 67, 135-144.

14. Pitchel J.R., Dick, W.A., and Sutton, P. (1994) Comparison of amendments and management practices for long-term reclamation of abandoned mine lands, *J. Environ. Qual.* 23, 766-772.
15. Peters, T.H. (1995) Revegetation of the Copper Cliff tailing area, in J.M. Gunn (ed), *Restoration and Recovery of an Industrial Region,* Springer-Verlag, New-York, pp. 123-134.
16. Tisch, B., Beckett, P., Okonski, A., Gordon, C., and Spiers, G. (2000) Remediation and revegetation of barren copper tailings using paper mill sludge; an overview. CLRA Annual meeting, Edmonton, Alberta Canada.
17. Alloway, B. (1995) *Heavy Metals in Soils.* 2nd edition, Blackie Academic and Professional, New-York.
18. Mench, M., Didier, V., Löffler, M., Gomez, A., and Masson, P. (1994) A mimicked in situ remediation study of metal-contaminated soils with emphasis on cadmium and lead, *J. Environ. Qual.* 23, 58-63.
19. Singh, B.R. and Osté, L. (2001) In situ immobilization of metals in contaminated or naturally metal-rich soils, *Environ. Rev.* 9, 81-97.
20. McBride, M.B. and Martinez, C.E. (2000) Copper phytotoxicity in a contaminated soil: Remediation tests with adsorptive materials, *Environ. Sci. Technol.* 34, 4386-4391.
21. Mench, M., Bussière, S., Boisson, J., Castaing, E., Vangronsveld, J., Ruttens, A., De Koe, T., Bleeker, P., Assunção, A., and Manceau, A. (2003) Progress in remediation and revegetation of the barren Jales gold mine spoil after *in situ* treatments, *Plant Soil* (in press).
22. Knox, A.S., Kaplan, D.I., and Hinton, T.G. (2001) Remediation of metals and radionuclides by phytoextraction and sequestration, in L. Evans (ed.), *Proc 6th ICOBTE Conference*, University of Guelph, Ontario, Canada, p. 314.
23. Unterköfler, J., Wenzel, W.W., Adriano, D.C., Wieshammer, G., Sommer, P., and Fitz, W. (2001) Integrated phytoextraction and (physico-)chemical immobilization – A new approach to remediate contaminated soils, in L. Evans (ed.), *Proc 6th ICOBTE Conference*, University of Guelph, Ontario, Canada, p. 411.
24. Boisson, J. (1999) *Réhabilitation de Sols Pollués en Eléments Traces par des Amendements Minéraux. Faisabilité et Durabilité d'après la Mobilité des Eléments et la Phytotoxicité du sol.* PhD thesis, Sciences agronomiques, Institut National Polytechnique de Lorraine, Nancy, France.
25. Young, S.D., Tye, A., Carstensen, A., Resende, L., and Crout, N. (2000) Alternative methods for the determination of labile cadmium and zinc, *European Journal Soil Science* 51, 129-136.
26. Swindoll, C.M. and Firth, M.J. (1998) Phytorestoration – Regulatory, industry, and public concerns, in J. Vangronsveld and S. Cunningham (eds.). *Metal-Contaminated Soils: In Situ Inactivation and Phytorestoration,* Springer-Verlag, Berlin and R.G. Landes Company, Georgetown, TX, pp. 249-259.
27. Knox, A.S., Seaman, J.C., Mench, M., and Vangronsveld, J. (2000) Remediation of metal- and radionuclides-contaminated soils by *in-situ* stabilization techniques, in I.K. Iskandar (ed.), *Environmental Restoration of Metals-Contaminated Soils*, CRC Press LLC, Lewis Publishers, pp. 21-60.
28. Mench, M., Vangronsveld, J., Lepp, N.W., and Edwards, R. (1998) Physico-chemical aspects and efficiency of trace element immobilisation by soil amendments, in J. Vangronsveld and S. Cunningham (eds.), *Metal-Contaminated Soils: In Situ Inactivation and Phytorestoration,* Springer-Verlag, Berlin and R.G. Landes Company, Georgetown, TX, pp. 151-182.
29. Axe, L. and Trivedi, P. (2001) Intraparticle diffusion of metal contaminants and their attenuation in amorphous aluminium, iron, and manganese oxides, in L. Evans (ed.), *Proc 6th ICOBTE Conference*, University of Guelph, Ontario, Canada, p. 107.
30. Lebeau, T., Jézéquel, K., Bagot, D., and Fabre, B. (2001) Control of the cadmium transfer in the soil-plant system by the inoculation of micro-organisms: a preliminary

study of cadmium biosorption by free and immobilised micro-organisms in soil extract medium, in L. Evans (ed.), *Proc 6th ICOBTE Conference*, University of Guelph, Ontario, Canada, p. 363.
31. Valls, M., Atrian, S., De Lorenzo, V., and Fernandez, L.A. (2000) Engineering a mouse metallothionein on the cell surface of *Ralstonia eutropha* CH34 for immobilization of heavy metals in soil, *Nature Biotechnology* 18, 661-665.
32. Brown, S.L., Henry, C.L., Compton, H., Chaney, R.L., and De Volder, P. (2000) Using municipal biosolids in combination with other residuals to restore zinc and lead contaminated mining areas, in Luo Y.M et al. (eds.), *Proc. Int. Conf. Soil Remediation SoilRem2000*, October 15-19, Hangzhou, China, pp. 285-289.
33. Lepp, N.W., Alloway, B., Penny, C., Warren, G., and Bochereau, F. (2000) The use of synthetic zeolites as in situ soil amendments to reduce metal transfer from soils to vegetables, in Luo Y.M et al. (eds.), *Proc. Int. Conf. Soil Remediation SoilRem2000*, October 15-19, Hangzhou, China, pp. 280-284.
34. Vangronsveld, J., Ruttens, A., Colpaert, J., and Van der Lelie, D. (2000) In situ fixation and phytostabilization of metals in polluted soils, in Luo Y.M et al. (eds.), *Proc. Int. Conf. Soil Remediation SoilRem2000*, October 15-19, Hangzhou, China, pp. 262-267.
35. Lebourg, A., Sterckeman, T., Ciesielski, H., Proix, N., and Gomez, A. (1996) Intérêt de différents réactifs d'extraction chimique pour l'évaluation de la biodisponibilité des métaux en traces du sol, *Agronomie* 16, 201-215.
36. Manceau, A, Marcus, M.A., and Tamura, N. (2002) Quantitative speciation of heavy metals in soils and sediments by synchrotron X-ray techniques, in P.A. Fenter, M.L. Rivers, N.C. Sturchio, and S.R. Sutton (eds.), *Applications of Synchrotron Radiation in Low-Temperature Geochemistry and Environmental Science*, Reviews in Mineralogy & Geochemistry 49, Geochemical Society and Mineralogical Society of America, pp. 341-428.
37. Vangronsveld, J., Mench, M., Lepp, N.W., Boisson, J., Ruttens A., Edwards R., Penny, C., and Van der Lelie, D. (2000) In situ inactivation and phytoremediation of metal- and metalloid contaminated soils: field experiments, in J. Wise, D. Trantolo, E. Cichon, H. Inyang, U. Stotmeister (eds.), *Bioremediation of Contaminated Soils*, Marcel Dekker Inc, New York, pp. 859-884.
38. Sappin-Didier, V. (1995) *Contrôle des Flux de Métaux dans les Agrosystèmes par Apport d'un Composé du Fer.* Thèse de Doctorat, Chimie analytique et Environnement, Université Bordeaux I, ENSCPB, Bordeaux, France.
39. Vangronsveld, J., Van Assche, F., and Clijsters, H. (1995) Reclamation of a bare industrial area, contaminated by non-ferrous metals: *in situ* metal immobilization and revegetation, *Environ. Pollut.* 87, 51-59.
40. Vangronsveld, J., Colpaert, J.V., and Van Tichelen, K.K. (1996) Reclamation of a bare industrial area contaminated by non-ferrous metals: physico-chemical and biological evaluation of the durability of soil treatment and revegetation, *Environ. Pollut.* 94, 131-140.
41. Hargé, J.C. (1997) *Spéciation Comparée du Zinc, du Plomb et du Manganèse dans des Sols Contaminés,* PhD thesis, Géophysique-Géochimie-Géomécanique, Univ. J. Fourier, Grenoble, France.
42. Verkleij, J.A.C., Karenlampi, S., De Koe, T., Mench, M., and Vangronsveld, J. (1999) Strategies for rehabilitation of metal polluted soils: in situ phytoremediation, immobilization and revegetation, a comparative study (PHYTOREHAB), Final report ENV4-CT95-0083, EU DGXII Environment & Climate programme, Vrije Universiteit Amsterdam, The Netherlands.
43. Semane, B. (2001) Evolution d'un sol pollué aux métaux lourds en mésocosme. Importance des amendements pour la phytostabilisation, Maîtrise Biologie des populations et des écosystèmes, Université Bordeaux I, Bordeaux, France.

44. Pannetier, S. (2000) Etude de l'incidence des amendements de béringite sur l'abondance des minéraux amorphes dans les sols: conséquences sur l'immobilisation des cations métalliques, IUP EGID, Bordeaux III, Bordeaux, France.
45. Hartley, W., Edwards, R., and Lepp, N.W. (2001) A study of novel methods for the *in situ* remediation of arsenic contaminated soils, in L. Evans (ed.), *Proc 6th ICOBTE Conference*, University of Guelph, Ontario, Canada, p. 38.
46. Boisson, J., Ruttens, A., Mench, M., and Vangronsveld, J. (1999) Immobilization of trace metals and arsenic by different soil additives evaluation by means of chemical extractions, *Communications Soil Science Plant Analysis* 30, 365-387.
47. Basta, N.T., Armstrong, F.P., and Hanke, E.M. (2001) Effect of chemical remediation of contaminated soil on arsenic mobility and gastrointestinal availability, in L. Evans (ed.), *Proc 6th ICOBTE Conference*, University of Guelph, Ontario, Canada, p. 40.
48. Brown, S.L., and Chaney, R. (1999) A rapid in-vitro procedure to characterize the effectiveness of a variety of in-situ lead remediation technologies, in W.W. Wenzel, D.C. Adriano, B. Alloway, H.E. Doner, C. Keller, N.W. Lepp, M. Mench, R. Naidu, and G.M. Pierzynski (eds), *Proc. Ext. Abs. 5th Int. Conf. Biogeochemistry Trace Elements*, Int. Soc. Trace Element Research, Vienna, pp. 419-420.
49. Orlova, A.O., Farfel, M.R., Lees, P.S.J., Chaney, R., and Ashley, P. (2001) Biosolid application for reducing urban soil lead hazards: baseline results, in L. Evans (ed.), *Proc 6th ICOBTE Conference*, University of Guelph, Ontario, Canada, p. 235.
50. Hettiarachchi, G.M., Pierzynski, G.M., and Ransom, M.D. (2001) In situ stabilization of soil lead using phosphorus, *J. Environ. Qual.* 30, 1214-1221.
51. Vangronsveld, J. (1998) Case studies in the field-industrial sites. Phytostabilisation of zinc-smelter contaminated site: The Lommel-Maatheide case, in Vangronsveld J. and S.D. Cunningham (eds.) *Metal Contaminated Soils: In Situ Inactivation and Phytorestoration*, Springer-Verlag, Berlin and R.G. Landes Company, Georgetown, TX, pp. 211-216.
52. De Koe, T., Bleeker, P.M., and Assunção, G.L. (1998) Field experiments at the Jales minespoil, in Verkleij J.A.C (ed), Strategies for rehabilitation of metal polluted soils: *in situ* phytoremediation, immobilization and revegetation, a comparative study (PHYTOREHAB), Progress report n°5 ENV4-CT95-0083, EU DGXII Environment & Climate programme, Vrije Universiteit Amsterdam, NL, pp. 44-67.
53. Goldberg, S., Lebron, I., Suarez, D.L., and Hinedi, Z.R. (2001) Surface characterization of amorphous aluminium oxides, *Soil Sci. Soc. Am. J.* 65, 78-86.
54. Boisson, J., Mench M., and Chartier, S. (1998) Limited soil-plant transfer of As by using immobilizing soil additives: a semi-field study. Proc. Int. Soil Science Society, Montpellier, 1998.
55. O'Reilly, S.E., Strawn, D.G., and Sparks, D.L. (2001) Residence time effects on arsenate adsorption / desorption mechanisms on goethite, *Soil Sci. Soc. Am. J.* 65, 67-77.
56. Lin, C.F., Lo, S.S., Lin, H.Y., and Lee, Y. (1998) Stabilization of cadmium contaminated soil using synthesized zeolite, *J. Hazard Mat.* 60, 217-226.
57. Sun, B., Zhao, F.J., Young, S.D., Tye, A. and McGrath, S.P. (2000) Availability and fixation of Zn and Cd in soils amended with metal sulphate, in Luo Y.M et al. (eds.), *Proc. Int. Conf. Soil Remediation SoilRem2000*, October 15-19, Hangzhou, China, pp. 354-365.
58. McLaughlin, M.J., Nardecchia, D., Maier, N.A., Smart, M.K., and Cozens, G.D. (2000) Remediation of cadmium-contaminated soils, in Luo Y.M et al. (eds.), *Proc. Int. Conf. Soil Remediation SoilRem2000*, October 15-19, Hangzhou, China, pp. 275-279.
59. Vacha, R., Podlesakova, E., and Nemecek, J. (2000) The use of organic and inorganic amendments for immobilization of trace elements in agricultural soils, in Y.M. Luo et al. (eds.), *Proc. Int. Conf. Soil Remediation SoilRem2000*, October 15-19, Hangzhou, China, pp. 315-318.

60. Wu, Q.T., Huang, G.F., and Xu, Z.L. (2000) Heavy metal pollution of soils in Guangdong and its remediation technologies, in Y.M. Luo et al. (eds.), *Proc. Int. Conf. Soil Remediation SoilRem2000*, October 15-19, Hangzhou, China, pp. 341-343.
61. Heeraman, D.A., Claasen, V.P., and Zasoski, R.J. (2001) Interaction of lime, organic matter and fertilizer on growth and uptake of arsenic and mercury by Zorro fescue (*Vulpia myuros* L.), *Plant Soil* 234, 215-231.
62. Meagher, R.B., Rugh, C.L., Kandasamy, M.K., Gragson, G., and Wang, N.J. (2000) Engineering phytoremediation of mercury pollution in soil and water using bacterial genes, in N. Terry and G.S. Bañuelos (eds.), *Phytoremediation of Contaminated Soil and Water*, CRC Press Inc, Boca Raton, FL, pp. 201-219.
64. Sopper, W.E. (1993) *Municipal Sludge Use for Land Reclamation*, Lewis Publishers, Ann Arbor, MI.
65. Compton, H., Brown, S., Henry, C., and Sprenger, M. (2001) Use of biosolids and lime to restore a metal affected ecosystem in Leadville, CO, in L. Evans (ed.), *Proc 6th ICOBTE Conference*, University of Guelph, Ontario, Canada, p. 35.
66. Svendsen, A.J., Brown, S.L., and Henry, C.L. (2001) Use of biosolids and lime to ameliorate subsoil acidity in mine tailings, in L. Evans (ed.), *Proc 6th ICOBTE Conference*, University of Guelph, Ontario, Canada, p. 36.
67. Brown, S.L., Henry, C.L., and Chaney, R.L. (2001) Restoration of large scale metal contaminated sites using biosolids and other residuals, in L. Evans (ed.), *Proc 6th ICOBTE Conference*, University of Guelph, Ontario, Canada, p. 282.
68. Christensen, B. and Brown, S. (2001). Assessment of methods for the in situ remediation of heavy metal contaminated soils, in L. Evans (ed.), *Proc 6th ICOBTE Conference*, University of Guelph, Ontario, Canada, p. 37.
69. O'Neill, P. (1995) Arsenic, in B. Alloway (ed.), *Heavy Metals in Soils*. 2nd edition, Blackie Academic and Professional, New-York, pp. 105-121.
70. De Koe, T. (1994) *Arsenic Resistance in Submediterranean Agrostis Species*. PhD thesis, Vrije Universiteit, Amsterdam, The Netherlands.
71. Mench, M., Guinberteau, J., and Recalde, N. (2003) Ectomycorrhizal fungi in the contaminated Jales soil after *in situ* treatment and phytostabilisation, in G. Gobran, et al. (eds.) *Ext. Abst. Proc. 7th ICOBTE*, Uppsala, Sweden (in press).
72. Eisler, R. (1994) A review of arsenic hazards to plants and animals with emphasis on fishery and wildlife resources, in Nriagu J.O. (ed), *Arsenic in the Environment, Part II Human Health and Ecosystem Effects*, John Wiley & Sons, New York, pp. 185-259.
73. Saur, E., Brechet, C., Lambrot, C., and Masson, P. (1995) Micronutrient composition of xylem sap and needles as a result of P-fertilisation in maritime pine, *Trees* 10, 52-54.
74. Bleeker, P.M., Assunção, G.L., Teiga, P.M., De Koe, T., and Verkleij, J.A.C. (2002) Revegetation of the acidic, As contaminated Jales mine spoil tips using a combination of spoil amendments and tolerant grasses, *Sci. Total Environ.* 300, 1-13.
75. Friesl, W., Lombi, E., Horak, O., and Wenzel W.W. (2001) Use of amendments to reduce trace elements mobility, in L. Evans (ed.), *Proc 6th ICOBTE Conference*, University of Guelph, Ontario, Canada, p. 217.
76. Lombi, E., Zhao, F.J., Zhang, G., and McGrath, S.P. (2001) In situ fixation of heavy metals by industrial co-products, in L. Evans (ed.), *Proc 6th ICOBTE Conference*, University of Guelph, Ontario, Canada, p. 42.
77. McGowen, S.L., Basta, N.T., and Brown, G.O. (2001) Use of diammonium phosphate to reduce heavy metal solubility and transport in smelter-contaminated soil, *J. Environ. Qual.* 30, 493-500.
78. Seaman, J.C., Arey, J.S., and Bertsch, P.M. (2001) Immobilization of nickel and other metals in contaminated sediments by hydroxyapatite addition, *J. Environ. Qual.* 30, 460-469.
79. Berti, W.R., Cunningham, S.D., and Cooper, E.M. (1998) Case studies in the field - In-place inactivation and phytorestoration of Pb-contaminated sites, in J. Vangronsveld and S.D. Cunningham (eds.), *Metal Contaminated Soils: In Situ Inactivation and*

Phytorestoration, Springer-Verlag, Berlin and R.G. Landes Company, Georgetown, TX, pp. 235-248.

80. Brown, S.L., Chaney, R., and Berti, B. (1999) Field test of amendments to reduce the in situ availability of soil lead, in W.W. Wenzel, D.C. Adriano, B. Alloway, H.E. Doner, C. Keller, N.W. Lepp, M. Mench, R. Naidu and G.M. Pierzynski (eds), *Proc. Ext. Abs. 5th Int. Conf. Biogeochemistry Trace Elements*, Int. Soc. Trace Element Research, Vienna, pp. 506-507.
81. Chen, M., Ma, L.Q., Singh, S.P., and Harris, W.G. (2001) Feasibility of using phosphate amendments to stabilize heavy metals in lead contaminated soils: field demonstration, in L. Evans (ed.), *Proc 6th ICOBTE Conference*, University of Guelph, Ontario, Canada, p. 477.
82. Shi, Z. and Erickson, L.E. (2001) Mathematical model development and simulation of in situ stabilization in lead-contaminated soils, *J. Hazard Mater.* 87, 99-116.
83. Peryea, F.J. and Kammereck, R. (1997) Phosphate-enhanced movement of arsenic out of lead arsenate contaminated topsoil and through uncontaminated subsoil, *Water Air Soil Pollut.* 93, 243-254.
84. Gulz, P. and Gupta, S.K. (2001) Is it possible to improve arsenic uptake in sunflower by adding phosphate to As contaminated soils ? in L. Evans (ed.), *Proc 6th ICOBTE Conference*, University of Guelph, Ontario, Canada, p. 319.
85. Seaman, J.C., Meehan T., and Bertsch, P.M. (2001) Immobilization of cesium-137 and uranium in contaminated sediments using soil amendments, *J. Environ. Qual.* 30, 1206-1213.
86. Lothenbach, B., Furrer, G., and Schulin, R. (1997) Immobilization of heavy metals by polynuclear aluminium and montmorillonite compounds, *Environ. Sci. Technol.* 31, 1452-1462.
87. Lothenbach, B., Furrer, G., Schärli, H., and Schulin, R. (1999) Immobilization of zinc and cadmium by montmorillonite compounds: Effects of aging and subsequent acidification, *Environ. Sci. Technol.* 33, 2945-2952.
88. Krebs, R., Gupta, S.K., Furrer, G., and Schulin, R. (1999) Gravel sludge as immobilizing agent in soils contaminated by heavy metals: a field study, *Water Air Soil Pollut.* 115, 465-479.
89. Oliver, D.P, Hanmam, R., Tiller, K.G., Wilhelm, N.S., Merry, R.H., and Cozens, G.D. (1994) The effects of zinc fertilization on cadmium concentration in wheat grain, *J. Environ. Qual.* 23, 705-711.
90. Van der Lelie, D., Tibarzawa, C., Corbisier, P., Vangronsveld, J., and Mench, M. (2000) Bacterial biosensors to quantify bioavailable concentration of heavy metals in polluted soils and to predict their risk of transfer to the food chain, Int. Conf. Heavy Metals, session 25, Bioadsorption and biomonitoring, Toronto.
91. Bouwman, L., Bloem, J., Römkens, P.F.A.M., Boon, G.T., and Vangronsveld, J. (2001) Beneficial effect of the growth of metal tolerant grass on biological and chemical parameters in copper- and zinc contaminated sandy soils, *Minerva Biotech.* 13, 19-26.
92. Geebelen, W., Adriano, D.C., Van der Lelie, D., Mench, M., Carleer, R., Clijsters, H., and Vangronsveld J. (2003). Selected bioavailability assays to test the efficacy of amendment-induced immobilization of lead in soils. *Plant Soil* (*in press*).
93. Osté, L.A., Lexmond, T., Dolfing, J., Ma, W-C. (2001) The effect of immobilization on Cd accumulation in earthworms is determined by the uptake route, in L. Evans (ed.), *Proc 6th ICOBTE Conference*, University of Guelph, Ontario, Canada, p. 335.
94. Conder, J.M., Lanno, R.P., and Basta, N.T. (2001) Assessment of metal availability in smelter soil using earthworms and chemical extractions, *J. Environ. Qual.* 30, 1231-1237.
95. Brown, S.L., Chaney, R.L., Sprenger, M., and Compton, H. (2002) Assessing impact to wildlife at biosolids remediated sites, *BioCycle* August, 50-58.
96. Jones, C.A., Inskeep, W.P., and Neuman, D.R. (1997) Arsenic transport in contaminated mine tailings following liming, *J. Environ. Qual.* 26, 433-439.

97. Brown, S.L., Chaney, R.L., Sprenger, M., and Compton, H. (2002) Soil remediation using biosolids, *BioCycle* June, 41-44.
98. Larison, J.R., Likens, G.E., Fitzpatrick, J.W. and Crock, J.G. (2000) Cadmium toxicity among wildlife in the Colorado Rocky Mountains, *Nature (London)* 406, 181-183.
99. Cooke, J. and Johnson, M. (2001) Sustainable restoration and residual chemicals in contaminated land, in L. Evans (ed.), *Proc 6th ICOBTE Conference*, University of Guelph, Ontario, Canada, p. 169.
100. Basta, N.T., Gradwohl, R., Snethen, K.L., and Schroder, J.L. (2001) Chemical immobilization of lead, zinc, and cadmium in smelter-contaminated soils using biosolids and rock phosphate, *J. Environ. Qual.* 30, 1222-1230.
101. Bodéman, F, Laperche, V., and Bény, C. (2001) Investigation of As trapping in sulfate-minerals, in L. Evans (ed.), *Proc 6th ICOBTE Conference*, University of Guelph, Ontario, Canada, p. 619.
102. Menzies, N. (2001) Redox processes and attenuation of metal availability in soils, in L. Evans (ed.), *Proc 6th ICOBTE Conference*, University of Guelph, Ontario, Canada, p. 108.
103. Matocha, C.J., Sparks, D.L., Amonette, J.E., and Kukkadapu, R.K. (2001) Kinetics and mechanism of birnessite reduction by catechol, *Soil Sci. Soc. Am. J.* 65, 58-66.
104. Salinger, Y., Sparks, D.L., Ghodrati, M., and Hendriks, G.J. (1994) Kinetics of ion removal from an iron-rich industrial co-product: I. Chloride, *J. Environ. Qual.* 23, 1194-1200.
105. Bhumbla, D.K., Sekhon, B.S., and Sajwan, K.S. (2001) Trace elements bioavailability in mine soils treated with sewage sludge and fly ash mixtures, in L. Evans (ed.), *Proc 6th ICOBTE Conference*, University of Guelph, Ontario, Canada, p. 368.
106. Ferguson, C.C. (1999) Assessing risks from contaminated sites: Policy and practice in 16 European countries, *Land Contamination Reclamation* 7, 33-54.
107. Rock, S., and Sayre, P.G. (2002) Regulatory considerations for phytoremediation, in I. Raskin and B.D. Ensley (eds.), *Phytoremediation of Toxic Metals: Using Plants to Clean Up the Environment*, John Wiley & Sons, Inc. New York, pp. 43-49.
108. Canadian Council of Ministers of the Environment (2001) Canadian Environmental Quality Guidelines, CCME Documents, Winnipeg, Canada.
109. Kärenlampi, S. (2002) Risk of GMO's: General introduction, political issues, social and legal aspect, in M. Mench and B. Mocquot (eds.), *Risk Assessment and Sustainable Land Management Using Plants in Trace Element-Contaminated Soils*, 4th WG2 Workshop, COST Action 837, Bordeaux'2002, INRA, Villenave d'Ornon, France, pp. 99-100.
110. Glass, D.J. (2000) Economic potential of phytoremediation, in I. Raskin and B.D. Ensley (eds.), *Phytoremediation of Toxic Metals: Using Plants to Clean Up the Environment*, John Wiley & Sons, Inc. New York, pp. 15-31.
111. Ruttens, A., Vangronsveld, J., and Mench, M. (2003) A lysimeter approach to evaluate metal immobilisation by soil additives in a metal-contaminated soil. Part I: Reductions in metal percolation and exchangeable soil metal contents, *Environ. Pollut.* (submitted).

CHAPTER 6

PHYTOMANAGEMENT OF RADIOACTIVELY CONTAMINATED SITES

H. VANDENHOVE
SCK-CEN, Radiation Protection Research Department
Boeretang 200, 2400 Mol, Belgium

Abstract

The application of nuclear energy and use of radionuclides for industrial, medical and research purposes have caused significant contamination of certain sites and their environment, which could result in health problems for several centuries if nothing is undertaken to remedy these situations.

Except for immediate environment of the facility, where decontamination activities may be feasible and affordable, the contamination often extents over a vast area and decontamination would be costly and could result in vast amounts of waste. Therefore, more realistic yet efficient remediation options should be searched for of which phytomanagement is among the potential options.

1. INTRODUCTION

The application of nuclear energy and the use of radionuclides for industrial, medical and research purposes have caused significant contamination of certain sites and their environment, which could result in health problems for several centuries if nothing is undertaken to remedy these situations. Except for the close environment of the facility, where decontamination activities may be feasible and affordable, the contamination often extents over a vast area and decontamination would result in considerable amounts of waste. Therefore, more realistic yet efficient remediation options should be searched for of which phytomanagement is among the potential options.

2. SCOPE AND LIMITATIONS

Below a short survey is given of the most important site and environmental radioactive contamination problems encountered. The potential role of different phytomanagement options is discussed and illustrated with

J.-L. Morel et al. (eds.), Phytoremediation of Metal-Contaminated Soils, 191–228.

examples. The phytomanagement options considered are: phytoextraction (including rhizofiltration), alternative land uses and phytostabilisation. The radionuclides considered are the fission products ^{137}Cs and ^{90}Sr and the natural radionuclide uranium.

3. MAJOR SOURCES OF SITE AND ENVIRONMENTAL RADIOACTIVE CONTAMINATION

3.1. Nuclear Weapons' Testing

Nuclear weapon testing started in 1945; the last atmospheric explosions took place in 1980. Most tests were performed by the USA, the former Soviet Union, China, France and UK. In addition to the contamination of the test sites and their surroundings, the atmospheric tests caused a significant global fallout, of which traces can be found back everywhere in the environment. The individual doses caused by the global fall-out are, however, negligible and remedial actions are therefore not feasible. Local contamination close to the test sites amounts to about 10% of the activity and significant local contamination and public exposures were experienced for example in the Altai region in Russia due to the atmospheric tests in Semipalatinsk. Today the doses to the public are low and no remediation actions are planned for the area (Zeevaert *et al.*, 1997).

3.2. Nuclear Weapons' Production

Information on environmental contamination linked with nuclear weapon production sites is scant, due to the general secrecy on all military activities. In Europe nuclear weapons have been fabricated in the FSU, France and UK. In the FSU nuclear materials were produced at seven sites, the most important of which were located close to Mayak (S. Ural), Tomsk and Kransnoyarsk. Some of these installations gave rise to considerable environmental contamination. At Mayak, an important surface was contaminated with ^{137}Cs and ^{90}Sr, through discharge of vast amounts of medium and high level liquid wastes into the Techa and Iset rivers and the explosion of a radwaste tank. A radiological rehabilitation plan was developed in 1982. The programme provided for a regulation of farming and forest use in the contaminated areas, primarily through optimised land use, in order to reduce the dose to the population (Romanov and Drozhko, 1996).

3.3. The Nuclear Fuel Cycle

The nuclear fuel cycle comprises mining and milling of uranium ore, conversion to nuclear fuel, fission of the fuel in the reactor, reprocessing of the used fuel and storage and disposal of the radioactive waste. Except for the process of uranium mining and milling, electricity produced by nuclear energy does not give rise to any significant health risk in routine operations. Throughout the years, accidents causing a release of radioactive material to the environment have occurred. Mostly these accidents were of a local concern.

Uranium mining and milling operations involve the removal and processing of large quantities of uranium ore and is one of the predominant sources of radioactive contamination in the nuclear power production cycle. In Europe uranium mining was mainly carried out in Germany, France, Spain, Bulgaria, Hungary, Romania and the Czech Republic. Most mining and milling sites in France and Spain are closed and remediated or remediation is in progress. There are also many abandoned sites in Eastern Europe; none are, however, remediated. Many of these sites pose serious problems because preservation and protection of the environment has often been neglected and because these sites are often situated close to human populations. The major atmospheric exposure in the vicinity of tailings comes from the inhalation of ^{230}Th and ^{222}Rn and ^{226}Ra, ^{238}U, ^{210}Po and ^{232}Th on dust and ingestion from ^{238}U leached to groundwater. For the remediation of the very low geomechanical stability of the fine-grained tailings there is no reasonable alternative than in-situ stabilisation and encapsulation. Heaps of mining debris are very seldom remediated.

The nuclear accident that so far released the highest amount of radioactivity and that was of global concern, was the accident in the Chernobyl reactor in Ukraine in April 1986. Apart for the first weeks, the major contributor to the ingestion dose and the external dose is radiocaesium. The total area of land contaminated at levels of 37 kBq m^{-2} ^{137}Cs or more was about 13,0000 km^2. About 10,000 km^2 were contaminated at 0.5 MBq m^{-2}. In these most contaminated regions the dose dropped from 20 mSv in 1986 to about 2 mSv in 1993.

Table 1 gives Figs. on radionuclide releases resulting from the major nuclear accidents.

Table 1. Radionuclide releases (PBq) resulting from the major nuclear accidents.

	Kyshtym	Windscale	Three-Mile Island	Chernobyl
Xe-133	-	14	370	6300
I-131	-	1	0.00005	1600
Cs-137	0.03	0.05		85
Sr-90	4	0.0002		8
Pu-239	-	0.002		0.03

Remediation actions applied in agricultural areas included physical operations (removal of contaminated soil and vegetation, deep an normal ploughing), soil improvement (application of Cs and Sr binders, potassium fertilizer liming, to decrease the soil-to-plant transfer) and reorientation of the agricultural industry (selection of crops (varieties) with a lower soil-to-plant transfer, cultivation of industrial (non edible) crops, from arable system to forestry). As such reduction factors (activity concentrations in alternative product or in product after treatment/ activity concentration in original product) ranging between 1.5 and 100 could be attained. The resulting dose reduction was between 50% and a factor 2.

3.4. Industrial Processes Involving Radionuclides

Of the primordial naturally occurring radionuclides ^{238}U, ^{235}U and ^{232}Th are particular since they are followed by a series of daughter products. Uranium and thorium are present in the Earth's crust at average concentrations of 4.2 and 12.5 mg kg^{-1}, respectively (corresponding to 50 Bq kg^{-1} for each of the radionuclides). However, levels of up to a few percent are found in particular ores.

During recent years, increasing interest has been paid to contamination linked with industries handling materials containing elevated levels of natural radionuclides (NORs). Ore extraction and processing may lead to enhanced levels of naturally occurring radionuclides (NORs) in products, by-products and waste and in surroundings and installations of the facility. The most contaminating industries are uranium mining and milling, metal mining and smelting and the phosphate industry. Radionuclide levels in products and/or waste products from the oil and gas extraction industry and of the rare earth and zirconium and ceramics industries may be particularly elevated, but waste streams are limited. Impact on the public from coal mining and power production from coal is commonly considered low (Vandenhove *et al.*, 2000, Vandenhove, 2000).

Remediation technologies applied to those sites may be divided in five categories: removal of sources; containment (capping and subsurface barriers) (capping may consist of multi-layer covers including vegetation cover); immobilisation (cement and chemical based solidification); separation combined with removal (soil washing, flotation and chemical/solvent extraction); and 'do nothing'. At many sites, remediation has not started.

4. POSSIBLE ROLE OF PHYTOMANAGEMENT

4.1. Phytoextraction

Phytoextraction is an approach to the clean-up of contaminated soils using accumulator plants. Phytoextraction requires that the target metal (radionuclide) is available for the plant root, absorbed by the roots and translocated from the root to the shoot and biomass production should be substantial. The metal (radionuclide) is removed from the site by harvesting of the biomass after which it is processed to either recover the metals or further concentrate the metal (thermal, microbial, chemical treatment) to facilitate disposal.

To maximise the metal content in the biomass, it is necessary to use a combination of improved soil management inputs (e.g. optimised soil pH and mineral nutrition, addition of agents which increase the availability of metals in soils), improved genotypes with optimised metal uptake, translocation and tolerance and improved biomass yield. The economics of the operation do not only depend on the phytoextraction efficiency but also on costs associated with crop management [soil management, crop establishment costs (yearly returning for annual crops), crop harvest], post-harvest biomass transport and biomass treatment and potential disposal.

Contrary to the research on heavy metals, the phytoextraction approach for radionuclides (RN) is rather new. Furthermore, most experiments conducted to test the phytoremediation approach for RN were done in hydroponics systems (Vasudev *et al.*, 1996; Ensley *et al.*, 1997).

4.1.1. The Potential for Phytoextraction

The soil to plant transfer of metal contaminants is a major process that has to be considered in the management of metal contaminated sites. These fluxes should either be minimised to reduce exposure of soil pollutants in the food chain or should be maximised in an attempt to remediate a soil by phytomining or phytoextraction. This part will give an overview of the soil-plant fluxes of metal contaminants. Examples will be given for three radiocontaminants that are considered important due to their toxicity and/or ubiquity. These elements are the natural radionuclide uranium, one of the predominant contaminants in the uranium mining and milling industry, the phosphate industry, and other NORM industries and the long-lived fission products radiocaesium (^{137}Cs) and radiostrontium ^{90}Sr.

The flux of an element from soil to plant is often referred to as the crop off take or removal of that element, i.e. the removal of the element from soil with the harvested part of the crop. The removal of a contaminant from the soil with the harvested biomass (in Bq ha^{-1}), is the product of the concentration in the plant (C_{plant}, in Bq kg^{-1}) and the yield of the harvested biomass (kg ha^{-1}):

$$\text{Crop Removal} = \text{Yield x } C_{plant} \quad (1)$$

The concentration of an element in the plant depends on its concentration in the soil, type of soil, plant type etc. It has been demonstrated that crop concentrations of non-essential trace elements, U, ^{137}Cs and ^{90}Sr, are proportional to their concentration in soil, for the same crop, soil type etc. The proportionality constant is defined as the Transfer Factor (TF, dimensionless)

$$TF = C_{plant}/C_{soil} \tag{2}$$

with C_{plant} the concentration of the radiocontaminant in the plant (Bq/g) and C_{soil} the concentration of the contaminant in the soil. The transfer factor is hence an important parameter determining the potential of phytoextraction. Table 2 gives a summary of transfer factors.

Table 2. TFs for Cs, Sr and U: total ranges, ranges of recommended values and upper (several sources)

	Total range	Range of recommended values	Upper limit	Comment on conditions for upper limit
Cs	0.00025–7.5	0.0038–0.29	7.5	Brassica, organic soil
Sr	0.0051–22	0.017–3.2	22	Green vegetables, sandy soil
U	0.000006–21.13	0.00075–0.02	21.13	Tubers, sandy soil

With Equations 1 and 2, the soil-to-plant transfer factor (TF), the percentage yearly reduction in soil activity can be calculated as follows:

$$\text{Annual removal (\%)} = \frac{TF \times Yield}{W_{soil}} \text{ x } 100 \tag{3}$$

with W_{soil} the weight of the contaminated soil layer (kg ha^{-1}). As made clear by Equation 3, the annual removal percentage increases with yield and TF. However, TF and yield values are not independent: high yield is often associated with lower TFs because of growth dilution effects.

Phytoextraction requires several years and the future trend in radionuclide concentration in the soil can be calculated according:

$$C_{soil,t} = C_{soil_{t=0}} \exp\left\{-\left(\frac{TF \times yield}{W_{soil}} + \frac{0.69}{t_{1/2}}\right) \times t\right\} \tag{4}$$

The second term in the exponent of Equation 4 was included to account for radioactive decay ($t_{½}$ is the half life of the radionuclide). Given the half life of ^{238}U, 4.5 10^9 year, this component will not affect the phytoextraction potential. For ^{137}Cs and ^{90}Sr, with half-lives of 30 years, the phytoextraction potential will be affected (2.33% yearly loss in activity merely through radioactive decay). Equation 4 assumes a constant bioavailability of the contaminant, i.e. a constant TF.

For a soil depth of 10 cm and a soil density of 1.5 kg dm^{-3}, soil weight is 1500000 kg. Table 3 shows the % annual removal for different crop yields and TFs.

Yields of more than 20 ton ha^{-1} and TFs higher than 0.1 (Table 2) may be regarded as average values or upper limits, except for Sr. This would result in an annual reduction percentage of 0.1% (excluding decay). In case of a TF of 1, annual reduction is about 1%.

Rearranging Equation 4 allows calculating the number of years needed to attain the required reduction factor as a function of annual removal percentage. Table 4 tabulates the years required to attain a reduction of the contaminant concentration up to a factor 100, given an annual extraction percentage or percentage reduction in radionuclide activity varying between 0.1 and 20%. With an annual removal of 0.1% it would take more than 2000 years to decontaminate a soil to 10% of its initial activity. With an annual removal of 1%, more than 200 years are required. It is hence clear that measures should be taken to increase the annual removal efficiency through crop selection, increasing the bioavailability with soil additives and through technical actions (e.g. decreased soil depth).

Table 3. Percentage yearly reduction of the soil contamination due to phytoextraction [and radioactive decay (t½ 30 years)] (soil depth 10 cm; soil density 1.5 kg dm^{-3}).

	Annual reduction % due to phytoextraction					Annual reduction % due to phytoextraction and radioactive decay				
Yield (t ha^{-1})	5	10	15	20	30	5	10	15	20	30
TF (g g^{-1})										
0.01	0.0033	0.0067	0.01	0.013	0.02	2.333	2.337	2.34	2.343	2.35
0.1	0.033	0.067	0.1	0.133	0.2	2.363	2.397	2.43	2.463	2.53
1	0.33	0.67	1.00	1.33	2.00	2.66	3	3.33	3.66	4.33
2	0.67	1.33	2.00	2.67	4.00	3	3.66	4.33	5	6.33
5	1.67	3.33	5.00	6.67	10.00	4	5.66	7.33	9	12.33
10	3.33	6.67	10.00	13.33	20.00	5.66	9	12.33	15.66	22.33

Table 4. Years required to decontaminate a soil (soil depth 10 cm; soil density 1.5 kg dm^{-3}) given a required (desired) reduction factor and a given annual removal percentage.

Desired reduction	% activity remaining $C_{soil.t}/C_{soil.t=0}$	Annual removal, % y^{-1} 20	15	10	5	3	2	1	0.1
1.5	66	2	2.5	4	8	13	20	40	400
2	50	3	4	7	14	23	34	69	693
3	33	5	7	11	22	36	55	110	1108
4	25	6	9	13	27	46	69	138	1609
5	20	7	10	15	31	53	80	160	1650
10	10	10	14	22	45	76	114	229	2301
20	5	13	18	28	58	98	148	298	2994
50	2	18	27	37	76	128	194	389	3910
100	1	21	28	44	90	151	228	458	4603

Decreasing the soil depth increases the removal percentage according to equation 3 and may intensify root-soil contact, and may result in an increased TF. The soil can be spread on geomembranes which impede roots from penetrating to deeper layers. These membranes will also limit contamination of the underlying clean soil, but a substantial area may be needed for treatment. In most cases of soil contamination, one has limited control on the depth of the contamination, though it may be feasible and advantageous to excavate and spread out the soil over the desired soil depth for phyto-decontamination purposes. Therefore, considering a soil depth of 10 cm may be a realistic assumption.

Decreasing the soil depth is radionuclide and plant unspecific. The other factors influencing radionuclide bioavailability (crop selection and increasing the bioavailability of the radionuclide of concern) will be discussed per radioelement. For each element a short description on the behaviour of the element in soil will be given.

4.1.2. Uranium

Behaviour of U in soil

Natural uranium exists as three isotopes, ^{238}U, ^{235}U and ^{234}U, with a relative abundance of 99.27, 0.720 and 0.0055%, respectively. The world average ^{238}U activity in soil is 40 Bq kg^{-1} (UNSCEAR, 1993). Uranium behaviour is similar to that of other heavy metals and its physiological toxicity, mimics that of lead. Uranium is chemically toxic to kidneys and insoluble U-compounds are carcinogenic (Hossner *et al.*, 1998).

The complex, pH-dependent speciation of U in soils makes the study of U uptake by plants difficult. U is present in soil primarily (80-90%) in the +VI oxidation state as the uranyl (UO_2^{+2}) cation (Ebbs, 1997). Under acidic conditions, UO_2^{+2} is the predominant U species in the soil. Hydroxide complexes form under near-neutral conditions, while carbonate complexes predominate under alkaline conditions. Generally, negligible amounts of UO_2^{+2} are present in available forms, due to the high solid-liquid distribution coefficient of uranium (K_D: ratio of the radionuclide in the solid soil fraction to the radionuclide concentration in the soil solution: Bq kg^{-1}/Bq dm^{-3}) of uranium (620 dm^3 kg^{-1}, Baes, 1982). Though extensive work has been done on U solubility in soils, there is comparatively little information regarding the uptake and translocation of U by plants as affected by soil properties. Work has focused primarily on either the U content of native plants growing in U contaminated environments (Whicker and Ibrahim, 1984; Ibrahim and Whicker, 1988, 1992) or U uptake by field and garden crops of importance to animals and humans (Frissel and van Bergeijck, 1989; Sheppard *et al.*, 1985; Sheppard *et al.*, 1989).

Uranium soil-to-plant transfer factors

The U soil-to-plant transfer factor (TF) varies with plant compartment and plant species. Roots incorporate much more uranium than stems, leaves and seeds (Apps et al., 1988; Lakshmanan and Venkateswarlu, 1988; Ebbs et al., 1998).

Leafy vegetables generally show higher U TFs, followed by root, fruit and grain crops (Table 5). TF values for the plants that have been studied rarely exceed a value of 0.1, except in plants grown on highly contaminated, often acidic, U-mining sites. A study of U accumulation by mixed grass species, forbs and big sagebrush (*Artemesia* spp.) growing near a U mine/mill complex indicated that plants growing on exposed tailings (low pH) had a TF value that approached unity while plants grown in proximity to the tailings pile had TF values ranging from < 0.1 to approximately 0.4 depending on the distance from the tailings and the wind direction (Ibrahim and Whicker, 1988). Sagebrush, the plant with the highest TF for natural conditions, is pH sensitive and does not grow at a pH below 4.

Table 5. Some specific U-TF to natural vegetation and garden crops.

Plant/crop/vegetation type		TF	Experimental conditions
Natural vegetation at U-milling			
site[1,2]		0.12	Background
Sagebrush		0.90	Edge of tailing pond
		0.07	Background
Mixed grasses		0.07	Edge of tailing pond
		0.69	Exposed tailings
		0.16	Background
Forbs		0.08	Edge of tailing pond
		1.1	Exposed tailings
Grasses[3]		0.003-0.18	Sandy and loamy soil
Vegetables			
Leaf	Lettuce[3]	0.025	Loam, pH 7
vegetables	Spinach[3]	0.033	Loam, pH 7
	Turnip green[1]	0.0058	Lake sediments, sandy, pH 4,
	Indian	0.007	1.5% OM
	mustard[5,6]	0.0057	Loam, OM 4.2%, pH 7.7
	Carrots[3]	0.019	Sandy loam, pH 7
Root crops	Potato flesh[3]	0.002	Loam, pH 7.3
	Potato flesh[3]	0.009-	Sand, pH 4.9
	Potato[4]	0.0009	Sand, pH 8.1 (TF for fresh
	Sugar beet[3]	0.01-0.06	weight)
	Red beet[3]	0.0024	Sand or Sandy loam, pH 7
	Turnip root[1]	0.00099	Loam, pH 5.1-7.0
Fruit	Tomato[3]	0.0005	Lake sediments, sandy, pH 4,
vegetables	Cucumber[3]	0.0009	1.5% OM
			Sandy loam, pH 7
			Sandy loam, pH 7
Cereals	Bush bean[1]	0.0018	Lake sediments, sandy, pH 4,
	Rice[3]	0.0005	1.5% OM
	Barley[3]	0.0021	Loam, pH 7.3
	Corn[3]	0.00021	Loam, pH 7.3
	Corn[1]	0.00021	Loam, pH 7.3
	Corn[5]	0.006-0.01	Lake sediments, pH 4, 1.5%
			OM
			Not specified
Amaranthus[7]		~0.007	

1. Whicker and Ibrahim, 1984;
2. Ibrahim and Whicker, 1992;
3. Frissel and van Bergeijck, 1989;
4. Lakshmanan and Venkateswarlu, 1988;
5. Mortvedt, 1994;
6. Ebbs *et al.*, 1998;
7. Huang *et al.*, 1998

Ebbs (1997) showed that free UO_2^{+2} is the U species most readily taken up and translocated by plants. Since this U species is only present at a pH of 5.5 or less, acidification of U-contaminated soils may be necessary for phytoextraction (Fig. 1). The uranyl cation also binds to the soil solids and organic matter, reducing the extent of plant uptake (Sheppard *et al.*, 1984). Therefore, in addition to acidification, soil amendments that increase the availability of U by complexation may also be required.

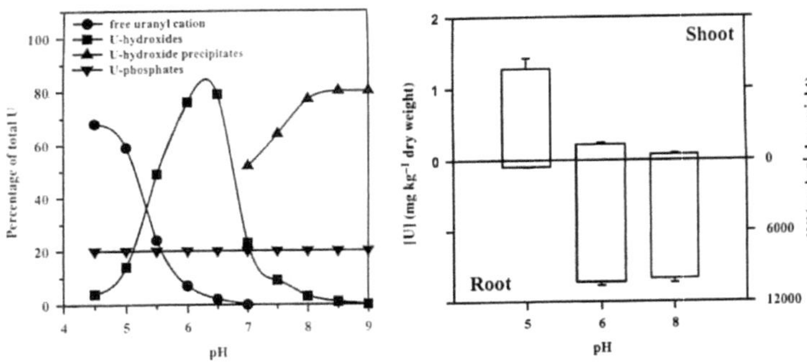

Fig. 1. pH –dependence of U-speciation and U-uptake (after Ebbs *et al.*, 1997)

Ebbs *et al.* (1998) and Huang *et al.* (1998) studied the role of acidification and chelating agents on the solubilisation of uranium. Of the organic acids tested (acetic acid, citric acid, oxalic acid, malic acid), citric acid was the most effective for increasing U in the soil solution. The results furthermore indicated that citric acid solubilised more U than simple acidification (HCl, SO, HNO_3). Chelating agents like EDTA and DTPA did not increase uranium solubility significantly. Compared to potassium citrate, citric acid was much more effective for increasing U solubility and hence accumulation in plants. With the addition of 20-25 mmol citric acid per kg soil, soil pH decreased 0.5 to 1.0 pH units, depending on initial soil pH. Following citric acid treatment (20 mmol kg^{-1}) the U-accumulation in Indian mustard (*Brassica juncea)* was increased 1000-fold (Huang *et al.*, 1998). Ebbs *et al.* (1998) and Ebbs (1997) observed a factor 10 increase in U-accumulation in beet (*Beta vulgaris)* after citric acid treatment (25 or 20 mmol kg^{-1}). Citric acid was always applied one week before harvest by spraying a solution over the soil surface. The citric-acid triggered U-accumulation in plants was observed within 24 h after application and a maximum was reached after 3 days. Thus, plants would maintain very low U-concentrations for most of their growth and start accumulating rapidly following after citric acid application after which it can be quickly harvested. Since citric acid is rapidly bio-degradable there should be no problems with residual levels of citric acid in soil.

The potential to phytoextract uranium (U) from a sandy soil contaminated at low levels was tested in the greenhouse by Vandenhove *et al.* (2001). Two soils were tested: a control soil (317 Bq ^{238}U kg^{-1}) and the same soil washed with bicarbonate (69 Bq ^{238}U kg^{-1}). Ryegrass (*Lolium perenne cv. Melvina)*, Indian mustard (*Brassica juncea cv. Vitasso*) and Redroot Pigweed (*Amarathus retroflexus)* were used as test plants. Plants were selected on the basis of their reportedly high transfer factors and relative important dry-weight production.

Citric acid addition resulted in a decreased dry weight production (all plants tested) and crop regrowth (in case of ryegrass) (Fig. 2).

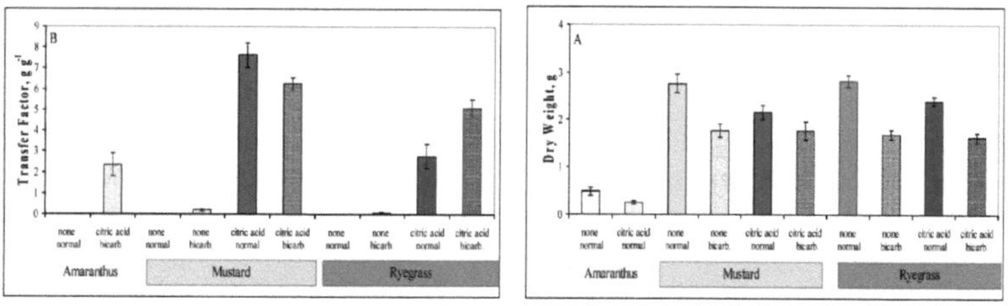

Fig. 2. Effect of citric acid addition one week before harvest on the dry weight production and uranium transfer factor of ryegrass and mustard and redroot pigweed

The annual removal of the soil activity with the biomass was less than 0.1 % for the control scenario. The addition of citric acid (25 mmol kg^{-1}) one week before the harvest increased U uptake up to 500-fold (Fig. 2). With a ryegrass and mustard yield of 15000 kg ha^{-1} and 10000 kg ha^{-1}, respectively, up to 3.5% and 4.6% of the soil activity could annually be removed with the biomass (Table 6). With the required activity reduction level of 1.5 and 5 for the bicarbonate-washed and control soil, respectively, it would take 10 to 50 years to attain the release limit.

A linear relationship was found between the plant ^{238}U concentration and the ^{238}U concentration in the soil solution of the control, bicarbonate washed or citric acid treated soil. This points to the importance of the soil solution activity concentration in determining U uptake and hence to the importance of solubilising agents to increase plant uptake.

Addition of citric acid is required to attain extraction levels that may make phytoextraction a feasible remediation option. There are important uncertainties, which affect the phytoextraction potential. (1) How does plant growth stage affect the efficiency of citric acid addition and what is the expected yield at the growth stage of maximal accumulation? (2) What is the optimal level of citric acid addition to attain the highest increase in U-TF and the lowest impact on yield? (3) What is the effect of continuous citric acid treatment on the soil? (4) What is the global cost of phytoextraction compared to other remediation options?

Table 6. Phytoextraction potential (annual removal, %) of untreated and citric acid treated soils (First Harvest) (Soil Depth 10 cm; Soil Density 1.5 kg dm^{-3})

Plant	Soil	Citric acid	Annual extraction %
Amaranthus	Control	No	0.0009 ± 0.0003
	Control	Yes	0.178 ± 0.058
Mustard	Control	No	0.010 ± 0.002
	Control	Yes	4.618 ± 0.384
	Bicarb.	No	0.103 ± 0.030
	Bicarb.	Yes	3.284 ± 0.250
Ryegrass	Control	No	0.007 ± 0.004
	Control	Yes	2.810 ± 0.689
	Bicarb.	No	0.052 ± 0.008
	Bicarb.	Yes	3.477 ± 0.474

Given the deleterious effect of citric acid on plant growth, one should possibly revert to other soil additives to increase the phytoavailability. However, several authors (Huang *et al.*, 1998; Ebbs *et al.*, 1998) have screened a number of additives, and only citric acid was found to be effective in increasing the TF. Though the experimental conditions of the latter authors were similar to the experimental conditions used by Vandenhove *et al.* (2001), they did not mention a significant decrease in dry weight production or a dying-off of the plants after citric acid treatment. Adding equivalent amounts of citric acid at pH 7 (adjusted with KOH) ryegrass regrowth was also hampered (Vandenhove and Van Hees, results not shown). This implies that dying off of plants following citric acid addition is not predominantly due to a low pH but rather to an increase in ion strength of the soil pore water when nutrients are solubilised under the action of citric acid.

Apart from application of soil additives to increase U export with the plant biomass, plant selection may also be important for improving the phytoextraction potential. However, the expected effects are small compared to the increase in uptake following soil amendments like citric acid addition. Huang *et al.* (1998) recorded a factor two difference in TF for four varieties of Indian mustard (*Brassica juncea).* They also demonstrated that plant species and cultivars differed significantly in response to citric acid treatment. Of the species tested, four showed significant potential in citric-acid triggered U accumulation: Indian mustard (*B. juncea),* Chinese mustard *(B. narinosa),* Chinese cabbage *(B. chinensis)* and Redroot Pigweed (*A. retroflexus)*. A two-fold difference between cultivars was also observed.

4.1.3. Caesium

Behaviour of Cs in soil and soil factors affection caesium availability

After the Chernobyl accident, the study of the fate of radiocaesium in the environment was of particular importance, given its relatively long half-life, its widespread contamination and its similarity with K favouring its uptake by plants (Lembrechts, 1993; Nisbet *et al.*, 1993; Shaw, 1993; Smolders & Merckx, 1993).

The soil K-status affects Cs-availability: in general the higher the soil K, the lower the TF. This K-effect is explained by the increased solution concentration of a cation, which competes with radiocaesium for uptake. There is now a general consensus that the solid/liquid partitioning of radiocaesium in the soils, and hence its soil bioavailability, is governed by the presence of micaceous, potassium bearing clay minerals (Evans *et al.*, 1983; Brouwer *et al.*, 1983; Cremers *et al.*, 1988). The process relates to the action of a small number of very selective sites, the so-called Frayed Edge Sites (FES) located at the edges of illite particles. In general, the higher the soil clay content, the lower the caesium availability.

In studies following the Chernobyl accident, lime was applied to reduce the radiocaesium TF through an increase in pH (extent of increase not reported) and transfer reductions of a factor 2 were obtained (Nisbet *et al.*, 1993; Konoplev *et al.*, 1993). Plant growth may however be reduced at lower pH and in reducing conditions (high water content). As such for example, the Cs-TF may be much higher on an organic soil, but yield may also substantially be reduced.

Enhanced uptake of caesium by plants is reported in the presence of increased amounts of NH_4 (Lembrechts, 1993; Bondar and Dutov, 1992; Belli *et al.*, 1995) or when K is depleted (Cline & Hungate, 1960; Shaw *et al.*, 1992; Smolders *et al.*, 1997). NH_4 additions increased the TF up to two-fold and also enhanced biomass production, both factors favouring the phyto-extraction potential (Vandenhove, 1999). Extremely low soil fertility with regard to potassium may increase the radiocaesium TF 10 to 100 fold but will also decrease plant growth. The rather high TF for ryegrass, 3.3 g/g, recorded by IUR (1989) may have been obtained for low-fertile soil.

Given the number of parameters which may possibly enhance the radiocaesium TF (low pH, high ammonium, low K), it may be concluded that increasing the TFs reported in literature two-to-three fold is a possibility worth to be tested.

Phytoextraction potential of Cs

In normal agricultural systems is the annual Cs-flux small compared to that present in soil (Table 7). The ^{137}Cs off take values are all less than 1% and the highest off take is found for grassland. The high sorption of ^{137}Cs in soil and the typical K levels in soil required for optimal plant growth both limit high off take values.

Table 7. The annual crop off take of ^{137}Cs for some agricultural crops, expressed as a fraction of total content in the plough layer (arable crops) or in the 0-12.5 cm layer (grassland). The range in dry weight based transfer factors (TF) represents a typical range. TFs of ^{137}Cs have been derived from Nisbet & Woodman (2000).

Crop	Yield ton dm /ha	TF g/g	Crop off take % of total in soil
Cereals (grain)	5-7	0.0004-0.25	0.0005-0.06
Potato tuber	6-10	0.003-0.89	0.0006-0.3
Leafy vegetables	5-10	0.008-1.7	0.001-0.6
Grassland	10-15	0.01-1.0	0.007-1.0

Table 8 presents radiocaesium TFs for sandy soils (generally higher TF) for some crops and natural vegetation. Most crops have a TF<< 1 g/g resulting in annual removal percentages of around 3%, decay included (Table 4).

Many crops may have a wide range of TFs, even for a specific soil type. For example, the range of TFs observed for ryegrass grown on sandy soil extends from 0.05-3.3 (IUR, 1992), meaning TFs substantially higher than one are within reach. Potential ryegrass yield is between 15 and 20 t/ha provided adequate fertilisation. This would mean a phyto-extraction ranging between 5.3 and 6.3%, decay included (Table 4).

Table 8. Average radiocaesium soil-to-plant transfer factors.

Crop	Transfer Factor, g/g
Winter wheat (leaves + grain)*	0.03
Summer wheat (leaves)*	0.05
Ryegrass*	0.03
Potato (tuber)*	0.09
Lettuce*	0.24
Yellow lupine (seeds)§	1.64
Yellow lupine (straw)§	1.02
Sunflower (seeds)§	0.15
Sunflower (straw)§	0.59
Sunflower (roots)§	2.88
Ryegrass°	1.2 (0.05-3.3)
Ryegrass!	1.87 (0.92-2.82)
Flax (leaves)$	0.66 (0.57-0.84)
Indian mustard^	0.4-0.5
Red Root Pigweed (Amaranthus species)^	2.2-3.2
Tepary bean^	0.2-0.3
Different amaranthus species¨	0.53-2.03

*: Lembrechts, 1993; §: Gopa, 1996: °: IUR, 1989; !: Vandenhove *et al.*, 1996;
$: Van Hees & Vandenhove, 1998 (unpublished); ^Lasat *et al.*, 1998;
¨Sorochinsky, 1998

Vandenhove (1999) tested the effect of ammonium addition on the phytoextraction potential of two crops with reportedly high TF and DW-production in a greenhouse experiment, using sandy Cs-contaminated soil (5 Bq/g) at a waste processing facility. Target was 1 Bq/g Cs. Ammonium addition increased DW-production with 20% and the TF with 80%, resulting in a TF of 0.8 g/g. With a realistic yield of 20 t/ha under field conditions this would result in an annual reduction of 3.3% (decay included). This would imply, 50 years of continued phytoextraction to reach the target of 1 Bq/g. Lasat *et al.* (1998) observed that ammonium addition did not significantly affect the Cs-TF for three species (Amaranthus, tepary bean, Indian mustard) (Fig. 3).

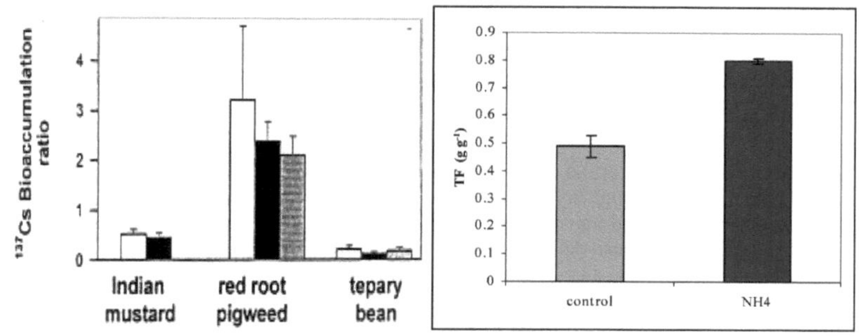

Fig. 3. Effect of ammonium addition on the Cs-TF of Indian mustard, Redroot Pigweed and tepary been (Lasat *et al.*, 1998) and ryegrass (Vandenhove, 1999)

Amaranthus species screened by Lasat *et al.* (1998) and Sorochinsky (1998) have TFs as high as 3.2 g/g and with a yield potential prospected at 30 t/ha/year (based on two harvests per year which is very improbable). The TF reported by Lasat *et al.* (1998) were derived for a sandy-loam from Hanford with contamination levels of 15 Bq/g soil and the annual removal, including decay would hence be 8.3% (Table 4). Their target was a 4-fold reduction in soil activity (to 4 Bq/g) which would require 16 years. In the more likely event of a yield of 15 t/ha, time required would be 28 years.

4.1.4. Strontium

Factors affecting the strontium TF

As radiocaesium, radiostrontium is among the most abundant radionuclides in the suite of nuclear fission products that are routinely or accidentally released. Its relatively long half-life (~30 years) and its metabolic similarity with Ca, favours its uptake in plants.

For the Sr-TFs, the TFs for each crop tend to be highest for sand and lowest for clay and organic soil types: the maximal differences between soil types for any one crop tend to be around an order of magnitude. The soil exchangeable Ca-content seems to affect the Sr-TF but not significantly. Sauras *et al.* (1999) found that lower Sr-TF were observed with increased soil solution Ca+Mg and increased CEC.

Phytoextraction potential of Sr

The off take of ^{90}Sr is higher than that of ^{137}Cs because the Sr availability is typically 10-fold above that of Cs. The TFs of ^{90}Sr in green vegetables and *Brassica*'s are typically around unity and upper levels are around 10 (Table 9).

With a yield of 10 ton/ha for leafy vegetables, and a TF of 10, 9% could be annually removed considering the contamination is restricted to 10 cm and accounting for decay. A five-fold reduction of the soil contamination would under those conditions require 15 years.

Phytoextraction of ^{90}Sr has not yet been investigated at field scale. The high off take values in agricultural crops (Table 9) suggest that this phytoextraction should be explored for this element.

Table 9. The annual crop off take of ^{90}Sr for some agricultural crops, expressed as a fraction of total content in the plough layer (arable crops) or in the 0-12.5 cm layer (grassland). The range in dry weight based transfer factors (TF) represents a typical range. The TFs of ^{90}Sr have been derived from Nisbet and Woodman (2000).

Crop	Yield ton dm /ha	TF g/g	Crop off take % of total in soil
cereals (grain)	5-7	0.02-0.94	0.0037-0.22
potato tuber	6-10	0.03-1.4	0.006-0.5
leafy vegetables	5-10	0.45-9.1	0.07-3.0

4.1.5. Conclusions for the Potential of Phytoextraction

Except for radiostrontium, TFs are generally too low to allow phytoextraction to be efficient without soil additives which increase bioavailability.

Following citric acid addition the uranium-TF may increase with a factor 500 resulting in annual activity reductions up to 3% but even then it would still require more than 50 years to reduce the soil contamination level with only a factor 5. Moreover, the effect of continuous citric acid treatment on the soil and plant growth and TF is not clear.

For Cs the same remark can be made. TFs are a factor 10-100 higher than the U-TF but additives will only increase the availability with maximally a factor 2 without really affecting the DW-production.

The high off take values in agricultural crops for Sr suggest that phytoextraction should be explored for this element.

Still remains the question about long-term effectiveness: will TF remain constant or will they decrease whilst radionuclide concentration decreases and when ageing processes occur.

There is the question of cost of waste treatment and site monitoring. With respect to this last issue, phytoextraction involves costs at different stages in the process. The soil will have to be prepared for crop establishment and the crop has to be well maintained. During crop establishment or before the harvest the soil may have to be treated with radionuclide specific amendments to improve crop yield. The treatment of 1 m^3 contaminated soil (10 m^2 if 1 dm soil layer) will annually result in about 10 to 20 kg biomass (~2-4 kg ashes) which will have to be harvested, transported and treated, entailing considerable costs. This scheme of action will have to be repeated on a yearly basis for several years.

4.2. Rhizofiltration

Rhizofiltration is the use of plants to accumulate compounds from aqueous solutions through adsorption on the roots or assimilation through the roots and eventual translocation to the aerial biomass (= phytoextraction). Rhizofiltration is being investigated for the removal of, radionuclides from aqueous waste streams, including groundwater and wastewater.

The removal of a radionuclide from an aqueous waste stream is governed by the plant dry weight production and the concentration factor, CF (ratio of Bq/g plant to Bq/ml water). Since absorption in (waste) water per volume is lower than in soil, the CF is higher than the TF. This becomes clear when considering the relationship between TF and CF which is:

$$TF = CF/K_D \qquad (5)$$

in which CF is the concentration factor, i.e. the ratio of the radionuclide activity concentration in plant shoots to that in soil solution, and in which K_D is the solid liquid distribution coefficient of the radionuclide ($dm^3\ kg^{-1}$) (i.e. the ratio of radionuclide activity concentration in the solid phase to that in the soil solution)

The K_D for Cs in soils ranges from 10-10^5 ml/g (Wauters, 1994), for Sr from 1-100, for U the geometric mean K_D per soil group range from 35-1600 (full range from 0.03-395100) (Thibault *et al.*, 1990) and for Ra the average K_D per soil group range from 490-36000 ml/g (full range from 1-1.810^7, IAEA 1994).

With the K_D for all radionuclides studied being generally substantially higher than 1, it is clear that the CF exceeds the TF with the same factor and that rhizofiltration is generally more effective that soil phytoextraction.

Vasudev *et al.* (1996) commented upon the removal of radionuclides from contaminated waste water and ground water. They found a 95% reduction in the U concentration in water through U-absorption by sunflowers (*Helianthus annuus*) after 24 h contact time.

In a comparable experiment using Chernobyl contaminated pond water, the Cs and Sr levels decreased with, respectively, 90 and 80% after 12 days contact time with 8-week-old sunflower plants replaced every 48 hours. Sunflowers showed higher removal rates than timothy, meadow foxtail, Indian mustard and peas (Cs-CF to shoots ranging from 400-2300 L/kg, Cs-CF to shoot from 860 -8600 L/kg; for Sr, respectively 650-2000 and 110-980 L/kg).

Dushenkov *et al.* (1997) tested a number of species for the removal of uranium from water in a greenhouse experiment. Beans and mustard were less effective than sunflower in U removal. The latter removed 95% of the U in the water in less than 24 hours. Practically all U was concentrated in the roots, almost none in the shoots (Fig. 4).

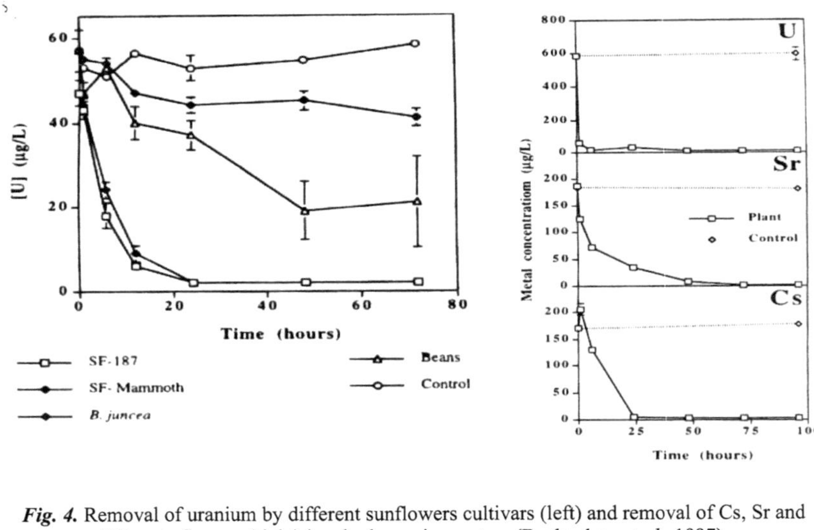

Fig. 4. Removal of uranium by different sunflowers cultivars (left) and removal of Cs, Sr and U by sunflowers (right) in a hydroponics system (Dushenkov *et al.*, 1997)

Concentration of uranium in the water did not affect the CF, but pH did: an average bioaccumulation coefficient for U in sunflower roots was 6624 and 3370 L/kg, respectively at pH 5 and pH 7. As mentioned before, U accumulation is affected by U speciation. The U-species most readily taken up by the plant is the uranyl form. This form predominates at pH below 5.5 (Ebbs *et al.*, 1998).

Uranium was removed much faster from the contaminated water compared to Cs and Sr.

In natural waters, uranium is usually complexed with carbonate, hydroxide, sulphate, phosphate. These complexes increase the solubility of U and make uranium precipitation more challenging. For this reason it seems U is one of the best candidates for a biological removal process.

Dushenkov *et al.* (1997), however, demonstrates that the rhizofiltration method has its limits. Setting up an experiment with rather highly contaminated waste water (1000 µg/L U; 20% above the upper limit of local processing waste water) and high flow rate (1.05 L/min), 95% of the U was removed by 6-week-old sunflowers grown for 2 weeks in the waste water, resulting in effluent concentrations of 40-70 µg/L, above the 20 µg/L drinking water limit.

4.3. Alternative Landuse: Non-Food Crop Production In Contaminated Areas

4.3.1. Introduction

Fifteen years after the Chernobyl accident, the question still remains how to manage large territories contaminated with a broad spectrum of radionuclides. An area of 29200 km^2 in Belarus, Ukraine and Russia was contaminated with levels exceeding 185 kBq ^{137}Cs m^{-2} (IAEA, 1996) and as much as 4300 km^2 of agricultural land had to be excluded from use. The Sr-contamination occurs mainly within a 70-km radius from the reactor although some significant contamination (37-74 kBq m^{-2}) can be found in the area northeast of Gomel (Rauret and Firsakova, 1995).

Belli and Tikhomirov (1996) reported that nine years after the accident the radiocaesium concentrations in plants grown in forests and on meadows did not significantly decline. In the more contaminated territories of Belarus, Ukraine and Russia, no lifting of the restrictions on land use is likely in the foreseeable future.

When selecting an appropriate countermeasure for a specific area the optimal solution to a given problem will apart from radiological criteria depend as much on economic, social and political factors as on sound scientific considerations (Segal, 1993).

Many studies have been targeting possible agricultural countermeasures, in response to too high concentration levels in food and agricultural crops. The removal of the contaminated soil surface layer is maybe the most effective measure but generates a large quantity of waste and is only applicable to small quantities of land (Vovk *et al.*, 1993). Moreover, the most fertile layer of the soil is removed in that process. Ploughing, in particular deep ploughing, dilutes the activity concentration at the root level and can be up to 80-95% effective in reducing the radiocaesium uptake by plants, i.e. a reduction with a factor 5 to 20 (Alexakhin, 1993; Maubert *et al.*, 1993; Vovk *et al.*, 1993). However, the average reduction factor was only 1.5 (Vovk *et al.*, 1993).

Potassium fertilisation will generally decrease the soil-to-plant transfer of radiocaesium due to competitive effects, certainly when the soil is poor in K. Reported reduction factors varied between studies but overall reduction factors ranging between 1.1 and 5.0 were obtained (Lonsjö *et al.*, 1990; Alexakhin, 1993; Konoplev *et al.*, 1993; Vovk *et al.*, 1993).

The behaviour of radiostrontium and its uptake by plants is dominated by its similarity to calcium. Many investigators have found a significant correlation between the Sr-TF and the reciprocal of the exchangeable Ca-content (Lembrechts, 1993). Consequently, much of the research and actions on reducing the strontium uptake by plants has centred on the use of lime as a soil-based countermeasure. The use of lime reduced the SR uptake by a maximum of 40%, limed compost by 60% (Lembrechts, 1993).

Chemical amendments such as zeolites, ammonium-hexacyano-ferrate (AFCF) or clay minerals also reduce radionuclide uptake by plants, since the radionuclides are trapped and so rendered less available for plants (Vandenhove *et al.*, 1996, 1998; Thiry *et al.*, 1997; Valcke *et al.*, 1997). A reduction factor of 4.6 in radiocaesium transfer was obtained for a sandy soil after addition of 1% zeolites (Valcke *et al.*, 1997) and a factor 25 applying 10 g AFCF per m^2 (Vandenhove *et al.*, 1996).

The presence of organic matter in soils is important because of its tendency to form co-ordinate linkages with Ca and Sr, which binds Ca and Sr more strongly than the ion exchange sites in mineral soils. Organic matter addition resulted in Sr-TF reductions with a factor 1.2 to 7 (the latter value obtained after addition of 15% OM to a sandy soil; Lembrechts, 1993).

Finally, countermeasures can also be based on the selection of crops which exhibit smaller radionuclide uptake, on food processing or choosing for non-food crops such that the products from the land are radiologically acceptable.(Alexakhin, 1993; Segal, 1993; Renaud and Maubert, 1997). Impact on dose to people and on the ecology and economy of the affected area may vary enormously: change in crop variety will have much smaller impact than more radical changes as the substitution of vegetables by cereals or changing from an arable or cattle system to forestry.

Many studies have hence been conducted to test the effect of different physical and chemical countermeasures. In contrast to this, information on long-term effect of countermeasures and especially the change to non-food crops is still limited.

When advocating an alternative crop, the principle questions to be asked are: (1) What is the fate of the radionuclide in the cultivation system and conversion routes and what is the expected radionuclide concentration in the end-products?; (2) How does the radionuclide behave during the biomass processing? (3) What is the dose acquired during biomass cultivation and processing? (4) How well are the crops adapted to the climate and soil conditions prevailing in the contaminated area? (5) What are the conclusions with regard to economic feasibility for the production and use of these alternative crops? (6) What are the perspectives for these various alternative crops, with emphasis on SRC, as alternative landuse for large contaminated surfaces?

In what follows, short rotation coppice is being commented upon in some detail, relying on literature and field data. Some other energy crops and reforestation are commented upon in less detail. Most information can be found in Vandenhove *et al.*, 1999, 2001, 2002a,b.

4.3.2. Feasibility of Willow Short Rotation Coppice for Energy Production for Revaluation of Contaminated Land

Coppicing is a method of vegetative forest regeneration by cutting trees at the base of the trunk at regular time intervals. Fast-growing species of the *Salix* genus (willows) are frequently used in a coppice system because of the ease of the vegetative reproduction and the large biomass production. Furthermore, they are tolerant to a wide range of climatic and edaphic factors (Ledin, 1996). Most limiting factors for growth are vegetation period and precipitation (water availability) (Perttu, 1999). The harvested biomass is converted into heat or power. As such this non-food industrial crop is a potential candidate for the valorisation of contaminated land with restrictive use. In case of revalorization of contaminated land, SRC may be preferred to traditional forestry since revenues come sooner and more regularly (every 3-5 years) after establishment. SRC yields are also high on good agricultural soils, and it is not a drastic change in land use (easy to apply and easy return to food crops). SRC may also be considered as complementary to forestry, given the different culture requirements of both vegetation systems (forests perform well on sandy soils whereas SRC requires soils with sufficient water retention capacity.

SRC has furthermore other potential advantages in a contamination scenario. Since it is a perennial crop, dispersion of radionuclides may be reduced. Harvest can be in winter when the soil may be covered by snow, resulting in protection for the people. Finally, SRC cultivation is not labour intensive, which is also advantageous with relation to exposure.

Willow SRC may be a suitable rehabilitation tool for highly contaminated land but only if the radionuclide levels in the wood are below the exemption limits for fuelwood, if the average yearly dose received during coppice cultivation and coppice wood conversion is acceptable and if SRC can be grown successfully in the contaminated territories (soils, climate) and if the cultivation of SRC is technically feasible and the SRC production and conversion are economically profitable.

In order to assess the incorporation of Cs in the wood, SRC field trials were established in Belgium (sandy and loamy soil, 4 willow clones), Belarus (sandy and peaty soil, 4 willow clones) and sampling was done in 14 existing willow SRC fields, planted on radiocaesium contaminated land in Sweden (Chernobyl deposition) (Vandenhove *et al.*, 1999, 2001).

Broadly we can say that in light textured soils with a low radiocaesium interception potential (RIP: measure for caesium fixation) and low soil K, as is the case for most soils in Belarus, the TF to wood varies between 0.5 10^{-3} and 2 10^{-3} m^2 kg^{-1} (Fig. 5). For soils with a medium to high RIP (finer textured soils) and sufficient K, as is the case for most Western European agricultural soils, the TF varies between 2 10^{-5} and 5 10^{-6} m^2 kg^{-1}. From experiments conducted in Sweden and Belgium, it could be concluded that appropriate K-fertilisation could bring down the Cs-TFs recorded with about a factor 50. No effect of culture variety and stand maturity on TF could be derived.

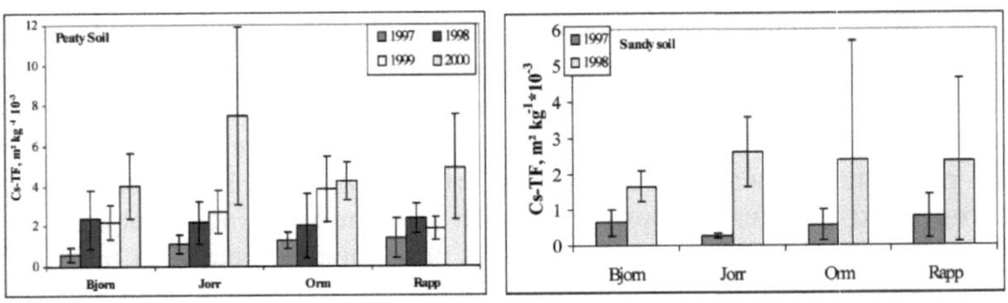

Fig. 5. Radiocaesium transfer factor to willow SRC wood (4 clones: Bjorn, Jorr, Orm, Rapp) at a peaty and a sandy soil at Savichy (Belarus) during a full rotation cycle for the peaty soil and two first years on the sandy soil

If the TF to SRC wood applies for coppice grown on relatively fertile soil with a moderate to high RIP, wood can be safely burnt and the ashes can be disposed off without concern. In case the high SRC-TFs for low-RIP soils pertain (2 10^{-3} m^2 kg^{-1}), which is potentially relevant for low fertile and low RIP soils, possibly acid and with a high ammonium concentration in the soil solution, wood burning would only be permitted when willow is grown on a soil contaminated with < 370 kBq m^{-2} ^{137}Cs (considering Belarus exemption limit for fuelwood of 740 Bq kg^{-1}). Given that TFs for common forestry (2 10^{-3} m^2 kg^{-1}) and for straw of winter wheat and oil seed rape (0.3 10^{-3} and 0.63 10^{-3} m^2 kg^{-1}) are comparable, the same applies for burning wood or straw for energy. At higher soil contamination levels, wood (straw) could potentially still be burnt in commercial electricity or heat plants but adequate exhaust filtering systems should be installed and appropriate disposal of the ashes should follow. In Belarus, waste and contaminated biomass from the Chernobyl accident can be incinerated (special regulations) if activity levels do not exceed 3.7 MBq kg^{-1} (Grebenkov, Personal communication).

Regarding the agrotechnical aspects affecting SRC production and feasibility, this crop has generally a high annual yield of about 12 t/ha. From the test sites in Belarus, it was seen that on peaty soils the yield ranged from

7.8 to 16.0 t ha^{-1} y^{-1}. Peaty soils are hence considered suitable for SRC production. On the sandy soils first-year yields were 0.25 t ha^{-1} y^{-1} and SRC died off two years after establishment.

According to their suitability for SRC cultivation (biomass yield) and their vulnerability to ^{137}Cs transfer, three main soil groups can be identified for Belarus and Ukraine:

- the sandy podzolic soils account for about 50% of the soils in the Chernobyl contaminated area in Belarus and Ukraine. Even with sufficient water supply, the SRC yield is below the 6 t ha^{-1} y^{-1} required for making SRC production feasible on economic grounds. These sandy textured soils which are very sensitive to water shortage could maybe better be allocated to pine forests, dry meadows or drought-tolerant grasses. On these soils caesium and strontium factors are high so that that exemption limits for fuel wood may be exceeded and valorisation into energy of the SRC wood produced may be hampered. Consequently, sandy soils should be classified as very likely not suitable for SRC.
- due to their high water reserve, the peaty soils are not sensitive to water shortage and SRC yields are high. They account for less than 10% of the farmland in Belarus. The high Cs-TF does theoretically not allow for the conversion into energy of the wood, since exemption levels may be exceeded but considering the high biomass potential and the low annual dose to workers cultivating coppice (SRC plantations are not labour extensive), these soils are classified as suitable for SRC. Moreover, adequate K fertilisation may substantially bring down the Cs-TF and Sr-TFs are low enough so that it is rather improbable that exemption limits will be exceed. At the conversion plant site, personnel working in vicinity of ash collectors and deposits should be radiation controlled.
- for the other soil types (sandy loam, loamy sand and loam), there is no sanitary problem for SRC wood conversion (low ^{137}Cs TFs). The SRC yields are intermediary between sandy and peaty soils (7-10 t ha^{-1} y^{-1}), but as these soils are generally not drought sensitive, these soils are classified as suitable for SRC. However, the competition with other potential food sources in a context of reduced soil resources can limit the establishment of SRC plantations, particularly on low contaminated soils.

If wood is burnt in a combustion unit, the personnel may be exposed. In case of high contamination levels in the wood (3000 Bq kg^{-1}, i.e. higher than the exemption level for fuelwood set at 740 Bq kg^{-1}) doses in vicinity of ash collectors are ash collectors may exceed the acceptable level of 1 mSv a^{-1} for a member of the general public. With appropriate work-rotation-schemes doses can most probably be brought down to less than 1 mSv a^{-1}, even in case of high(er) contamination levels. Contributions from other possible exposure pathways are negligible (external exposure during culturing and transport, inhalation dose in the combustion plant and to the public following wood burning).

The economic sustainability of cultivation and conversion of SRC was evaluated for Belarus by Vandenhove *et al.* (2002). Several scenarios (scales/modes of production and conversion) were screened. At the production site, a minimal yields of 6 t ha^{-1} y^{-1} is required for the Belarus production cost conditions and 12 t ha^{-1} y^{-1} for W. European production cost conditions, if all other parameters are optimal. Hence, since the maximal SRC biomass production on (non-irrigated) sandy soils is only about 5 t $ha^{-1}y^{-1}$, SRC production should not be advised on these lands in Belarus. Since about 50% of the soils most soils in Belarus are of a sandy nature, only a small percentage of soils in Belarus are apt for SRC cultivation. Heat schemes may be a viable option for wood conversion in Belarus whereas electricity schemes are not. In Europe, subsidies are required to make wood conversion economically feasible.

From this study it was also concluded that the existence of a contamination scenario does not necessarily hamper the economic viability of the energy production schemes studied. Radiological exposure is limited are can be kept minimal and associated costs are low. The ash which is left after gasification or combustion should in case of severe contamination of wood, be considered as waste. The cost associated with disposal of contaminated ashes was estimated as less than 1% of the biofuel cost (Vandenhove *et al.*, 1999, 2002) and will not affect the economic feasibility.

4.3.3. Feasibility of Other Potential Alternative Land Uses: Liquid Biofuel Crops and Forestry

Forests

Soil-to-wood TFs to coniferous and deciduous wood range from average $6.7*10^{-4}$ m^2 kg^{-1} to 2.8 10^{-3} m^2 kg^{-1} (Zadbudko *et al.*, 1995) and are hence comparable with the TFs to willow wood observed in the Belarus test sites. They are high compared with the TFs observed for willow in finer textured soils and soils with an adequate potassium status. The TFs recorded to bark are a factor 5-10 higher.

Comparison between the willows grown on the Belgian test sites and a 17 year-old-forest in Belarus (4 years-old at the time of the accident) has shown that the net annual radiocaesium accumulation is about 35 times higher in the forest standing biomass than in coppice. Moreover, annual biomass increase is only 6 t ha^{-1} for forests and may attain 12 t ha^{-1} for SRC grown on soils with an adequate water reserve and fertility status. On these types of soil, SRC may hence be a more promising land-use option than traditional forestry. On soils with low water reserve (e.g. sandy soil) willow yield without irrigation is maximally about 5 t ha^{-1}, and here forestry may be the preferred option (Vandenhove *et al.*, 2001)

The TF value obtained for the newly planted pine forest on contaminated arable land (sandy soil) was $2.7*10^{-4}$ and hence a factor 5-10 lower than the TFs recorded for forests in place at the time of the accident.

With the fuel wood exemption limit for ^{137}Cs of 740 Bq kg^{-1} and an average soil-to-wood TF of 2 10^{-3} m^2 kg^{-2}, maximal soil contamination level is 370 kBq m^{-2} for regular use, respectively. For the newly planted forests, soil contamination levels may be 10 times higher for safe fuel use.

Liquid biofuel crops

From the literature, TFs were assembled for the potential biofuel crops considered and for some other crops (Table 10). TF to the useable product (rape seed, wheat grains, beet root, barley, potatoes and winter rye) for the other biofuel crops considered range between 0.01 and 0.5 10^{-3} m^2 kg^{-1} and the liquid biofuels are almost free from activity. No information was found on the TF of radiocaesium to the by-products of the liquid biofuel cycles. Also radiocaesium levels in the waste-products is generally of no concern. In case of high deposition (~ 1000 kBq m^{-2}), Cs-levels in the oil cake from oil seed rape (~2000 t ha^{-1}) and the pulp and vines from sugar beet (~ 4000 t ha^{-1}), exceed the exemption level for use as animal fodder and for incineration and should be disposed off. Since amounts to be disposed off are substantial, this involves a high disposal cost (up to 20 and 60% for OSR and SB, respectively). Possibly this waste could also be burnt to obtain a mass reduction by about 90%.

Table 11. Cs and Sr TF to different plant parts of some potential biofuel crops.

Crop	PlantComponent	Cs-TF, 10^{-3} m^2 kg^{-1}
Spring wheat	Leaves	0.23-0.42
	Straw	0.23-0.36
	Grain	0.08-0.16
Cereals	Grain	0.0048 / 26(2.6-260)
	Grain (peaty)	83(8.3-830)
Winter wheat	Straw	0.27-0.44
	Seeds	0.08-0.18
Spring rape seed	Stems and leaves	0.12 (0.46**)
	Straw	0.017 - 1.4**
	Seeds	0.03-0.04 (0.0006 - 0.5**)
Brassicaceae	Seeds	0.037-3.4*
Legumes	Seeds	94(12-750)*
Sugar beet	Root	0.43 #
Root crops	Root	0.025 / 11(1.1-110)*
Green vegetables	Leaves	0.07 - 4.86*
	Leaves (peaty)	260(25-2700)*

Data for Belarus from Grebenkov (1997), * from Nisbet *et al.* (1999) (average and 95% confidence limits), ** from Gopa 1996 and # from IAEA, 1994.

All energy crops considered can be grown successfully in Belarus if soil conditions are appropriate to attain potential yields. The cost of liquid biofuels is about 3 to 4 times the cost of fossil fuels and hence price subsidy is needed to compete with fossil fuels and therefore the cultivation of crops for liquid fuel production cannot be recommended.

4.3.4. Conclusions

Energy production from SRC in contaminated area is a feasible remediation option on radiological, technico-agricultural and economic grounds under certain conditions. On sandy soils, it is advisable to install forests or drought resistant grasses instead of SRC or to apply irrigation since SRC is clearly not adapted to dry sandy soil conditions.

For the other energy crops considered, there is generally no concern on radioecological grounds, but conversion to liquid biofuels is not profitable and therefore the cultivation of crops for liquid fuel production cannot be recommended.

4.4. Phytostabilisation

Phytostabilisation reduces the risk presented by un- or sparsely-vegetated contaminated soil by the use of plants and/or soil amendments to establish a stable vegetation cover which may progressively reduce the soil labile metal pool. This technology does not achieve a clean up of the soil, but changes the mobility of potentially toxic elements by either reducing concentrations in the soil water and other freely exchangeable sites within the soil matrix or by reducing re-entrainment of toxic particulates following the development of a stable and permanent vegetation cover. Both processes alter the speciation of soil metals, reducing potential environmental impact. These technologies draw upon fundamental plant and soil chemical processes as well as established agricultural practices. The development of a stable and self-perpetuating ecosystem as a result of this type of treatment may be a further beneficial process, as in some circumstances, plant root activity may change metal speciation (changes in redox potential, secretion of protons, chelating agents); the microflora associated with their root systems may produce similar effects. Further spreading of the contamination from the site is prevented. The technique also has significant implications for "polishing" less contaminated soils and composts.

In these notes, application of the phytostabilisation approach will be discussed for two different contamination scenarios: (1) the phytorehabilitation approach with willow studied in the Dnieper catchment contaminated by the

Chernobyl accident and; (2) the application of a vegetation cover (as the final layer of a multi-layer cover) for a radioactive contaminated site resulting from NOR material extraction or processing, e.g. a uranium tailing pond or the use of a vegetation cover for the stabilisation of uranium mining debris.

4.4.1. Phytomanagement with Willow Vegetation Systems in the Chernobyl Exclusion Zone

Most radioactive fall-out following the Chernobyl accident was deposited within the Dnieper catchment system, which adjoins the site of the Chernobyl nuclear power plant (ChNPP). This area and adjacent drainage basins form an extensive area from which contaminated water and sediments flow downstream through the Pripyat and Dnieper rivers across the Ukraine to the Black Sea.

Remedial actions are still being investigated, to control the radionuclide efflux from the Chernobyl Exclusion zone. Phytorehabilitation technologies (in general less expensive measures) could in this context also be considered as remedial options. In this regard, willow plantations are put forward given their acknowledged phytorehabilitation properties and their already successful application as environmental plantings. A willow plantation can function as a windbreak or a buffer strip along a stream. It can stabilise river borders and hill slopes and as such decrease erosion. Wind erosion can also be effectively diminished by planting willow in the concerned areas. Alternatively, this kind of plantations are used to recycle sewage or industrial sludge and wastewater. These different uses of willow can have their application in the contaminated territory, near the Pripyat River.

Three phytorehabilitation approaches with willow vegetation systems for the Dnieper catchment system in the vicinity of the Chernobyl Nuclear Power Plant were studied (Vandenhove *et al.*, 2001) (Fig. 6):

1. the potential effect of willow vegetation on vertical migration of radionuclides
2. the potential effectiveness of SALIMATs (willow mats) for the stabilisation of the Chernobyl Cooling Pond sediments
3. the potential effectiveness of willows for the horizontal erosion control in case of inundation of the Pripyat floodplain

Experimental sites in the Chernobyl Exclusion Zone were assigned and a number of screening tests with willow were carried out on the contaminated floodplains of the Pripyat river and on small artificial and natural ponds. Survival of new willow plantation was tested at several locations. Most information can be found in Vandenhove *et al.*, 2002.

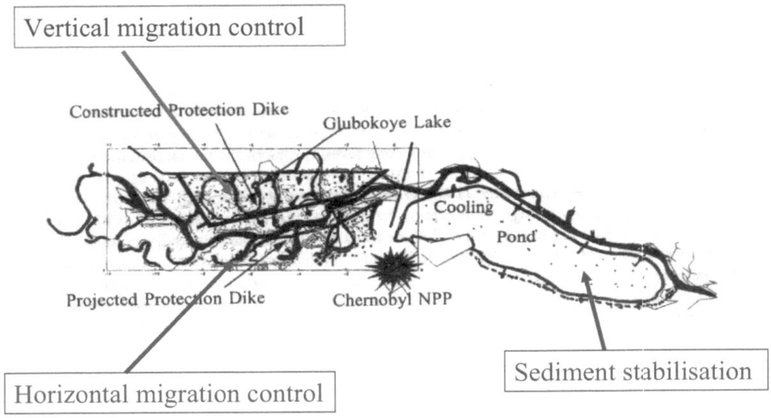

Fig. 6. Phytostabilisation approaches at the contaminated area of the Dnieper close to the Chernobyl NPPP

Vertical migration control with willow

The area of interest to study the vertical migration control by willow was an extremely contaminated zone of 16 km^2 at the left bank of the Pripyat (between 3.7 and 18.5 TBq km^{-2} ^{90}Sr and ^{137}Cs and 0.37 TBq km^{-2} Pu) which is partly protected from the spring floods by a dam. It was investigated if the radionuclide migration can be minimised by reduced downward water flow by the high evapotranspiration of willow and partly by removal of the radionuclides by fixation in the plants.

Willows seem to naturally develop after inundation and sedimentation of the Pripyat river floodplains but biomass production is extremely low (low soil nutrient status and limited water availability after spring). New plantations of willows are difficult to establish due to the extreme nutrient and water shortage on most of the soil types in the 30-km zone (mostly sandy soil or sandy alluvial, some bog soils). Most of the willow clones developed for high biomass production have also high nutrient and water requirements and unless the soil would be irrigated and fertilised, the potentially high biomass will never be obtained. Thus it is very unlikely that it is advisable to establish willow SRC plots in this area.

Through modelling exercises it was shown that, due to their high evapotranspiration rate, willow SRC stands are expected to decrease the groundwater table level by 100-200 cm in fertilised stands. Without fertilisation, only a decrease of groundwater table level of less than 50 cm was predicted. Since the immobilisation potential of ^{137}Cs and ^{90}Sr in the willow wood is limited (< 0.1% for ^{137}Cs and 1.9-7.9% for ^{90}Sr for 10 years old willows), the influence of plant uptake on migration remains low.

Stabilisation of the Chernobyl Cooling Pond Sediments with the SALIMAT technology

Following the effective closure of the ChNPP, the water level of the cooling pond (22.5 km^2; depth between 1.5 and 15 m; with about 111 TBq ^{137}Cs and 37 TBq ^{90}Sr) will drop with 4-7 m and 15 km^2 of the sediments will come to surface and may be in need for stabilisation.

In this respect the SALIMAT (SALIMATs consists of a roll of willow rods (stems) rolled around a central disposable tube. Biodegradable string is used to tie the rods in place. The SALIMATs are unrolled (horizontally) by dragging them across the lagoon) option was investigated to see if it was a feasible and justifiable option for the stabilisation and reclamation of the cooling pond sediments. Small tests have demonstrated that willow placed as SALIMATs on contaminated pond sediments establish well and produce a full vegetation cover already during the second year following establishment. Willow clones selected for increased growth potential are better candidates than the local willow varieties. The approximated cost of the phytostabilisation option ranges from 0.8 to 1.9 10^6 EUR for the reclamation of 15 km^2 of sediments and this is a low cost compared with the prospected cost of removal of the sediments (6000000 $, transport and disposal costs not included) or the maintenance of the present water level (200000 $ per year). Therefore, phytostabilisation of the sediments is possibly the optimal 'minimal' strategy to limit resuspension (wind erosion) of the contaminated sediments.

Horizontal erosion control with willows

The projected area is the right bank of the Pripyat river which was significantly less contaminated than the left bank, yet not protected with a dam. After inundation, part of the activity is eroded and transported to the Pripyat with the withdrawing water. It was calculated that even in case of extremely high flooding, dense willow plantation will effectively decrease the horizontal soil erosion and concomitant transport of radionuclides into the Dnieper river system.

The costs and benefits of the latter two operations should also be thoroughly calculated through. Dose reduction is the potential benefit one can expect from the operation. However, the present radiation exposure from contaminated sediments entering the Dnieper Cascade is very limited and far below the dose limit. Under these conditions it is difficult to make a proper cost-benefit analysis. But the fact that the measures may result in an important reduction in environmental loading is an important one which may find cost-effective application in other contamination situations, e.g. where the population lives closer to the contaminated site.

4.4.2. Uranium Mining Tailings and Debris Heaps

The final step in closing out an impoundment for uranium mill tailings is the design and placement of a cover which will give long-term stability and control to acceptable levels of radon emanation, gamma radiation, erosion of the cover and tailings and infiltration and precipitation into the tailings and heaps. Surface vegetation can be effective in protecting tailings or a tailing cover from water and wind erosion. Factors affecting the effectiveness of surface revegetation on impoundments can be broadly classed into climatological and agrobiological factors. Plants should be chosen to match the local climate conditions. Concerning agrobiological factors, the nature of the ore and the mill processes will largely determine the tailing characteristics from their point of view of sustaining growth. Considerable efforts to correct adverse characteristics, such as low or high pH values and low plant nutrient content, will usually be required before tailings can sustain growth. An adequate soil cover may need to be established.

Water and wind erosion are the primary causes of erosion of tailings or tailing cover material. A vegetation cover may decrease the erosion hazard. However, vegetation surfaces may raise concerns: the vegetation can promote radon emanation by drying out the tailing and tailing; tree roots may penetrate the contaminated material and break the cover integrity.

Given the increased evapotranspiration rate following vegetation establishment, the vegetation cover alters the water household of the tailings and may decrease seepage; decreased water content (less precipitation water entering tailing; higher evaporation).

The effect of a vegetation cover on the radionuclide dispersion through an alteration of the water balance but also potentially, because of the effect of plant roots on the physico-chemical characteristics of the tailing material (biologically driven acidification of the tailing material), has not been studied intensively so far.

At an 35-year old reclaimed site on a uranium mining dump near Schlema (Saxony, Germany), Thiry and co-workers performed an exhaustive sampling campaign: 4 full-grown pine trees were sampled for biomass. The different plant compartments (stem, branches, needles, roots) and soil compartments (barren rock, 3 organic layers) were analysed for U and nutrient content and some other factors. They concluded that the biomass could reduce infiltration by 40-60% due to interception by the canopy (25-40%) and increased transpiration. They further found that of the 165000 g/ha U in soil (30 cm depth), only 4 g was in the above-ground plant parts and 510 g/ha in the below ground plant parts. Ninety percent of the uranium which is taken up during the growing season is recycled (returned) with the needles. U-dispersion by uptake through vegetation is hence minimal. It may be concluded from these preliminary results (unpublished) that a forest vegetation may reduce infiltration rate and will defavorise radionuclide dispersion.

5. CONCLUSIONS

Different phytomanagement options may be applied for the remediation of radioactive contaminated sites. For vast contaminated surfaces, only the phytostabilisation option and alternative land use options seem feasible options. The effect of the vegetation cover on the radionuclide dispersion and the subsequent dose should be thoroughly studied. Similarly, the effectiveness of alternative landuse should not be studied on mere radiological grounds only. Socio-economic factors are equally important.

Phytoextraction for soil-clean-up seems only reasonable for limited surfaces and will only be effective under very specific conditions. Side effects (costs, treatment of contaminated biomass, potential for ground water contamination, ...) should be evaluated in depth.

REFERENCES

1. Absalom, J.P., S.D. Young, N.M.J. Crout, A.F. Nisbet, R.F.M. Woodman, E. Smolders and A.G. Gillet, 1999. Predicting soil to plant transfer of radiocaesium using soil characteristics. *Environmental Science and Technology,* 33, 1218-1223.
2. Alexakhin, R.M. 1993. Countermeasures in agricultural production as an effective means of mitigating the radiological consequences of the Chernobyl accident. *Science of the Total Environment,* 137, 9-20.
3. Allen, P.G., J.M. Berg, C.J. Chisholm-Brause, S.D. Conradson, R.J. Donohoe, D.E. Morris, J.A. Musgrave and C.D. Tait, 1994. Determining uranium speciation in contaminated soils by molecular spectroscopic methods: Examples from the uranium in soil integrated demonstration. Los Alamos, Los Alamos National Laboratory.
4. Apps, M.J., M.J.M. Duke and L.G. Stephens-Newsham, 1988. A study of radionuclides in vegetation on abandoned uranium tailings. *Journal of Radioanalytical and Nuclear Chemistry,* 123(1): 133-147.
5. Baes, C.F. 1982. Environmental transport and monitoring: Prediction of radionuclide Kd values from soil-plant concentration ratios. *Transuranic American Nuclear Society,* 42: 53-54.
6. Belli, M. and Tikhomirov, F. 1996. *Behaviour of radionuclides in natural and semi-natural environments.* ECSC-EC-EAEC 145.
7. Belli, M., Sansone, U., Ardiani, R., Feoli, E., Scimone, M., Menegon, S. & Parente, G. 1995. The effect of fertilizer applications on ^{137}Cs uptake by different plant species and vegetation types. *Journal of Environmental Radioactivity,* 27, 75-89.
8. Biewinga, E.E. and van der Bijl, G. 1996. Sustainability of energy crops in Europe: A methodology developed and applied. Centre for Agriculture and Environment, Utrecht, CLM-234.
9. Bondar, P.E. and Dutov, A.I. 1992. Parameters of radiocaesium transfer into oats harvest on lime soils in connection with the application of mineral fertilizers and chemical

ameliorants. In *Collections of Scientific Works - Problems of Agricultural Radiology.*, Kiev, Ukraine. 125-132

10. Brackhage, C., Dienemann, H. Dannecker and E.G. Dudel, 1999. Radionuclide and heavy metal distribution in soil and plants from a 35 year old reclaimed uranium mining dump site, Proc. 5th Intern. Conf. On The Biogeochem. of Trace Elements. Vienna, p. 1174-1175.
11. Brouwer, E., Bayens, B., Maes, A. & Cremers A. 1983. Caesium and rubidium ion equilibria in illite clays. *Journal of Physical Chemistry,* 87, 1213-1219.
12. Cline, J.F. & Hungate, F.P. 1960. Accumulation of potassium, caesium-137 and rubidium-86 in bean plants grown in nutrient solution. *Journal of Plant Physiology,* 35, 826-829.
13. Cremers, A., A. Elsen. and P. De Preter. 1988. *Nature,* 335, 247-249.
14. Dushenkov, S., Vasudev D., Kapulnik, Y., Gleba, D., Fleisher D., Ting K.C. and Ensley, B. 1997. Removal of uranium from water using terrestrial plants. *Environment, Science and Technology,* 31: 3468-3474.
15. Ebbs, S.D. 1997. Identification of plant species and soil amendments that improve the phytoextraction of zinc and uranium from contaminated soil. *Faculty of Graduate Studies.* Michigan, Cornell University: 174.
16. Ebbs, S.D., W.A., Norvell and L.V. Kochian, 1998. The effect of acidification and chelating agents on the solubilisation of uranium from contaminated soil. *Journal of Environmental Quality,* 27: 1486-1494.
17. Ensley, B.D., Raskin, I., and Salt D.E. 1997. Phytoremediation applications for removing heavy metals contamination from soil and water. *Biotechnology in the Sustainable Environment,* 6, 59-64.
18. Ettenhuber, E. (1996) Environmental restoration plans and activities in Germany. RER/9/022 TC Regional project on Environmental restoration in Central and Eastern Europe, Romania, 4-8 November 1996.
19. Evans, D.W., Alberts, J.J. and Clack, R.A. 1983. Reversible ion-exchange fixation of cesium-137 leading to mobilization from reservoir sediments. *Geochimica et Cosmochimica Acta* 47, 1041-1049.
20. Frissel, M.J. and van Bergeijck, K.E. (Eds.) 1989. VIth Report of the IUR working group soil-to-plant transfer factors: Report of the working group meeting in Guttannen Switzerland, 198924-26, 1989. Bilthoven, RIVM: 240.
21. Goor, F., V. Davydchuk and J.-F. Ledent. 2001. Assessment of the potential of willow SRC plants for energy production in areas contaminated by radionuclide deposits: methodology and perspectives. *Biomass & Bioenergy*, 21, 225-235.
22. GOPA, 1996. Belarus: Study on Alternative Biodiesel Sources in Relation with Soil Decontamination. Project No TACIS/REG93
23. Grebenkov, A. (1997) Internal deliverables for RECOVER
24. Hossner, L.R., Loeppert, R.H., Newton, R.J. and Szaniszlo, P.J. 1998. Literature review: Phytoaccumulation of chromium, uranium, and plutonium in plant systems. Springfield, VA, Amarillo National Resource Centre for Plutonium: 51.
25. Huang, J.W., Blaylock, M.J., Kapulnik, Y. and Ensley, B.D. 1998. Phytoremediation of uranium contaminated soils: Role of organic acids in triggering hyperaccumulation in plants. *Environ. Sci. Technol.* 32: 2004-2008.
26. IAEA (1996) One decade after Chernobyl: the basis for decision. *IAEA Bulletin* 38 Nr 3, 10 pp.
27. IAEA (International Atomic Energy Authority) 1992. Current practices for the management and confinement of uranium mill tailings. Technical report series No 335. Printed by IAEA in Austria, PP. 140.
28. IAEA (International Atomic Energy Authority), 1994. Handbook of Parameter Values for the Prediction of Radionuclide Transfer in Temperate Environments, Technical Report Series No 364, IAEA, Vienna
29. Ibrahim, S.A. and Whicker F.W. 1992. Comparative plant uptake and environmental

behaviour of U-series radionuclides at a uranium mine-mill. *Journal of Radioanalytical and Nuclear Chemistry,* 156(2): 253-267.
30. Ibrahim, S.A. and Whicker, F.W. 1988. Comparative uptake of U and Th by native plants at a U production site. *Health Physics* 54(4): 413-419.
31. IUR, 1989, VIth Report of the IUR working group soil-to-plant transfer factors: Report of the working group meeting in Guttannen Switzerland, 198924-26, 1989. Bilthoven, RIVM: 240.
32. IUR, 1992, VIIIth Report of the Working Group soil-to-plant transfer factors, Report of the Working Group Meeting in Madrid, Spain, June 1-3, 1992, IUR Pub R-9212-02, IUR Technical Secretariat, Balen, Belgium.
33. Jackson, W.A., Craig, D. and Lugo, H.M. 1965. Effects of various cations on cesium uptake from soils and clay suspensions. *Soil Science*, 99, 345-353.
34. Konoplev, A.V., Viktorova, N.V., Virchenko, E.P., Popov, V.E., Bulgakov, A.A. and Desmet, G.M. 1993. Influence of agricultural countermeasures on the ratio of different chemical forms of radionuclides in soil and soil solution. *Science of the Total Environment*, 137, 147-162.
35. Kopp P., O. Oestling and W. Burkart. Availability and uptake by plants of radionuclides under different environmental conditions. *Toxicology and Environmental Chemistry,* 23, 53-63.
36. Lakshmanan, A. R. and Venkateswarlu V.S. 1988. Uptake of uranium by vegetables and rice. *Water, Air and Soil Pollution,* 38: 151-155.
37. Lasat, M.M., Fuhrmann, M., Ebbs, S.D., Cornish, J.E. and Kochian, L.V. 1998. Phytoremediation of a radiocaesium contaminated soil: evaluation of caesium-137 accumulation in the shoots of three plant species. *Journal of Environmental Quality,* 27, 165-169.
38. Ledin, S .1996. Willow wood properties, production and economy. *Biomass and Bioenergy*, 11: 75-83
39. Lembrechts, J. 1993. A review of literature on the effectiveness of chemical amendments in reducing the soil-to-plant transfer of radiostrontium and radiocaesium. *Science of the Total Environment*, 137, 81-88.
40. Lönsjö, H. Haak, E. and Rosén, K. 1990. Effects of remedial measures on the long term transfer of radiocaesium from soil to agricultural products as calculated from Swedish field experimental data. Proceedings of a Symposium on Environmental Contamination Following a Major Nuclear Accident, Vienna, Austria. STI/PUB/825, IAEA, 151-162.
41. Maubert, H. Vovk, I., Roed, J., Arapis, G. and Jouve A. 1993. Reduction of the soil-plant transfer factors: mechanical aspects. *The Science of the Total Environment*, 137, 163-167.
42. Minotti, P.L., Craig, D. and Jackson, W.A. 1965. High caesium uptake in wheat seedlings cultured with ammonia. *Soil Science Society Proceedings: Notes*. 220-221.
43. Mortvedt, J.J. 1994. Plant and soil relationship of uranium and thorium decay series radionuclides - A review. *Journal of Environmental Quality,* 23: 643-650.
44. Nisbet A.F. and Woodman R.F.M. (2000). Soil-to-plant transfer factors for radiocaesium and strontium in agricultural systems. Health Physics 78(3): 279-288
45. Nisbet, A.F. 1993. Effect of soil-based countermeasures on solid-liquid equilibria in agricultural soils contaminated with radiocaesium and radiostrontium. *Science of the Total Environment,* 137, 99-118.
46. Nisbet, A.F., Konoplev, A.V., Shaw, G., Lembrechts, J.F., Merckx, R., Smolders, E., Vandecasteele, C.M., Lönsjo, H., Carini, F. and Burton, O. 1993. Application of fertilisers and ameliorants to reduce soil-to-plant transfer of radiocaesium and radiostrontium in the medium to long term - a summary. *Science of the total Environment*, 137, 173-182.
47. Nisbet, A.F., Woodman, R.F.M. and Haylock, R.G.E. 1999. Recommended soil-to-plant transfer factors for radiocaesium and radiostrontium for use in arable systems. National Protection Board, NRPB-R304, Didcot, UK, pp 50.
48. Perttu, K.L. 1998. Environmental justification for short-rotation forestry in Sweden. *Biomass and Bioenergy,* 15, 1-6.

49. Perttu, K.L. 1999. Environmental and hygienic aspects of willow coppice in Sweden. *Biomass and Bioenergy,* 16, 291-297.
50. Rauret, G. and Firsakova, S. 1996. Transfer of radionuclides through the terrestrial environment to agricultural products, including the evaluation of agrochemical practices. Final report for the Experimental Collaboration Project n°2 (ECP-2, EU 16528), 1996, pp. 179.
51. Renaud, P. and H. Maubert. 1997. Agricultural countermeasures in the management of the post-accidental situation. *Journal of Environmental Radioactivity,* 35, 53-69.
52. RIR. 1990. The basic laws of radionuclide accumulation in agricultural plants in the territory contaminated by radionuclides and scientific base to develop methods to reduce the radionuclide levels in plants under the ChNPP-program. Report RIR, Gomel, 71p.
53. Romanov, G.N. and Drozhko, Y.G. (1996) Ecological consequences of the activities at the Mayak plant. In *Proceedings of the NATO Advanced Study Institute on Radio-active Contaminated Site Restoration,* Zarechny, Russia, 19-28 June, 1995, 45-56.
54. Sauras, T.Y., Vallejo, V.R., Valcke E., Colle C., Förstel H., Millan R. and Jouglet H. 1999. 137 Cs and 90Sr root uptake prediction under close-to-real controlled conditions. *Journal of Environmental Radioactivity,* 45: 191-217.
55. Segal, M.G. 1993. Agricultural countermeasures following deposition of radioactivity after a nuclear accident. *The Science of the Total Environment,* 137, 31-48.
56. Sennerby-Forsse, L. 1994. The Swedish energy frestry programme. In *Willow vegetation filters for municipal wastes and sludges: A biological purification system,* Uppsala, Sweden, SLU Info/Repro, Uppsala, 19-22.
57. Shaw, G. 1993. Blockade of fertilizers of caesium and strontium uptake into crops: effects of root uptake process. *Science of the Total Environment,* 137, 119-133.
58. Shaw, G., Hewamanna, R., Lillywhite, J. and Bell, J.N.B. 1992. Radiocaesium uptake and translocation in wheat with reference to the transfer factor concept and ion competition effects. *Journal of Environmental Radioactivity,* 16, 167-180.
59. Sheppard, M.I., Sheppard, S.C. and Thibault, D.H. 1984. Uptake by plants and migration of uranium and chromium in field lysimeters. *Journal of Environmental Quality,* 13(3): 357-361.
60. Sheppard, M.I., Thibault, D.H. and Sheppard S.C. 1985. Concentrations and concentration ratios of U, As and Co in Scots Pine grown in waste site soil and an experimental contaminated soil. *Water, Air and Soil Pollution* 26: 85-94.
61. Sheppard, S.C., Evenden, W.G. and Pollock, R.J. 1989. Uptake of natural radionuclides by field and garden crops. *Can. J. Soil Sci.* 69: 751-767.
62. Simon, S.L. and S.A. Ibrahim (1987) The soil/plant concentration ratio for calcium radium, lead and polonium: Evidence for non-linearity with reference to substrate concentration. *Journal of Environmental Radioactivity,* 5: 123-142.
63. Simon, S.L. and S.A. Ibrahim (1988) Biological uptake of radium by terrestrial plants: e review. In *Radium Monograph,* IAEA.
64. Smolders, E., Van den Brande, K. and Merckx, R. 1997. Concentrations of ^{137}Cs and K in soil solution predict the plant availability of ^{137}Cs in soils. *Environmental Science and Technology,* 31, 3432-3438.
65. Smolders, E. and Merckx, R. 1993. Some principles behind the selection of crops to minimize radionuclide uptake from soil. *Science of the Total Environment,,* 137, 135-146.
66. Smolders, E., Sweeck, L., Merckx, R. and Cremers A. 1996. Cationic interactions in radiocaesium uptake from solution by spinach. *Journal of Environmental Radioactivity,* 34, 161-170.
67. Sorochinsky, B., 1998. Application of phytoremediation technologies in real conditions in the Chernobyl zone. In *Proceedings of the Chernobyl Phytoremediation and Biomass Energy Conversion Workshop.,* Feb. 23-25, Slavutych, Ukraine. 229-234.
68. Sweeck, L., Th. Zeevaert, G. Volckaert and C.M. Vandecasteele, 1998, Geologische berging van geconditioneerd langlevend hoogradioactief afval: Lange termijn performantiestudies. Biosfeerparameters in peformantie and veiligheidsanalyses Deel III.:

R-3194, SCK•CEN, Mol, Belgium.
69. Thibault, D.H., Sheppard, M.I. and Smith P.A. 1990. *A critical compilation and review of default soil solid/liquid partition coefficients, Kd, for use in environmental assessments.* Atomic Energy Canada, AECL,-10125, Whiteshell Nuclear Research Establishment, Mnitoba, Canada, 1990
70. Thiry, Y., Vandenhove H., Van Hees, H. and Vandecasteele, C.M. 1997. Prospects of alternatives in agro-chemical site remediation. In *Biotechnology for Waste Management and Site Restoration,* eds. C. Ronneau and O. Bitchaeva. Kluwer Academic Publisher, The Netherlands, 213-219.
71. UNSCEAR, 1993. Sources, effects and risks of ionising radiation. United Nations Scientific Committee on the effects of atomic radiation. Report to the General Assembly, United Nations, New York.
72. Valcke, E., Elsen, A. and Cremers, A. 1997. The use of zeolites as amendments in radiocaesium- and radiostrontium contaminated soils: A soil chemical approach. Part IV: A potted soil experiment to verify laboratory-based predictions. *Zeolites*: 18: 225-231.
73. Van Hees, M. and Vandenhove, H. 1998. Unpublished data
74. Van Netten, C. and Morley, D.R. 1982. Differential uptake of U, Mo and Cu by roots, stalks and seeds of oats grown on uranium-rich soils. *Int. J. Environ. Studies* 19: 259-262.
75. Vandenhove, H. 1999. Phyto-extraction of low-level contaminated soil: Study of Feasibility of the Phyto-extraction Approach to Clean-up ^{137}Cs Contaminated Soil from the Belgoprocess Site; Part 2: Transfer factor screening test: discussion of results. Internal SCK•CEN report R-3407.
76. Vandenhove, H. 2000. European sites contaminated by residues from ore extraction and processing industries. In *International Symposium on Restoration of Environments with Radioactive Residues*, Arlington, Virginia, USA, 29/11-3/12/1999. IAEA, STI/PUB/1092, Austria, Vienna, pp. 61-89.
77. Vandenhove, H., M. Van Hees and S. Van Winckel. 2001. Feasibility of the phytoextraction approach to clean-up low-level uranium contaminated soil, *International Journal of Phytoremediation*, 3(3), 301-320.
78. Vandenhove, H., F. Goor, S. Timofeyev, V. Podalyak, A. Timofeyev, V. Drobyshevskaja, V. Ageyets, O. Voitsekovitch, N. Victorova, B. Sorochinsky, A. Mikheev, J. Burema and M. Van der Perk. 2002a. PHYTOR: Evaluation of Willow Plantations for the Phytorehabilitation of Contaminated Arable Land and Flood Plane Areas. Final Report, EC-DG XII project under contract ERB IC15-CT98 0213, SCK•CEN, BLG 909
79. Vandenhove, H., Th. Zeevaert, A. Bousher, D. Jackson, B. Lambers and P. Hedemann Jensen. 2000. Investigation of a possible basis for a common approach with regard to the restoration of areas affected by lasting radiation exposure as a result of past or old practice or work activity - CARE, Final report for EC-DG XI-project 96-ET-006, Radiation Protection 115, Luxembourg, Office for Official Publication by the European Communities, printed in Belgium. pp. 238.
80. Vandenhove, H., Thiry, Y., Gommers, A., Goor, F., Jossart, J.M., Holm, E., Gävfert, T. and Roed, J. (1999). RECOVER-Relevancy of short rotation coppice vegetation for the remediation of contaminated areas, Final report, EC-DG XII-project FI4-CT0095-0021c, SCK•CEN, BLG 826
81. Vandenhove, H., Van Hees, M., and Vandecasteele, C.M., 1998. Effectiveness of immediate and delayed AFCF application in reducing radiocaesium transfer to ryegrass. *Journal of Environmental Radioactivity,* 41, 47-63.
82. Vandenhove, H., Van Hees, M., de Brouwer, S. and C.M. Vandecasteele, C.M., 1996. Transfer of radiocaesium from podzol to ryegrass as affected by AFCF concentration. *The Science of the Total Environment,* 187, 237-245.
83. Vandenhove, H., Y. Thiry, A. Gommers, F. Goor, J.M. Jossart, E. Holm, T. Gäufert, J. Roed, A. Grebenkov and S. Timofeyev. 2001. Short rotation coppice for re-valuation of contaminated land. *Journal of Environmental Radioactivity,* 56, 157-184.

84. Vandenhove, H., F. Goor, S. Timofeyev and S. O'Brien. 2002b. Economic viability of short rotation coppice for energy production for refuse of caesium contaminated land in Belarus. *Biomass and Bioenergy*,22, 421-431.
85. Vasudev, D., Ledder T., Dushenkov, S., Epstein, A., Kumar, N., Kapulnic Y. and Ensleay, B. 1996. Removal of radionuclide contamination from water by metal-accumulating terrestrial plants. Prepared for presentation at Spring National Meeting, New Orleans, LA: *In situ soil and sediment remediation*. Unpublished.
86. Vovk, I.F., Blagoyev, V.V., Lyashenko, A.N. and Kovalev, I.S. 1993. Technical approaches to decontamination of terrestrial environments in the CIS (former USSR). *The Science of the Total Environment,* 137, 49-63.
87. Wauters, J. 1994. Radiocaesium in aquatic sediments: sorption, remobilization and fixation. PhD-dissertation 246, Faculty of Agronomy and Applied Biological Sciences, KULeuven, Belgium, pp. 110.
88. Whicker, F.W. and Ibrahim, S.A. 1984. Radioecological investigations of uranium mill tailings systems. Fort Collins, Colorado State University: 48.
89. Willebrand, E. Ledin S. and Verwijst T. 1993. Willow coppice systems in short rotation forestry: effects of plant spacing, rotation length and clonal composition on biomass production. *Biomass and Bioenergy*, 4, 323-331.
90. Zeevaert, Th., Vanmarcke H. and Govaerts P. (1997) Status of the restoration of contaminated sites in Europe, Radiation Protection 90, DGXI, Environment, Nuclear Safety and Civil Protection, Luxembourg.
91. Zabudko A.N., Petrov E.E., Kozmin, G.V., Svetove, V.A. 1995, Comprehensive radiological investigations of forests in Kaluga, Tula and Or'ol regions, IAEA Technical Meeting on "Clean-up Criteria for Forests and Forestry Products Following a Nuclear Accident" 30 Oct-3 Nov 1995, unpublished.

CHAPTER 7

PHYTOREMEDIATION OF INDUSTRIALLY-CONTAMINATED SITES USING TREES

N.M. DICKINSON
Liverpool John Moores University
School of Biological and Earth Sciences, JMU, Byrom Street, Liverpool, L3 3AF, England

Abstract

Phytoextraction of Cd using *Salix* is likely to be suitable only for urban and agricultural soils with low- level contamination; unrealistically long time scales make this unsuitable for mine spoils or heavily contaminated Industrial sites. Under short-rotation coppice, willows can be expected to have a productive lifecycle of about 30 years. After this period the root bole could be relatively easily removed. It would also be important to harvest foliage. Short-rotation coppice systems in northern Europe generally rely on harvesting the trees during winter, often when the ground is frozen to provide easier access for agricultural machinery. However, in terms of metal removal, a substantial proportion of Cd is relocated to the foliage; leaf fall in autumn may actually increase the concentration and availability of the metal in the surface and would require earlier harvesting along with the woody stems.

1. INTRODUCTION

Previous research at the universities in Liverpool has focussed on the accumulation of metals in food chains (Hunter, et al., 1987), evolution of metal tolerance (Bradshaw and McNeilly, 1982; Bradshaw, 1984) and ecosystem impacts of metal contamination (Dickinson, et al., 1996; Watmough and Dickinson, 1995). More recent experimental studies on the responses of trees to metal contamination (Turner and Dickinson, 1993; Watmough and Dickinson, 1996; Watmough, et al., 1996; Dickinson, 1996; Dickinson, et al., 1991; Wisniewski and Dickinson, 2003) have led to an interest in the use of trees for phytoremediation (Royle, et al., 2003; Putwain, et al., 2003; Dickinson, 2000). In this paper, it is considered whether *Salix* or other fast-growing trees meet this requirement, particularly to deal with widespread and ubiquitous contamination of soils with cadmium at brownfield sites and in urban environments.

J.-L. Morel et al. (eds.), Phytoremediation of Metal-Contaminated Soils, 229–240.

2. URBAN REGENERATION IN NORTH-WEST ENGLAND

Urban renaissance is currently receiving considerable attention from UK government. Regeneration of urban brownfield land for environmental health and economic regeneration aims to transform post-industrial and derelict land into a green and more pleasant landscape. Fundamental to this environmental policy was the creation of 12 Community Forests in England in the mid-1990s; an initiative to introduce multi-purpose forestry to increase lowland forest cover in urban and urban fringe areas. Creation of more naturalistic habitats of value to conservation is particularly highly favoured. Thus, economic and political changes are driving the remediation of brownfield land and are starting to address the legacy of contamination from an earlier industrial history. 'Low-cost' restoration to 'soft end-uses' is prioritised together with building development. The North-West of England possesses 25% (30,000 ha) of England's derelict land and currently has forest cover of only 5%, about a quarter of the European average.

Reclamation of Brownfield Land

Regeneration of urban brownfield land requires innovative, low-cost, ecologically-sensitive and effective techniques for clean-up of soils contaminated with heavy metals, avoiding expensive 'dig and dump' reclamation. Current problems associated with reclamation of brownfield land fall into two categories: (i) disposal of existing contaminated soil coverage and, (ii) import of new topsoil. Both are prohibitively expensive and this type of 'hard' reclamation lacks environmental sensitivity. Landfill disposal costs are currently at least €35 per t^{-1}, an amount that may represent just 1 m^2 of land surface. Purchase of topsoil typically costs €13-20 m^{-3}. Put together these translate to €65-100,000 ha^{-1} (plus significant associated costs) to achieve a minimal soil depth for tree planting. Furthermore, adding topsoil often is a difficult way to establish diverse semi-natural vegetation. Physical structure, fertility and integrity of soil are greatly affected by disturbance and stockpiling and, in any case, many topsoils are simply too fertile to support a diverse plant community. In many ways it is easier, and certainly more cost-effective, to work *in situ* with existing contaminated soil and to attempt to restore a healthy soil. For these reasons, natural attenuation is often favoured for low value brownfield land, sometimes combined with use of appropriate additives or amendments (Vangronsveld and Cunningham, 1998; Vangronsveld, et al., 2000). Phytoremediation through tree planting may also be suited to this task (Lepp and Dickinson, 1998; Dickinson, 2000).

3. PHYTOREMEDIATION AND CADMIUM

The possibility of exploiting plants for removal of metals from soil through uptake and harvest (phytoextraction) may meet these criteria is considered in other papers in this textbook and elsewhere (Meagher, 2000; US-EPA, 2000; US-EPA, 2001). To be successful phytoextraction requires high yielding plants that readily transport targeted metals from the soil into harvestable plant tissues.

3.1. Hyperaccumulators and Cadmium

As described elsewhere in this textbook, the extraordinary metal uptake traits of hyperaccumulator plants may offer real potential to extract nickel, zinc, selenium and some other metals but, unfortunately, only a few small shrubby species have been found (*Thlapsi caerulescens* and *Arabidopsis halleri*) that accumulate more than 100 mg Cd kg^{-1} into their tissues (McGrath et al., 2001; Kramer, 2000). Until other species are found, if indeed they exist, there is an opportunity to consider other forms of remediation including those that involve tree planting (Dickinson, 2000).

The use of woody plants to enhance the degradation of organic contaminants in soil and to protect groundwater is already a well-proven example of phytoremediation (Burken and Schnoor, 1997; Jones, et al., 1999; Wang, et al., 1999; Nzengung, et al., 1999; Newman, et al., 1997). Many complex and toxic organic molecules are naturally degraded by microbes in the soil into simpler harmless compounds. Microorganisms in the rhizosphere appear to be primarily responsible for degradation and, therefore, this process is generally enhanced in the presence of plants; either through uptake and volatilisation or simply by plant roots improving conditions for degradation in the soil. Metals cannot be degraded in the same way; they can only be treated by using physical or chemical stabilization to restrict their mobility, or else by separation and removal from soil. There is a large variation between plants in uptake of cadmium, and some trees appear to readily accumulate fairly high concentrations of this metal in their tissues. Potentially, selection of trees with high rates of uptake opens up the possibility of phytoextraction of cadmium.

3.2. Cadmium Toxicity

Cadmium is a non-essential element and one of the most hazardous of the trace elements or heavy metals, frequently exceeding established guidelines and thresholds in urban and agricultural soils. Mean background UK soil levels are 0.53 mg kg^{-1} in urban gardens and 0.37 mg kg^{-1} in agricultural soils, with current UK Soil Guideline Values for 'residential land use with plant uptake' are set at total concentrations of between 1-8 mg Cd kg^{-1}, depending on soil acidity (DEFRA, 2002). Cadmium is particularly mobile in biological systems and is considered to be primarily zootoxic, thus presenting potential risks to human health; ingestion is the main route of transfer. It is particularly significant that plants only tend to be sensitive to the metal when exposed to much higher concentrations than is the case for animals, thus increasing the risk of food chain transfer to humans and other animals.

In terms of current guidelines and risk assessment, measurements of total metal concentrations currently provide a very conservative regulatory assessment of the risks of contaminated soils to the environment and to human health (Dickinson, et al., 2000). With cadmium, soluble low molecular weight chelates are formed in soil solution that keep the free metal activity low but greatly increase soluble metal concentrations. A limited proportion of metals in soil are bioavailable, but simple techniques to describe the kinetics of metal speciation in soils and inherent risks are currently lacking (Nolan, et al., 2003). This lack of scientific certainty means that types of remediation that attempt to chemically stabilise the metal in soil are unlikely to be acceptable for cadmium. It is reasonable to be concerned that changed environmental conditions in the medium- to long-term may alter the rate of replenishment of labile pools of the metal.

3.3. Requirements for Phytoextraction

In biological terms, it is likely there is a trade-off in plants between (i) diverting metabolic energy into stress tolerance, that allows high uptake of toxic metals, and (ii) rapid growth that produces high yields. Some optimum combination of these traits is required for plants to be used for removal of metals from soils. Hyperaccumulators are plants that achieve high uptake, but generally have slow growth rates and poor biomass acquisition. Some tree species have high growth rates and are good biomass crops, but tissue metal concentrations are much lower than in hyperaccumulators. What evidence is there that trees are potential candidates for phytoextraction of cadmium?

4. CASE STUDIES

4.1. Salix and Cadmium Uptake

In fast-growing short-rotation crops of selected species and clones of *Salix* (willows and osiers), there is undoubtedly significant uptake of cadmium from soil to roots and then a ready translocation of the metal to aerial tissues when compared to other metals and to other plants (Stoltz and Greger, 2002; Sander and Ericsson, 1998). *Salix* preferentially accumulates

Cd in the roots, but transport of the metal to above-ground tissues varies between and within species. Higher shoot concentrations are found when plants are grown in hydroponics compared to field grown plants, but it appears that clones with higher rates of metal transfer to shoots can be readily identified using hydroponics (Pulford and Watson, 2003; Greger, and Landberg, 1999). Greger and Landberg (1999) argued that uptake properties of clones are stable, although work by others showed a clear ability of individual plants to acclimate to metal stress (Punshon and Dickinson, 1997).

Salix clones have been found to accumulate 100 – 250 mg Cd kg^{-1} in aerial tissues in hydroponics and pot experiments, leading to some exaggerated claims of hyperaccumulation. In field plots, uptake is generally lower. In *Salix* that colonised dredged river sediment landfills in Flanders, cadmium was taken up in proportion to sediment concentrations of the metal (Vandecasteele, et al., 2002). Cadmium uptake from land contaminated by sewage sludge application was more than 6 times higher in *Salix* than any of five other tree species in twigs and 13 x higher in foliage (Rosselli, et al., 2003). On mine spoils with 255 mg Cd kg^{-1}, *Salix cinerea* was sampled with 44 mg Cd kg^{-1} in foliage and 76 mg Cd kg^{-1} in woody stems (Punshon and Dickinson, 1997). In the same work, *S. caprea* was grown for 12 months on industrially-contaminated soil (14 mg HNO_3-extractable Cd kg^{-1}) and mine spoil (117 mg Cd kg^{-1}), recording above-ground woody tissue concentrations of >25 and >125 mg Cd kg^{-1} respectively, although productivity was particularly poor in the mine spoil. In the industrial spoil, where better growth was achieved, concentrations factors (above-ground tissues : soil) above 5 were recorded. Felix (1997) found a transfer coefficient of 3.4 for *S. viminalis*. Pulford et al. (2002) found stem Cd concentrations in 3 – year old stands of 20 clones of *Salix* that were up to 10 x higher than soil concentrations.

There does not appear to be a correlation between metal resistance and uptake in the biomass willow clones of *S. viminalis* (Greger and Landberg 1999), although some willows not commonly used as biomass clones are more resistant to metal toxicity and accumulate higher metal concentrations in wood (Watson, et al., 2003). Cadmium uptake is variable even within single clones of *Salix* (Punshon and Dickinson, 1997; Punshon and Dickinson, 1999). Lanberg and Greger (1994) found that uptake of to shoots from 1 μM Cd hydroponic solutions after 20 days varied by a factor of 80 between 94 clones of *S. viminalis* and *S. dasyclados*.

Pulford et al. (2002) suggested that the clone *S. burjatica* Germany may be suitable for removing significant quantities of Cd and other metals from soil. *S.* x *calodendron* Wimm., a hybrid of *S. viminalis*, *S. caprea* and *S. cinerea*) was favoured by Punshon and Dickinson (Punshon and Dickinson, 1999; Punshon and Dickinson, 1997). Differences in susceptibility of plant species to Cd toxicity almost certainly involve differences in structure, function and transporter systems in root plasma membranes, involving different biochemical pathways (Das, et al., 1997).

In summary, uptake of cadmium into *Salix* is dependent on the amount of cadmium in soil and some species and clones have higher uptake than others. It appears that tissue concentration in double Figs. (above 10 mg Cd kg^{-1}) can be achieved, with a bioconcentration factor up to 10, but it is unlikely that productive clones will be found with tissues concentration as high as 100 mg Cd kg^{-1}.

4.2. Mass Balance Calculations

In terms of mass balance equations, a *Salix* harvest of 15 t ha^{-1} annum^{-1} containing 25 mg Cd kg^{-1} would reduce soil concentrations in the surface 20 cm by 5 mg Cd kg^{-1} in about 27 years (Table 1). This would appear to represent (i) an uptake ratio that is currently and realistically achievable, (ii) a time period that may be considered appropriate for low-value urban-fringe derelict and underused land and (iii) the approximate length of the productive lifecycle of short-rotation coppice.

Table 1. Number of years to reduce soil Cd concentrations by 5 ppm (5 mg kg^{-1}) by planting Salix to different soil depths, calculated according to rates of metal uptake into above-ground plant tissues.

Depth	Plant tissue concentration (mg kg^{-1})			
(cm)	10	25	50	100
10	33	13	7	3
20	67	27	14	7
40	133	53	28	14

Calculation assumes (i) yields of 15 ODT ha^{-1}, (ii) bulk density of 1 and (iii) consistent Cd uptake throughout period. Fertilization and irrigation may be required to achieve these yields.

Currently, however, there is very little direct evidence that this can be achieved. The amount of Cd removed from field grown *Salix* at *un*contaminated sites has been estimated as 36 and 61g Cd ha^{-1} annum^{-1}, with and without leaf harvest (Greger and Landberg, 1999). This equates to an approximate reduction of 1 mg Cd kg^{-1} in soil (0-30 cm depth) after about 50 years. The questions remains as to whether this can be increased to 5 mg Cd kg^{-1} in 27 years when *Salix* is planted in contaminated soils.

Possibly the best evidence that soil concentrations have declined after repeated *Salix* harvest was reported by Eriksson and Leding (Eriksson and Ledin, 1999). They compared eight agricultural sites with marginally elevated Cd (< 0.3 mg kg^{-1}), derived from phosphate fertiliser, where *Salix viminalis* had been planted for 10 years or more. Although there were deficiencies with their experimental design, they found that $CaCl_2$-

extractable or exchangeable Cd (but not total Cd) in 8 soils taken from *Salix* stands and nearby unplanted reference soils were 30-40% lower in the planted soils. Few other studies have accumulated data beyond three years of growth.

4.3. Additional Possibilities

4.3.1. Removal from Root Bole

Concentrations of cadmium in roots are invariably substantially higher than in above-ground tissues. Although there are many fewer studies of root biomass and yield, willow stands in Quebec have been found to amount to 8-11 ODT ha^{-1} (www.reap-canada.com; www.esf.edu/willow). At the end of a period of remediation, it would appear that removal of the root bole for redevelopment of the land would probably remove an amount of metal almost to equivalent to that already harvested. This would be a simple operation for standard agricultural machinery.

4.3.2. Allowing for Hotspots of Cadmium Contamination

At brownfield sites, uniform contamination of the land surface is unusual. Soil contamination invariably varies in both vertical and horizontal profiles (Fig. 1). Tree roots activity forage avoid highly metal contaminated soil (Turner and Dickinson, 1993; Watmough and Dickinson, 1995) but, for cadmium-enriched soil, they are unlikely to be sensitive to soil concentrations < 15 mg Cd kg^{-1}, and probably much higher (Dickinson, et al., 2000). This spatial variability probably explains why lower plant tissue concentrations tend to be recorded in field studies than in pot experiments where soils are uniformly mixed. In a study where short-rotation coppice has been planted at several brownfield sites in Merseyside (including those in Fig. 1) stem : soil bioconcentration factors ranged from 1.4 to 13.3. Values towards the upper end of this range add evidence to arguments that phytoextraction of cadmium is achievable.

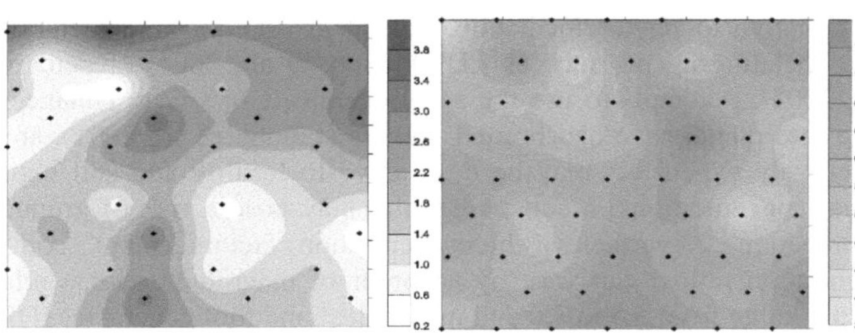

Fig. 1. Spatial profiles of HNO_3-extractable soil cadmium concentration at two brownfield sites in North-West England within 900m sampling plots (Putwain, et al., 2003)

For practical reasons it is unlikely that most field sites are mapped in such detail that hotspots could be routinely targeted, although improved techniques such as portable XRF may make this possible in the future. Whilst blanket coverage of *Salix* grown as short-rotation coppice for biomass is obviously preferred, achieving high concentrations of metals in tissues of the whole crop may not be necessary.

4.3.3. Bioavailability, Mycorrhizae and Soil Fauna

Other factors of importance to the success of phytoextraction relate to longer-term bioavailability of metals in soils or processes affecting uptake. For example, there is some evidence that mycorrhizae reduce uptake of metals to shoots (Dixon, 1988). Trees themselves may alter the bioavailability of metals in soils, and willows were found to increase EDTA-extractable Cd in sludge-treated soils.

It has long been suspected that earthworms are likely to alter the solubility of heavy metals in soil. Ireland (Ireland, 1975) found that *Dendrobaena rubida* altered the solubility of Pb, Zn and Ca in reclaimed Welsh mining spoils, reporting a 50% increase in availability of water-soluble Pb in faeces. Devliegher and Verstraete (1996, 1997) found that copper, chromium and cobalt availability was raised by between 6-30% in the soil by *L. terrestris*. Ma *et al.* found that species of *Pheretima* increased the mobility of metals in Pb/Zn mine tailing diluted with uncontaminated soil. They suggested that earthworms may benefit attempts to use plants for phytoextraction, by increasing the amount of metal in the soil available for plant uptake. The presence of *L. terrestris* in microcosms increased the concentration of Cd, Cu and Zn into roots and shoots (and of Pb and Fe into roots) of *Lolium perenne* seedlings (Rida, A.M.M.A., 1996).

Clearly these are additional factors of relevance to the phytoextraction process that we do not yet properly understand.

4.3.4. Chemical Enhancement of Phytoextraction

Alternative to phytoremediation include *ex situ* washing methods with strong chelating agents such as EDTA (Sun, et al., 2001) or acids (Chen, et al., 2003). Attempts to use the same chemicals *in situ* to enhance metal uptake into plants have had limited success (Lombi, et al., 2001), are often prohibitively expensive, may be destructive to fertility and soil biota, and residual concentrations in soil pore water may pose a risk of groundwater contamination. Nevertheless, bioconcentration factors (plant: soil metal concentration ratios) can vary by an order of magnitude over a relatively short pH range from 5.5 to 7.0 and in both leaf and root vegetables (DEFRA, 2002), and Cd uptake into *Salix* has been found to be highly pH dependent in field stands (Eriksson and Ledin, 1999). It may be possible to enhance Cd uptake with low-cost organic or inorganic acid soil amendments.

5. CONCLUSIONS

Although there is still a lack of sufficient field evidence, increased uptake of cadmium in *Salix* clones has been identified. The combination of high yield and rates of cadmium accumulation that realistically can be achieved suggest this is a viable technology. We could aim to reduce soil contamination by 5 mg Cd kg^{-1} in less than the crop lifecycle of 30 years. This could be substantially improved by removal of the root bole on final harvest. A failure to quantify the spatial dispersion of metals at contaminated sites has probably made the use of *Salix* look less favourable than is fair. Targeting of hotspots would provide more accurate quantification of the remediation process.

Phytoextraction of Cd using *Salix* is likely to be suitable only for urban and agricultural soils with low- level contamination; unrealistically long time scales make this unsuitable for mine spoils or heavily contaminated industrial sites. Under short-rotation coppice, willows can be expected to have a productive lifecycle of about 30 years. After this period the root bole could be relatively easily removed. It would also be important to harvest foliage. Short-rotation coppice systems in northern Europe generally rely on harvesting the trees during winter, often when the ground is frozen to provide easier access for agricultural machinery. However, in terms of metal removal, a substantial proportion of Cd is relocated to the foliage; leaf fall in autumn may actually increase the concentration and availability of the metal in the surface and would require earlier harvesting along with the woody stems.

This type of tree planting provides short- to medium-term aesthetic improvement of derelict and underused land. It also provides a crop with short-term economic benefits; woody residues can be used for renewable biomass energy, fibreboard manufacture and handicrafts. Agronomic practices from propagation to harvesting are well established and straight-forward for *Salix,* using either labour or machinery, avoiding the need to bring new crops into cultivation. What to do with the harvested willows and inherent risks associated with this form of soil clean-up has been discussed at length elsewhere (Dickinson, et al., 2000; Punshon and Dickinson, 1997; Greger and Landberg, 1999; Pulford and Watson, 2003).

Finally, there is a broader knowledge deficit. Accurate and realistic ecosystem models of metal fluxes, and an improved knowledge of soil ecology provide the key to using phytoremediation for the restoration of healthy soils on contaminated brownfield land. Metal mobility is significantly influenced by soil conditions created by plant growth, rhizosphere processes and earthworms, but this is poorly understood. Plants themselves affect metal mobility, and metals may be returned to the surface through uptake and leaf fall. Improved understanding is required of longer-term temporal changes in bioavailability in relation to the rate of replenishment of labile pools of the metal in contaminated soils.

REFERENCES

1. Bradshaw, A.D. (1984) Adaptations of plants to soils containing toxic metals - a test for conceit, in D. Evered and G.M. Collins, (eds.), Origins and Development of Adaptation, Pitman: London. pp. 4-19.
2. Bradshaw, A.D. and McNeilly, T. (1982) Evolution and Pollution, Edward Arnold, London.
3. Burken, J.G. and Schnoor, J.L. (1997) Uptake and metabolism of atrazine by poplar trees, Environ. Sci. Technol., 31, 1399.
4. Chen, Y.X., Lin, Q., Luo, Y.M., He, Y.F., Zhen, S.J., Yu, Y.L., Tian, G.M., and Wong, M.H. (2003) The role of citric acid on the phytoremediation of heavy metal contaminated soil, Chemosphere, 50, 807-811.
5. Das, P., Samantaray, S., and Rout, G.R. (1997) Studies on cadmium toxicity in plants: a review, Environ. Pollut., 98, 29-36.
6. DEFRA (2002) Soil guideline values for cadmium contamination, Environment Agency, Bristol. d. Vol. R&D publications SGV 3.
7. Devliegher, W. and Verstraete, W. (1996) Lumbricus terrestris in a soil core experiment: effects of nutrient enrichment processes (NEP) and gut-associated processes (GAP) on the availability of plant nutrients and heavy metals, Soil Biol. Biochem., 28, 489-496.
8. Devliegher, W. and Verstraete, W. (1997) The effect of Lumbricus terrestris on soil in relation to plant growth: Effects of nutrient-enrichment processes (NEP) and associated processes (GAP), Soil Biol. Biochem., 29, 341-346.
9. Dickinson, N.M. (1996) Metal resistance in trees, in I. Glimmerveen, (ed.), Heavy metals and trees, Institute of Chartered Foresters (ICF): Edinburgh. pp. 85-92.
10. Dickinson, N.M. (2000) Strategies for sustainable woodlands on contaminated soils, Chemosphere, 41, 259-263.
11. Dickinson, N.M., MacKay, J.M., Goodman, A., and Putwain, P. (2000) Planting Trees on Contaminated Soils: Issues and Guidelines, Land Contamination and Reclamation, 8, 87-101.
12. Dickinson, N.M., Turner, A.P., and Lepp, N.W. (1991) How do trees and other long-lived plants survive in polluted environments?, Funct. Ecol., 5, 5-11.
13. Dickinson, N.M., Watmough, S.A., and Turner, A.P. (1996) Ecological impact of 100 years of metal processing at Prescot, North West England, Environ. Rev., 4, 8-24.
14. Dixon, R. (1988) Response of ectomycorrhizal Quercus rubra to soil cadmium, nickel and lead, Soil Biol. Biochem., 20, 555-559.
15. Eriksson, J. and Ledin, S. (1999) Changes in phytoavailability and concentration of cadmium in soil following long term Salix cropping, Water Air and Soil Pollution, 114, 171-184.
16. Felix, H. (1997) Field trials for in situ decontamination of heavy metal potlluted soils using crops of metal-accumulating plants, Z. Pflanzenernahr Bodenk, 160, 525-529.
17. Greger, M. and Landberg, T. (1999) Use of willows in phytoextraction, International Journal of Phytoremediation, 1, 115-123.
18. Hunter, B.A., Johnson, M.S., and Thompson, D.J. (1987) Ecotoxicology of copper and cadmium in a contaminated grassland ecosystem. II. Invertebrates, J. appl. Ecol., 24, 587-599.
19. Hunter, B.A., Johnson, M.S., and Thompson, D.J. (1987) Ecotoxicology of copper and cadmium in a contaminated grassland ecosystem. I. Soil and vegetation contamination, J. appl. Ecol., 24, 573-586.

20. Ireland, M.P. (1975) The effect of the eartthworm Dendrobaena rubida on the solubility of lead, zinc and calcium in heavy metal contaminated soil in Wales., J. Soil Sci., 26, 313-318.
21. Jones, S.A., Lee, R.W., and Kuniansky, E.L. (1999) Phytoremediation of trichloroethene (TCE) using cottonwood trees, in A. Leeson and B.C. Alleman, (eds.), Phytoremediation and innovative strategies for specialized remedial applications, Battelle Press: Columbus. pp. 101-108.
22. Kramer, U. (2000) Cadmium for all meals - plants with an unusual appetite, New Phytol., 145, 1-3.
23. Landberg, T. and Greger, M. (1994) Heavy metal tolerant clones of Salix: possible vegetation filter on heavy metal contaminated land. in Willow vegetation filters for municipal wastewaters and sludges: a biological purification system. Swedish University of Agricultural Science, Uppsala.
24. Lepp, N.W. and Dickinson, N.M. (1998) Woody plants: role in remediation, in J. Vangronsveld and S.D. Cunningham (eds.) Metal-Contaminated Soils: In Situ Inactivation and Phytorestoration R.G.Landes Company, Georgetown., Texas.
25. Lombi, E., Zhao, F.J., Dunham, S.J., and McGrath, S.P. (2001) Phytoremediation of heavy metal-contaminated soils: natural hyperaccumulation versus chemically enhanced phytoextraction, J. Environ. Qual., 30, 1919-1926.
26. Ma, Y., Dickinson, N.M. and Wong, M.H. (2002) Toxicity of Pb/Zn mine tailings to the earthworm Pheretima and the effects of burrowing on metal availability, Biol. Fert. Soil 36, 72-78.
27. McGrath, S.P., Lombi, E., and Zhao, F.-J. (2001) What's new about cadmium hyperaccumulation?, New Phytol., 149, 2-2.
28. Meagher, R.B. (2000) Phytoremediation of toxic elemental and organic pollutants, Current Opinion in Plant Biology, 3, 153-162.
29. Newman, L.A., Strand, S.S., Choe, N., Duffy, J., Ekuan, G., Ruszaj, M., Shurtleff, B.B., Wilmoth, J., Heilman, P., and Gordon, M.P. (1997) Uoptake and biotransformation of trichloroethylene by hybrid poplars, Environ. Sci. Technol., 31, 1062-1067.
30. Nolan, A.L., Lombi, E., and McLaughlin, M.J. (2003) Metal bioaccumulation and toxicity in soils - why bother with speciation?, Aust. J. Chem., 56, 77-91.
31. Nzengung, V.A., Wang, C., Harvey, G., McCutcheon, S., and Wolfe, L. (1999) Phytoremediation of perchlorate contaminated water: laboratory studeies, in A. Leeson and B.C. Alleman, (eds.), Phytoremediation and innovative strategies for specialized remedial applications, Battelle Press: Columbus. pp. 133-138.
32. Pulford, I.D. and Watson, C. (2003) Phytoremediation of heavy metal contaminated land by trees - a review. Environment International, in press.
33. Pulford, I.D., Riddell-Black, D., and Stewart, C. (2002) Heavy metal uptake by willow clones from sewage sludge-treated soil: the potential for phytoremediation, International Journal of Phytoremediation, 4, 59-72.
34. Punshon, T. and Dickinson, N.M. (1997) Acclimation of Salix to metal stress, New Phytol., 137, 303-314.
35. Punshon, T. and Dickinson, N.M. (1997) Mobilization of heavy metals using short rotation coppice, Aspects of Applied Biology, 49, 285-292.
36. Punshon, T. and Dickinson, N.M. (1999) Heavy metal resistance and accumulation characteristics in willows, International Journal of Phytoremediation, 1, 361-385.
37. Putwain, P.D., Rawlinson, H.A., French, C.J., Dickinson, N.M., and Nolan, P. (2003) The Mersey Forest Brownfield Research Project, in H.M. Moore, H.R. Fox, and S. Elliott, (eds.), Land Reclamation: Extending the Boundaries, A.A. Balkema: Lisse, The Netherlands. pp. 165-172.
38. Rida, A.M.M.A. (1996) Trace-element concentrations and growth of earthworms and plants in soils with and without cadmium, copper, iron, lead and zinc contamination: interactions of soil and earthworms, Soil Biol. Biochem., 28, 1029-1035.

39. Rida, A.M.M.A. (1996) Trace-element concentrations and growth of earthworms and plants in soils with and without cadmium, copper, iron, lead and zinc contamination: interactions of plants, soil and earthworms, Soil Biol. Biochem., 28, 1037-1044.
40. Rosselli, W., Keller, C., and Boschi, K. (2003) Phytoextraction capacity of trees growing on a metal-contaminated soil, Plant Soil, in press.
41. Royle, A., Dickinson, N.M., Putwain, P.D., Rawlinson, H.A., Kng, R., and Gray-Jones, E. (2003) Phytoremediation of a contaminated dredegd canal sediment, in H.M. Moore, H.R. Fox, and S. Elliott, (eds.), Land Reclamation: Extending the Boundaries, A.A. Balkema: Lisse, pp. 281-284.
42. Sander, M.L. and Ericsson, T. (1998) Vertical distributions of plant nutrients and heavy metals in Salix viminalis stems and their implications for sampling, Biomass & Bioenergy, 14, 57-66.
43. Stoltz, E. and Greger, M. (2002) Accumulation properties of As, Cd, Cu, Pb and Zn by four wetland plant species growing on submerged mine tailings, Environ. Exp. Bot., 47, 271-280.
44. Sun, B., Zhao, F.J., Lombi, E., and McGrath, S.P. (2001) Leaching of heavy metals from contaminated soils using EDTA, Environ. Pollut., 113, 111-120.
45. Turner, A.P. and Dickinson, N.M. (1993) Copper tolerance of Acer pseudoplatanus L. (sycamore) in tissue culture, New Phytol., 123, 523-530.
46. Turner, A.P. and Dickinson, N.M. (1993) Survival of Acer pseudoplatanus L. (sycamore) seedlings on metalliferous soils, New Phytol., 123, 509-521.
47. US-EPA (2000) Introduction to phytoremediation, Report EPA/600/R-99/107, United States Environmental Protection Agency, Washington D.C. p. 94.
48. US-EPA (2001) Brownfield Technology Primer: Selecting and Using Phytoremediation for Site Cleanup. Report EPA 542-R-01-006, United States Environmental Protection Agency, Washington DC. p. 35.
49. Vandecasteele, B., De Vos, B., and Tack, F.M.G. (2002) Cadmium and Zinc uptake by volunteer willow species and elder rooting in polluted dredged sediment disposal sites, The Science of The Total Environment, 299, 191-205.
50. Vangronsveld, J. and Cunningham, S.D., Eds. (1998) Metal-contaminated soils: in situ inactivation and phytorestoration. R.G. Landes Company: Georgetown, Texas.
51. Vangronsveld, J., Ruttens, A., Mench, M., Boisson, J., Lepp, N.W., Edwards, R., Penny, C., and van der Lelie, D. (2000) In situ inactivation and phytoremediation of metal- and metalloid-contaminated soils: field experiments, in D.L. Wise and D.J. Trantolo, (eds.), Bioremediation of Contaminated Soils, Marcel Dekker, Inc.: New York. pp. 859-885.
52. Wang, X., Newman, L.A., Gordon, M.P., and Strand, S.E. (1999) Biodegradation of carbon tetrachloride by poplar trees: results from cell culture and field experiments, in A. Leeson and B.C. Alleman, (eds.), Phytoremediation and innovative strategies for specialized remedial applications, Battelle Press: Columbus. pp. 133-138.
53. Watmough, S.A. and Dickinson, N.M. (1995) Dispersal and mobility of heavy metals in relation to tree survival in an aerially-contaminated woodland ecosystem, Environ. Pollut., 90, 139-142.
54. Watmough, S.A. and Dickinson, N.M. (1996) Variability of metal resistance in Acer pseudoplatanus L (sycamore) callus tissue of different origins, Environ. Exp. Bot., 36, 293-302.
55. Watmough, S.A., Gallivan, C.C., and Dickinson, N.M. (1995) Induction of zinc and nickel resistance in Acer pseudoplatanus L. (Sycamore) callus cell lines, Environ. Exp. Bot., 35, 465-473.
56. Watson, C., Pulford, I.D., and Riddell-Black, D. (2003) Screening of willow species for tolerance to heavy metals: comparison of the performance in a hydroponics system and field trials, International Journal of Phytoremediation, in press.
57. Wisniewski, L. and Dickinson, N.M. (2003) Toxicity of copper to Quercus robur (English Oak) seedlings from a copper-rich soil, Environ. Exp. Bot., 50, 99-107.

CHAPTER 8

FACTORS LIMITING EFFICIENCY OF PHYTOEXTRACTION AT MULTI-METAL CONTAMINATED SITES

C. KELLER
Swiss Federal Institute of Technology (EPFL)
ENAC – ISTE – Laboratory of Soil Science
Bât. GR, Ecublens, 1015 Lausanne, Switzerland

Abstract

The paper is based mostly on results from 3 field experiments and pot experiments with additional references to the literature. The three field experiments of reference will be first presented. Then the site characteristics (including the local climatic and edaphic conditions as well as the nature and extent of the contamination), the plant characteristics and the aspects related to the legislative background will be presented successively. It should be remembered that all these aspects may interact and thus are not truly independent parameters.

1. INTRODUCTION

Phytoextraction has been proposed as a suitable alternative to destructive techniques used so far to clean up soils contaminated with heavy metals. Indeed, the use of plants to remove metals from soils is environmental friendly and its cost is low compared to engineering-based techniques such as soil capping, soil washing, vitrification, landfilling etc... Additionally, it is an *in situ* and solar generated technique that could help to rehabilitate large areas of agricultural soils contaminated mostly in the upper layer, maintaining or even restoring soil fertility. It also produces less waste because the biomass may be recycled for both energy and metals. Unlike phytostabilisation - its counterpart among the phytoremediation techniques - phytoextraction potentially resolves most of the legislative requirements that ask for metal removal down to a given threshold.

The general drawbacks of phytoextraction are well known (e.g. Vangronsveld and Cunningham, 1998) and its optimisation still requires a lot of research: phytoextraction is a slow process in comparison to the above mentioned techniques (10 years is the goal on which most people agree on to

J.-L. Morel et al. (eds.), Phytoremediation of Metal-Contaminated Soils, 241–266.

recognize phytoextraction as economically acceptable). Its application is limited because of the lack of established methods and successful completed remediation case studies, the lack of recognised economic performance and the risk of food chain contamination. Additionally it may not be able to remove 100% of the contaminants and its efficiency has been proved for some contaminants only. Alternatively, the plants might not be able to withstand highly toxic concentrations of the contaminants.

However, it is not the aim of this paper to review these drawbacks in detail: a preliminary comparison with the other available techniques should determine whether these negative aspects are limiting or not for choosing phytoextraction for a given site. Once it has been decided to use phyto-extraction as a clean up technique, many pitfalls may appear that will limit its efficiency; that is, its ability to reach the goal set in a reasonable time and with limited negative side effects. They may be due to the specificities of the site, soil, plants and contamination characteristics but also to external constraints like legislation. Highlighting these factors will help to define the boundaries conditions for the use of phytoextraction and thus our ability to offer an efficient technique to landowners or local authorities either as main remediation tool or as part of a "remediation package".

This paper is based mostly on results from 3 field experiments and pot experiments with additional references to the literature. The three field experiments of reference will be first presented. Then the site characteristics (including the local climatic and edaphic conditions as well as the nature and extent of the contamination), the plant characteristics and the aspects related to the legislative background will be presented successively. It should be remembered that all these aspects may interact and thus are not truly independent parameters.

2. SITE CHARACTERISTICS

2.1. Experimental Sites in Switzerland

Site characteristics include the climatic conditions, the nature of the soil that is contaminated, the extent of the soil volume concerned (surface and depth), the nature and the extent of the contamination, the metals to be removed, their concentrations, their availability to plants, their toxicity, the presence of other contaminants (both inorganic or organic), and the degree of heterogeneity of the contamination.

The three sites chosen to illustrate these different aspects and their impact on the final efficiency of phytoextraction are presented in Table 1. Two are agricultural soils and the third is a landfill that has been closed and revegetated with trees. They are moderately contaminated with metals, mostly Cu, Zn, Cd and Pb with some additional metals in lower concentrations but no significant organic contamination.

The Dornach site, NW Switzerland has been described by Kayser, et al., 2000. The source of the heavy metal contamination was a nearby brass-smelter emitting Cu, Zn and Cd in particulate form until the mid 1980's. The soil has been classified as calcaric regosol (Geiger, et al., 1993). The site is currently included in a crop rotation.

The Caslano site in southern Switzerland was contaminated with sludges from septic tanks spread on the site for 20 years until 1980. This has led to enrichment of the topsoil of the fluvisol with both organic matter and heavy metals (Cd, Cu, and Zn). Currently, the site is used as a private garden and a meadow for fodder production.

The third experimental site is located on the former landfill "Les Abattes" (Le Locle, Switzerland), in the Swiss Jura mountain chain. The 2-ha landfill was used for inert waste material deposits (lime and marl from nearby building areas). Heavy metal concentrations in these materials were very low (Dubois, 1991), but the landfill was capped in 1990 with a final top layer (0.05-0.60 m depth) made up by a mixture in varying proportions and thickness of contaminated compost originating from contaminated sewage sludge (mostly Cd, Cu and Zn) and uncontaminated local calcareous soil and gravel. Thus, heavy metal concentrations in the final top layer were spatially highly variable as a consequence of the heterogeneous mixing of the components (Rosselli, et al., 2003).

Phytoextraction experiments have been performed on the two agricultural sites with crop plants (tobacco, maize, Indian mustard, sunflower, bioenergy plants (*Salix viminalis*) and hyperaccumulators (*Thlaspi caerulescens, Alyssum montanum, Iberis intermedia*) on mini-plots. Both experiments have been conducted according to agronomic standards. Detailed descriptions of the experiments can be found elsewhere (Kayser, et al., 2000; Hammer and Keller, 2003). On the landfill, various tree species have been planted in groups of 10 plants of the same species, including pioneer (and bioenergy) trees like *S. viminalis* (Rosselli, et al., 2003). Additionally two clones (a local one and a Swedish clone n°78980) of *S. viminalis* were planted either as cuttings in 1997 or as one-year-old plants (1-m high) in 1998 in 30 groups of 12 plants with a random distribution of the 30 groups on a 30-m-side grid. The whole field was maintained as a typical forest plot (no weed removal but grass was cut once per year around the young plants). Tables 2 and 3 give an overview of the main characteristics of the experiments as well as some results obtained with *T. caerulescens* and *S. viminalis* that will be presented later on. In the case of Les Abattes only the results of the one-yr-old plants are given. Additionally, *S. viminalis* and *T. caerulescens* were tested in hydroponics as well as in pots in order to measure their extraction potential in controlled conditions.

Table 1. Site descriptions (after Kayser, et al., 2000; Hammer and Keller, 2003; Keller, et al., 2003 and Keller, et al., unpublished data). Soil parameters are average values with standard deviations given in brackets.

	Les Abattes	Dornach	Caslano
Location	Jura chain	Jura edge	Southern Alps
Altitude in m	1000	300	280
Mean temperature in °C	5,3	10	11,7
Mean rainfall in mm	1470	800	1860
Soil type	Calcaric Anthroposol	Calcaric Regosol	Acidic Fluvisol
Nb samples for analysis 0-0.2m	49	20	4
pH	7.4 (0.1)$^{(1)}$	7,2	4,9
% clay	43 (13)$^{(1)}$	30$^{(2)}$	12
% organic carbon	5.6 (2.1)$^{(1)}$	2.5$^{(2)}$	5,2
Total Zn in mg $kg^{-1(3)}$	245 (179)	645 (81)	1158 (216)
Total Cu in mg kg^{-1}	188 (158)	525 (62)	264 (43)
Total Cd in mg kg^{-1}	1.2 (0.7)	2.0 (0.4)	2.8 (0.7)
Other metals	Ni, W, Hg, Pb	-	As, Hg, Sn, Pb
Soluble Zn in mg $kg^{-1(4)}$	bdl$^{(5)}$	0.08 (0.03)	7.4 (5.9)
Soluble Cu in mg kg^{-1}	bdl$^{(5)}$	0.7 (0.2)	0.4 (0.1)
Soluble Cd in mg kg^{-1}	bdl$^{(5)}$	2.0 (0.6)	13 (11)

$^{(1)}$: 11 samples; $^{(2)}$ 3 samples; $^{(3)}$: Total = Extraction HNO_3 2M [8]
$^{(4)}$: Soluble = Extraction $NaNO_3$ 0.1M [8]; $^{(5)}$: bdl = below detection limit

Table 2. Overview of *Salix viminalis* (clone 78980) results (after Kayser, et al., 2000; Keller, et al., 2003; Hammer, et al., 2003; and Keller et al., unpublished data). Average yield and concentrations of 4 (Dornach and Caslano) or 30 plots (Les Abattes) are given with standard deviations in brackets.

	Les Abattes	Dornach	Caslano
Type of experiment	Forest rehabilitation	Phytoextraction	Phytoextraction
Plant density ha^{-1}	10000	12800 (+20000)	40000
Type of planting	one-yr-old plant	cuttings	cuttings
Yield after 2 years in t DM ha^{-1}	0.9 (0.7)[(1)]	6.5 (3.8)	15.8 (6.0)
Yield after 5 years in t DM ha^{-1}	-	32 (2.5)	-
Yield after 2 years in t DM ha^{-1} corrected for plant density	0,9	5,08	3,95
Concentrations after 2 years in mg k^{-1} DM			
Zn in whole plants	252 (77)	240 (46)[(2)]	785 (202)
Zn in leaves	475 (181)	665 (50)	1700 (465)
Zn in stems	183 (42)	120 (10)	430 (85)
Cu in whole plants	5.2 (0.7)	10 (1)	13 (1)
Cu in leaves	5.6 (0.7)	13 (1)	13 (1)
Cu in stems	5.1 (0.8)	9 (1)	13 (1)
Cd in whole plants	1.3 (0.5)	3.3 (0.5)	2.6 (0.3)
Cd in leaves	1.5 (0.5)	5.1 (0.7)	3.6 (0.7)
Cd in stems	1.2 (0.5)	2.3 (0.1)	2.2 (0.4)

(1): plants not cut the first year

(2): analyses performed on 3-yr-old plants for Zn, Cu and Cd

Table 3. Overview of *Thlaspi caerulescens* results (after Kayser, et al., 2000; Hammer and Keller, 2003; Keller, et al., 2003 and Keller et al., unpublished data). Average yield and concentrations of 4 plots are given with standard deviations in brackets. Ganges (St-Laurent-Le-Minier, France) and Prayon (Belgium) refer to two *T. caerulescens* populations.

	Les Abattes	Dornach	Caslano
Type of experiment	Forest rehabilitation	Phytoextraction	Phytoextraction
Plant density ha^{-1}	none	1000000	1000000
Type of planting	-	plantlets/seeds	plantlets/seeds
Yield for one harvest yr^{-1} in t DM ha^{-1} (sown)		0.6 (0.3)	1.0 (0.7)
Yield for one harvest yr^{-1} in t DM ha^{-1} (planted)		0.9 (0.3)	2.1 (0.2)
Concentrations in shoots of the first harvest (2000, planted, Ganges) in mg kg^{-1} DM			
Zn	-	3419 (335)	9787 (1345)
Cu	-	70 (17)	29 (5)
Cd	-	124 (21)	271 (50)
Concentrations in shoots of the first harvest (1998, planted, Prayon) in mg kg^{-1} DM			
Zn	-	2011 (629)	-
Cu	-	53 (4)	-
Cd	-	6.4 (2.2)	-

3. LIMITING FACTORS: SITE CONDITIONS

3.1. Climatic Conditions and Soil types

3.1.1. The Climate

Climatic conditions (mostly temperature and rain) play a role directly on biomass production and indirectly on metal concentrations (that are influenced by plant health). The results obtained for *S. viminalis* and presented in Fig. 1 are typical of the impact of the climatic conditions on the growth of plants: in a pot experiment no significant difference was measured in the biomass production of *S. viminalis* (clone 78980) grown 3 months in the Les Abattes, Dornach and Caslano soils (Hammer and Keller, 2002; Hammer, 1997), whereas in the field, the yield per plant in Les Abattes was lower due to a lower average temperature and high rainfall (Tables 1 and 2). Also, Van Splunger et al., 1996 have found that drought affects root distribution of Salicaceae by increasing root length and rooting depth and thus their ability to forage and take up nutrients from soil.

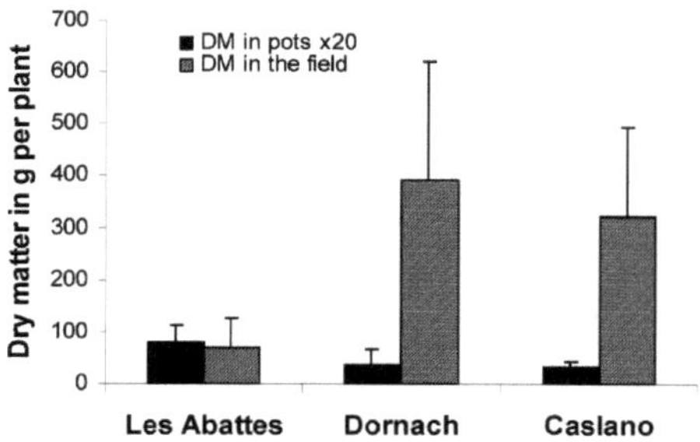

Fig. 1. Comparison of the biomass produced by plants of *Salix viminalis* (clone 78980) grown in 3 different soils in pots (after 3 months) and in the field (second growing season). Ratios between dry biomass produced in pot and in the field are 18, 118, and 190 for Les Abattes, Dornach and Caslano resp. After Hammer and Keller, 2002 and Table 1.

3.1.2. Soil Characteristics

Soil characteristics (pH, % C, % clay, nutrient availability, water table, compaction) are responsible for the general soil fertility and have a large impact on plant biomass and composition (e.g. Li, Y.-M., et al., 2003b). The type of soil also influences the root depth prospecting as shown in Table 4, where root depth may vary from 40 to 450 cm for the same species growing in different soils. For willows, it has been shown that root development varies according to soil texture that drives plant-available soil water and the maximum capillary rise: for example sand retains too little water and silt is usually too compact and oxygen-deficient to allow root growth at depth whereas sandy loam or clay loam allow deep root system (Danfors, et al., 1998).

Like the three sites studied that are located within the boundaries of towns, many contaminated soils in urban areas are fertile, especially kitchen gardens. But unfortunately, many other contaminated sites are, additionally to the contamination, degraded in a broader sense (Genske, 2003): they are poor in nutrients, have low water retention and more generally heterogeneous biological, chemical and physical properties that are not favourable for plant growth and may induce deficiencies Agronomic management, fertilisation and various additives may have to be used intensively to obtain optimal results and thus add to the costs of phytoextraction. Indeed, soil water content was found to be positively correlated to biomass production and metal accumulation in *Thlaspi, Alyssum and Berkheya* in both uncontaminated

Table 4. Examples of average root depth in cm for several tree species and according to soil type (modified from Polomski and Kuhn, 1998).

Soil type	*Acer pseudoplatanus*	*Populus* spp	*Populus tremula*	*Betula pendula*
With skeleton, carbonates, average depth Ex: Rendzina	–	–	–	40-60
Deep, without skeleton, silty clay Ex: Cambisol	70-500	120-140	90-150	100-150
Non hydromorphic sand, sandy soils	110-140	100-260	30- –	150-450
Different soils, slightly hydromorphic	130-140	110	–	–
Hydromorphic soils Ex: Pseudogley/redoxisol	40-50	–	–	90-130
Presence of groundwater table Ex: Gley/reductisol	40	80	–	–

and Ni-contaminated soils (Angle, et al., 2003). Thus irrigation may be necessary to reach an optimal yield. Additionally, whereas some plants proposed for phytoextraction have a high nutrient requirement that must be satisfied to obtain an efficient metal removal (e.g. N for *Salix* (Polomski and Kuhn, 1998; Ericsson, 1981), nitrogen fertilisation has also given positive results with hyperaccumulators, increasing the biomass without decreasing metal concentrations in the plants (Schwartz, et al., 2003; Li, Y.-M., et al., 2003). Iron deficiency may also occur and require the use of an Fe fertilizer to reduce chlorosis as observed on *S. viminalis* and *T. caerulescens* at Dornach and Caslano (Hammer and Keller, 2003; Hammer, et al., 2003).

Although the origin of the contamination influences the speciation of the metals present in the soil, soil characteristics have an impact on metal speciation in both the soil matrix and soil solution. This is best exemplified in unpolluted environments (e.g. Iyengar, et al., 1981; Keller and Védy, 1994) but also occurs in polluted soils (Hammer and Keller, 2002). This has an impact on the availability of metals and ultimately on metal uptake. For example, at Les Abattes, the large C content combined with a high pH and clay content resulted in non detectable soluble Cd, Zn and Cu as compared to the Dornach and Caslano soils (Table 1). Fig. 2 further shows the relation between pH and $NaNO_3$-extractable Zn and Cd (defined as the soluble metal fraction according to the Swiss law (OIS, 1998) in two soils (Kayser, 1999; Kayser, et al., 2001) amended with increasing amounts of elemental sulphur in order to decrease soil pH and consequently metal bioavailability to plants

(Kayser, et al., 2001). The increase in $NaNO_3$-extractable Cd and Zn was more pronounced in the Rafz soil (initial $pH_{CaCl2} = 6.8$, HNO_3-extractable Zn = 813 mg kg^{-1} and Cd = 0.9 mg kg^{-1}) than in the Dornach soil (pH 7.4) with an increase in Rafz of resp. 26- and 13-fold for Cd and Zn in Rafz and in Dornach of 15 and 11-fold with the highest S_8 addition. This illustrates that the changes in soil pH (either artificially induced or through natural acidification) may affect differently metal availability in different soils. Additionally, the effect on Cd and Zn concentrations in *S. viminalis* was 4.6-fold for Zn and 1.6-fold for Cd compared to the control plants grown on the non amended Rafz soil and 2.4- and 1.5-fold for Zn and Cd resp. *in S. viminalis* grown on the Dornach soil (Kayser, et al., 2001): it shows that there is no strict relationship between plant uptake and metal bioavailability assessed by chemical extractants. The question of bioavailability is developed extensively elsewhere and thus will not be further discussed.

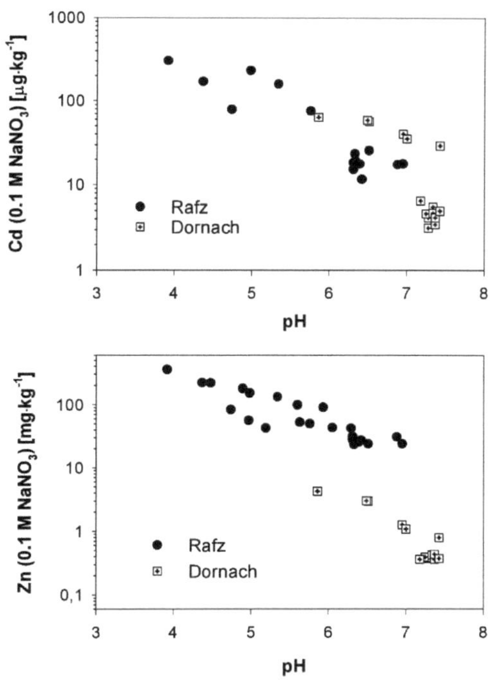

Fig. 2. Concentration of $NaNO_3$-extractable Zn and Cd in two Swiss soils (Rafz and Dornach) as a function of pH following application of increasing amounts of elemental sulphur (S_8) in order to decrease pH in a pot experiment (Kayser, 1999; Kayser, et al., 2001).

3.2. Nature and Extent of the Contamination

3.2.1. Toxicity of the Contaminants

An overall evaluation of the particular contamination issue and site must be made before starting a phytoextraction program. Indeed, sites to be remediated are very seldom contaminated with just single metals. On the contrary, multi-contamination with both organic and inorganic pollutants (and sometimes with physical constraints, such as compacted soil) are common and may induce toxicity to plants that are not tolerant to all the contaminants present. Plants are potentially sensitive to a broad range of elements if their concentrations in the soil are above a given threshold. However, the most toxic metals for both higher plants and certain microorganisms are Hg, Cu, Ni, Pb, Co, Cd and possibly Ag, Be and Sn (Kabata-Pendias, 2001). The effect of organic pollutants (toxicity) on plants and their fate in plants is poorly known. Attempts have been made to use plants to reduce their concentrations in soils. However, their uptake by roots, translocation to shoots, volatilisation and/or degradation seem to be highly plant, soil and component dependent (Siciliano and Germida, 1998; Bhadra, et al., 1999; Burken and Schnoor, 1998). However, it is clear that the presence of herbicide residues may reduce or even prevent plant establishment.

In general, critical limits in plant can be quickly reached when a large amount of a metal contaminant is in a form that is available in the soil for plant uptake, especially when it is combined with a large soil-plant transfer coefficient (Sauerbeck, 1989; Baker and Walker, 1990; Reeves and Baker, 2000) or when the threshold is low. For example, copper is a wide spread contaminant and so far there has been no reported example of successful extraction of Cu from contaminated soil. The remediation efforts are mostly directed towards Cu immobilization through the use of additives (e.g. McBride and Martinez, 2000). However its toxicity has been found to reduce phytoextraction efficiency because for most of the plants, the toxic level is already reached when the Cu concentration in the plant is above 15-20 mg kg^{-1} DM (Sauerbeck, 1989) or when the concentration in the soil solution reaches 0.02 to 0.06 mg L^{-1} (Pahlsson, 1989). This is also valid for hyperaccumulators like *T. caerulescens* that is known to be sensitive to copper (McLaughlin and Henderson, 1999). As a result of Cu toxicity, the biomass is reduced and the metal uptake is impaired (McLaughlin and Henderson, 1999; Ebbs and Kochian, 1997; Lombi, et al., 2001). Thus, soils with undesirable contaminants might not be easily decontaminated and a soil pre-treatment might be necessary to remove or inhibit the toxic compounds. Additionally, metals may have antagonistic or synergic effects on their toxicity and uptake as demonstrated for Cd by Costa and Morel (Costa and Morel, 1994).

3.2.2. Limited Availability of Metals

From a different perspective, low availability of a metal to be remediated may reduce the potential for its phytoextraction (Tables 1, 2, and 3). Although the metal available fraction in soil may be determined by different methods, it is indeed usually better correlated to plant uptake than total metal content in soil. For example, Table 5 shows that the correlation between plant uptake and $NaNO_3$-extractable soil Cd (soluble) is always better than the correlation between plant uptake and HNO_3-extractable ("pseudo-total") soil Cd. In Fig. 3, the low Cd uptake by *S. viminalis* is explained by the low soluble (s) Cd content in the soil compared to hydroponics that show the uptake potential by this plant: similar Cd concentrations in solution in both cases give similar Cd concentrations in the shoots of *S. viminalis* (clone 78980). Additionally it shows that the higher Cd soluble content in the Caslano soil is not responsible for the lower biomass production in this soil compared to the Dornach soil.

Table 5. Pearson correlation coefficients of the log-transformed concentrations of Cd measured in the dry matter of plants grown successively on the same soil that has been amended with increasing amounts of sludge and HNO_3-extracts resp. $NaNO_3$-extracts of a soil ($n = 20$),***$p < 0.001$, ** $p < 0.01$ (modified from Keller, et al., 2000).

	$NaNO_3$	HNO_3
Maize leaves	0.84***	
	0.77***	
Maize cobs	0.84***	
	0.79***	
Sugar-beet leaves	0.89***	
	0.85***	
Sugar-beet roots	0.89***	
	0.81***	
Potato leaves	0.97***	
	0.80***	
Potato tubers	0.74***	
	0.64**	
Lettuce	0.90***	
	0.75***	
Spinach	0.76***	0.39

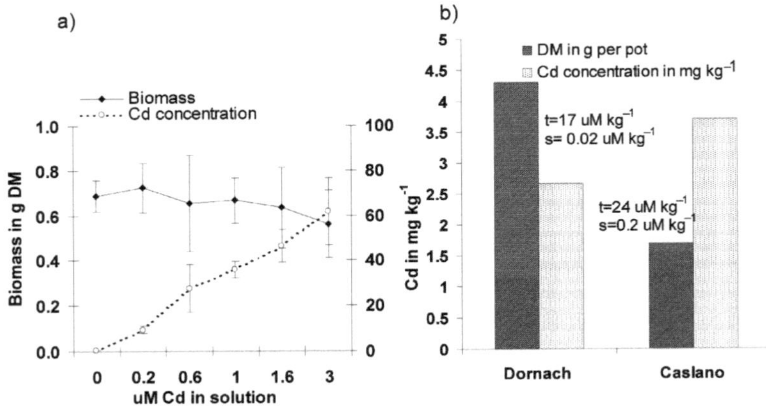

Fig. 3. Cadmium concentration and biomass production: comparison between a) *Salix viminalis* (clone 78980) grown in hydroponics with increasing Cd concentrations and b) the same clone grown in pots in the Dornach and Caslano soils. S = $NaNO_3$-extractable Cd (soluble) and t = HNO_3-extractable (pseudo-total). In Fig. 3b, the scale is for both the biomass (DM in g) and the Cd concentration (in mg kg^{-1}). After Hammer and Keller, 2002; Keller, et al., unpublished data.

A low metal uptake due to low metal availability increases the number of years necessary for metal removal. It may lead to the use of additives to increase bioavailability (Kayser, et al., 2000; Cooper, et al., 1999; Elkhatib, et al., 2001; Wenger, et al., 2002). EDTA, NTA, sulphur, and other chelating agents have been tested, but results have not been conclusive so far. EDTA is often preferred because it enables large metal mobilisation, especially for Pb phytoextraction (Cui,Y., et al., 2003; Liphadzi, et al., 2003; Huang, et al., 1997). However, the main drawback is its persistence and the risk of groundwater contamination through uncontrolled leaching (Wenzel, et al., 2003) and high cost. Its use should therefore be limited where leaching can be controlled or prevented and thus it may not be applicable at any site.

3.2.3. Heterogeneity of the Contamination

Contamination is highly heterogeneous at polluted sites: this includes spatial variation in composition and concentration (Genske, 2003; as well as variation with depth (Keller, et al., 2003).

An example of surface and at depth heterogeneity is displayed on Fig. 4 for Zn soil concentration at Dornach. The soil, contaminated from atmospheric deposition, has undergone disturbances that have resulted in a contaminated layer with a thickness varying between 0.2 and 0.7 m within a 400 m^2 area. In that case high heterogeneity prevents an optimal root colonisation of the contaminated layer (See Table 9, first column) because phytoextraction can only be achieved when roots are within the contaminated layer. In fact, various cases may occur that prevent an optimal uptake. The question of root colonisation will be developed in Section 3.

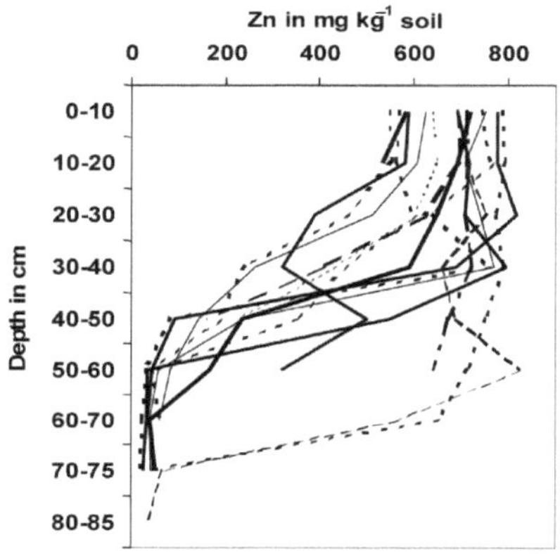

Fig. 4. Profiles of HNO_3-extractable Zn at Dornach. The 15 profiles have been collected within a 20x20m area with the Humax technique that allows for sampling of undisturbed soil cores.

4. LIMITING FACTORS: PLANT CHARACTERISTICS

The choice of the species must take into account the metals to be removed, the other contaminants present, the soil and the climate characteristics as well as the necessity of deriving a benefit from the land during the process of phytoextraction, and at last the final disposal method of the plants. However, the choice of the plant(s) will be always a compromise because there is no perfect plant for a given situation. Ideally the plant should be contaminant tolerant, adapted to the site conditions, quickly produce a large biomass with high metal concentrations in the shoots, be easy to manage, dispose of and/or recycle, and able to be grown continually or repeatedly. The use of the whole plant or parts of it for an additional profit (i.e. the use of fibres from lax or seeds from sunflowers) would be beneficial in reducing the cost of phytoextraction. There are no plants available that possess all these characteristics but three groups of plants are possible contenders: hyperaccumulators, crop plants and bioenergy plants. The latter two groups accumulate lesser concentration of metals in tissues but they are much higher yielding plants. The choice will depend on the final total extraction obtained multiplying the yield with the metal concentration in the shoots. This will require preliminary studies in order to estimate these two values.

4.1. The Hyperaccumulator Plants

Hyperaccumulators are metal specific and are usually wild plants with a small biomass and a slow growth under temperate climate. Their agronomic requirements are poorly known and this may result in large variations of yield and metal concentrations variations annually and in different soils. Thus, there may be poor control of the efficiency of the phytoextraction process. Table 3 gives the results obtained with *T. caerulescens* at Dornach and Caslano and Table 6 gives additional information on the variability of the results obtained with *T. caerulescens* at Dornach. It appeared that metal concentrations varied up to a factor of 30 depending on the population, the method used (planted or sown) and the time of the year. Presently, there are few species available for field application (lack of nurseries or seed producers) and within these species very few cultivars or populations so that the use of hyperaccumulators might have to be dismissed for remediation of large areas. Furthermore, methods to harvest the plants have to be adapted from crop species techniques and might not be always straightforward.

4.2. High Biomass Plants

The crop and energy plants have opposite characteristics to hyperaccumulators: they are well known plants with large yields and fast growth, but they are usually not metal specific and they accumulate below the hyperaccumulation thresholds. They have been selected for rather rich and fertilized agricultural soils and may thus not adapt easily to contaminated sites (see above). There exists a large choice of species and cultivars that can be tested according to site specifications and metal tolerance. However, the same management as for the usual production may not be efficient for phytoextraction. For example, plants used for bioenergy, like *Salix* spp., are perennial plants that are left 20 to 30 years on site. From data obtained at Dornach, Cd and Zn concentrations measured in both leaves and stems of *S. viminalis* decreased with time (Hammer, et al., 2003) most probably because roots had progressively extended downwards, outside the contaminated layer (Keller, et al., 2003) thus reducing the extraction efficiency with time. It may be necessary to remove the plants every 2 or 3 years in order to keep an optimal extraction.

Table 6. Variations of the Cd, Cu and Zn concentrations in shoots of *Thlaspi caerulescens* grown at Dornach, according to the population, the technique of propagation, the year and the time of the year or position within the rotation (after Hammer and Keller, 2003 and unpublished data).

Year Rotation	**Population**	**Method**	**Concentrations in mg kg^{-1} in DM** Cd	Cu	Zn	**Variability**
1997	Prayon	planted	7.1 (0.5)	75 (9)	2051 (315)	
1998	Prayon	planted	6.3 (2.2)	53 (4)	2011 (629)	
1999	Ganges	planted	184 (38)	53 (3)	5265 (612)	
2000/1	Ganges	planted	124 (21)	70 (17)	3419 (335)	
2000/2	Ganges	planted	67 (38)	91 (26)	2646 (1296)	
2000/3	Ganges	planted	49 (6)	43 (1)	3539 (313)	
2000/1	Ganges	sown	127 (47)	118 (17)	2551 (327)	
2000/2	Ganges	sown	77 (15)	52 (10)	3851 (558)	
2001	Ganges	sown	76 (24)	81 (2)	3836 (332)	
Factor Ganges 1999/Prayon 1998			*29*	*1.0*	*2.6*	*Inter-populations*
Factor Ganges 2000/1 /Ganges 2000/3			*2.5*	*1.6*	*1.0*	*Inter-harvests*
Factor Ganges 2000/1p /Ganges 2000/1s			*1.0*	*0.6*	*1.3*	*Inter-methods*

Additionally, as for hyperaccumulators (Table 6), the choice of the planting technique and crop management may have a long term effect on phytoextraction efficiency (Table 7): at Les Abattes, the field management was conducted as a forest management with limited weed control and both cuttings and one-year-old plants (1m height) were planted. At Dornach and Caslano only cuttings were used and weeds were systematically removed. In the first case cuttings did not survived whereas in the second one a good survival percentage was obtained and maintained.

Table 7. Percentage survival of cuttings and one-yr-old plants of *Salix viminalis* according to the field management at the three field experimental sites.

Site	Field management	Type of plantation	Months after plantation 5	18	30	42
Les Abattes	grass cut	cuttings	84	71	-	11
		1-yr-old plants	99	-	99	83
Dornach	hand weeding	cuttings	88	88	88	88
Caslano	hand weeding	cuttings	100	-	-	80

4.3. Concentrations in Plant Parts

Both hyperaccumulators and high biomass plants may be more or less tolerant to a range of contaminants depending on the contaminant, the species and even the cultivars or clones. This is particularly evident for *Salix* spp. that presents a large plasticity in Cd tolerance and accumulation (Landberg and Greger, 1996). But some critical limits may also be quickly reached for some elements like boron, which has a narrow deficiency/toxicity range varying widely between plant species (Kabata-Pendias, 2001).

In addition to having various accumulation capabilities (not detailed here), plants do not react the same way to increasing concentrations in the soil (Baker, 1981). In general their efficiency will decrease with increasing soil concentrations as shown by a number of crop plants in Fig. 5. Also, the bioaccumulation factor (shoot/soil concentration ratio) calculated for *T. caerulescens* decreased with increasing soil concentration for both Cd and Zn regardless of the population used (Zhao, et al., 2003).

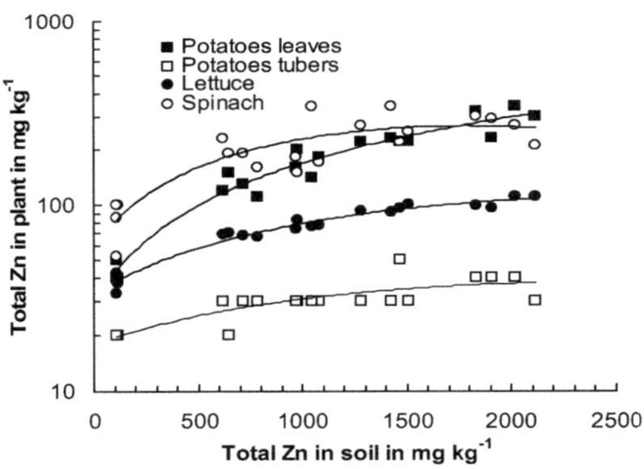

Fig. 5. Relationship between HNO_3-extractable Zn in the soil and Zn concentrations in 3 plants grown in mini-plots amended with increasing metal concentrations (Salez experiment described in (Keller, et al., 2000).

Another point is the distribution of metals inside the plants: metal concentration in the plant varies with the organ and the age of this organ for both hyperaccumulators and non-hyperaccumulators (Sauerbeck, 1989; Sander and Ericsson, 1998; Perronnet, et al., 2003). This has consequences on the time of the harvest and the parts to be harvested. In Fig. 6 concentrations of Cd and Zn are larger in leaves of *S. viminalis* than in stems or roots (with one exception). It means that, while only stems are harvested for energy production, leaves will have also to be collected if *Salix* is used for phytoextraction. Another difficulty is the extrapolation of the results from one metal or one soil to another: on Fig. 6 the ratios calculated between concentrations in leaves, stems and roots vary with the metal and the soil. Pot experiments may thus be necessary prior applying phytoextraction to a full scale field experiment.

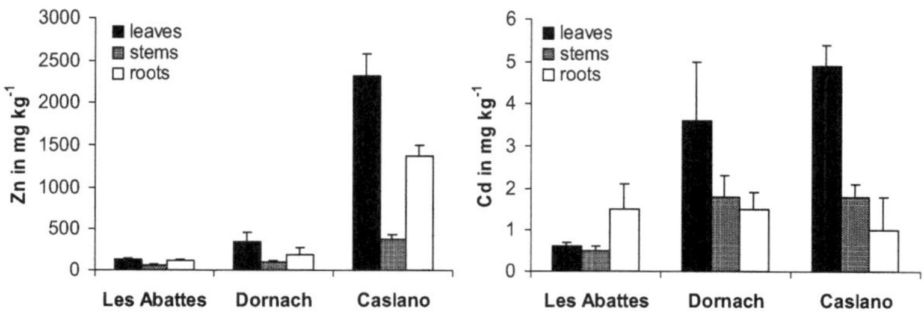

Fig. 6. Cd and Zn concentrations in the different parts of *Salix viminalis* (clone 78980) grown during 3 months in pots in three different soils. After Hammer and Keller, 2002 and Hammer et al., unpublished data.

For the hyperaccumulator *T. caerulescens* it has also been shown that most of the Cd and Zn in shoots are soluble in water and as a consequence may increase the bioavailable metal pool in soils if leaves are left on the ground (Hammer and Keller, 2003; Perronnet, et al., 2000). Owing to the high metal concentrations in *T. caerulescens* shoots, a thorough removal of the leaves (and probably also roots) will have to be performed.

4.4. Root Prospection

As demonstrated above, the extent of the root systems varies with the site conditions. However, for a given soil, plants develop very different root systems (Polomski and Kuhn, 1998) as well as different root:shoot ratios. Such characteristics have been measured at Dornach for various species and the results are presented in Table 8. Maize had the largest root length density and thus was able to prospect efficiently the whole soil profile whereas *T. caerulescens* had the largest root:shoot ratio (together with a relative larger proportion of fine roots) expressing an ability for increased element uptake (Keller, et al., 2003).

Table 8. Cumulative root density (L_A) 0 - 0.2 m depth and L_A 0 - 0.20 m / shoot biomass ratio calculated for 5 crops grown at Dornach (after Keller, et al., 2003). Standard deviations are in brackets.

Plants	I. Mustard	Tobacco	Maize	Willows[(1)]	*Thlaspi*
Zn concentrations in shoots in mg kg^{-1}	124 (13)	146 (21)	129 (30)	294 (38)	5265 (612)
Cu concentrations in shoots in mg kg^{-1}	20 (2)	38 (6)	10 (1)	14 (1)	53 (3)
Cd concentrations in shoots in mg kg^{-1}	1.0 (0.1)	3.5 (0.8)	0.6 (0.2)	3.4 (0.4)	184 (38)
Cumulative root density L_A 0-0.2 m depth in km m^{-2}	2.7 (0.7)	3.7 (0.8)	6.6 (1.8)	3.5 (0.9)	2.6 (1.1)
L_A 0-0.20 m / shoot biomass	4	3	4	3	28

[(1)] 2- and 3-year-old plants

Root colonisation of the soil is mostly driven by plant type and soil characteristics and may not match properly the metal distribution. Fig. 7 displays several of the possible situations that lead to optimal (I) or sub-optimal (II) metal extraction. Because of the localisation of the contaminated layer and the size of their root system plants may either 1) not colonize the whole layer, 2) not be able to reach the contaminated layer if it is at depth, 3) grow deeper than the contaminated layer or 4) avoid the contaminated spots (Breckle and Kahle, 1992; Schnoor, et al., 1995). On the other hand, *T. caerulescens* actively colonizes specific metal hot spots (Schwartz, et al., 1999) and root systems may be manipulated to a certain extent to force the root to reach the contaminated layer (Chappell, 1998).

Practically it may result in a decrease in extraction efficiency as shown in Table 9: for each miniplot soil cores were taken down to 0.75 m and metal concentrations of the soil were measured every 0.10 m as well as total root length for the same layer. A coefficient taking into account the depth of contamination and the maximal root depth (with root length density > 5000 mm dm^3) was calculated. The same calculation was made for a hypothetical steady shallow (0.20 m) or deep 0.70 m) contamination. In general *T. caerulescens* fitted better a shallow contamination because of its shallow root system, whereas maize and willows were more suitable when the contaminated layer was thick. In the specific case of Dornach, none of the species was optimal and either a mix culture or a rotation with plants with various root colonisation would be best. These results can probably be extended to other contaminated sites and emphasize the necessary of an appropriate and site-specific plant management.

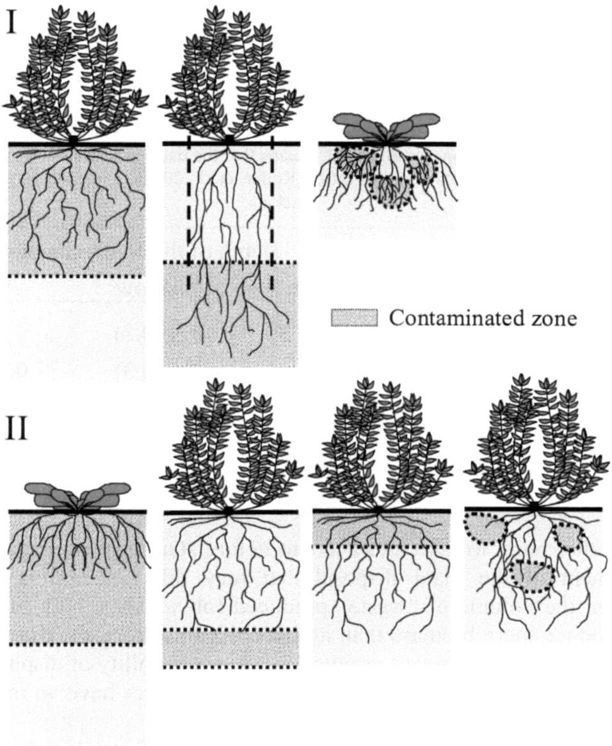

Fig. 6. Various possible combinations of root distribution and contaminated layers. I Optimal situations where the root system develops within (and only within) the contaminated layer and II Situations where the root system distribution does not match the contamination. *Salix* and *Thlaspi* have been taken as example as reported in the literature.

Additionally, total root length in the contaminated layer (and thus the extent of the root prospecting), and the root length/shoot biomass ratio may have an impact on metal concentrations in the biomass (Table 8). At Dornach, *T. caerulescens* (Ganges) had a distinctly larger L_A/ shoot biomass ratio in spite of having the lowest cumulative root density. Metal concentrations in its above ground biomass were also well above concentrations measured in the other crops. Within a single species (maize), Van Noordwijk et al., found a positive correlation between Cd uptake and root length.

Table 9. Matching of roots and Zn soil contamination expressed as the ratio between the maximal depth with a root length density > 5000 mm dm^3 and the soil contamination depth according three scenarios: contamination as measured in the soil profile of each plot, hypothetical shallow contamination (0.2 m), and hypothetical deep contamination (0.7 m). b) and c) depths are extreme values found in the field experiment at Dornach. Depth of contamination is calculated after removing 150 mg kg^{-1} Zn (Swiss guide value (OIS,1998) from the total soil Zn concentration. ">" means that the deepest root sampling gave a root length density above 5000 mm dm^3 (after (Keller, et al., 2003). Standard deviations are in brackets.

	Depth of the contamination		
	True	Shallow	Deep
Indian mustard	0.9 (0.3)	1.9 (0.8)	>0.5
Tobacco	0.5 (0.2)	1.3 (0.3)	0.4 (0.1)
Maize	1.4 (0.6)	>3.3	>0.9
Willows	1.5 (0.7)	>3.2	0.9 (0.2)
Thlaspi (Ganges)	0.4 (0.1)	1.0 (0.0)	0.3 (0.0)

Another point to take into account before starting phytoextraction, is an assessment of the plant disposal that may differ according to the site location, the volume of "waste" produced (obviously a high biomass plant will produce more biomass than an hyperaccumulator), the time of the year and duration of the biomass production, the accessibility of disposal facilities and the legislation (these parameters may themselves have an impact on the plant species to be chosen).

Finally, the choice of the species (one or several) needs a thorough study of the plant's potential and its suitability for the site. As it has been shown, the results are not easily predictable and pre-experiments in pots or even in mini-plots will be necessary before applying phytoextraction to the site.

5. LEGISLATION AND TIME REQUIRED

Models have been proposed to evaluate the time needed to decontaminate a given soil with a given plant (Kayser, et al., 2000; Zhao, et al., 2003; Robinson, et al., 2003; Klang-Westin and Eriksson, 2003). This time varies with the target value, the depth of contamination but also with the type of function (linear or exponential) used to describe the decrease of metal concentrations in soil. Indeed, both concentrations in the plants and yields may not be constant with time and efficiency may thus decrease because the available pools decrease but also because the roots are going deeper or because perennial plants need a certain time to establish themselves.

Unfortunately there are not enough long-term data sets available to validate either of these models.

The target values are also highly variable and depend mostly on legislations and the perception of a clean (or uncontaminated) soil. Some countries have already well developed legislations that give a framework for soil remediation, others do not. Table 10 gives the guidelines for Cd in three European countries: thresholds are different and based on different chemical extractions. From these data it is obvious that in the Netherlands the clean up of a soil below the Cd intervention value will take more time than the same procedure in Switzerland, and phytoextraction may be an option in the second case, while it might take too long in the first one. It has also been proposed to use phytoextraction as a stripping technique to remove the bioavailable pool only. Whereas the extent of the replenishment of the metal available pool is difficult to assess in the first place, it appears also that in some countries like Switzerland, both total and soluble contents have to be reduced below a given threshold and the BSM approach may just not be suitable for this reason.

Additionally to the total or soluble metal content, targets values can theoretically be based on the sum of metals, on functional criteria and toxicological aspects, exposure pathways or the idea of restoring soil multi-functionality. This will lead to a large range of concentrations that may have an impact on the technical and economical feasibility of phytoextraction.

Table 10. Three European policies for cadmium in soil (modified from Adriano, 2001).

Switzerland	Germany	Netherlands
Cleanup values	**Action values**	**Intervention values**
2M HNO_3 agricultural use 30 ppm private use 20 ppm children's playground 20 ppm 0.1 M $NaNO_3$ 0.1 ppm	"total" aqua regia children's playground in prep.	"total" aqua regia 12 ppm
Trigger values	**Trigger values**	**Intermediate values**
food crops 2 ppm fodder crops 2 ppm oral/inhaled/dermal uptake 10 ppm "soluble" 0.1 M $NaNO_3$ food crops 0.2 ppm fodder crops 0.2 ppm oral/inhaled/dermal uptake -	"total" aqua regia agricultural use 1-2 ppm children's playground 6 ppm residential areas 12 ppm parks 30 ppm industrial areas 60 ppm	"total" aqua regia 6.4 ppm food crops **Signal values** "total" aqua regia food crops 0.5-1.0 ppm grassland 2-3 ppm fodder crops 0.5-1.0 ppm ornemental plants 5-10 ppm
Guide values	**Preventive values**	**Target values**
"total" 2M HNO_3 0.8 ppm "soluble" 0.1 M $NaNO_3$ 0.03 ppm	(reference values derived from background values) various extractants 0.3-1.5 ppm	"total" aqua regia +/-0.8 ppm

So far, calculations made from preliminary experiments or, at most, from a few years field experiments have led to various results (results obtained with addition of large amounts of chelating agents like EDTA are not taken into account), predicting either remediation within few years, decades or centuries (Kayser, et al., 2000; Hammer and Keller, 2003; Hammer, et al., 2003; Klang-Westin and Eriksson, 2003; Robinson, et al., 2003). Obviously, all the factors presented above added to obtain such a large range of results. It emphasises again the necessity to choose carefully the sites to be remediated by phytoextraction as well as the planning of the procedure to be applied.

6. CONCLUSION

Phytoextraction has a great potential for cleaning soils contaminated with heavy metals, especially in cases where conventional technologies are not efficient, not possible or too expensive. From the early results obtained in the lab and in the field it is clear that phytoextraction could be applied either as the main remediation tool or more likely as a "polishing" technique combined with other conventional and "bio" techniques. There is obviously no single phytoextraction technique and each site will need a tailor-made scheme. The main difficulty in doing so is the optimisation of the efficiency. Indeed, it is difficult to spot the limiting factors and even more to try to find a hierarchy among them because of their large number and their interdependency. In order to help practitioners choose or discard phytoextraction technology, decision trees have been proposed (see for example the US-EPA phytoremediation decision trees at: www.clu-in.org/techdrct/techpubs.asp), which include some of these factors. However once phytoextraction has been selected as the appropriate technique, much effort will have to be devoted to reach a full control of the technique and to obtain a maximal efficiency, in particular on the long term. Research has a key position on that aspect and is responsible for the successful development of phytoextraction.

7. ACKNOWLEDGEMENT

This research was part of the project n°ENV4-CT97-0598/n°OFES97.0362 funded by the Swiss Federal Agency for Teaching and Research and was also financed by the Swiss Federal Institute of Technology, Zürich. I thank all the people involved in these researches including D. Hammer (EPFL), C. Cosio (EPFL), D. Vainstein (EPFL), and A. Kayser (ETHZ) for their work on the projects, U. Zimmermann, A. Grünwald, J.-D. Teuscher, E. Rossel, and K. Truninger (EPFL and ETHZ) for the help in plants and soil analyses, W.

Attinger (ETHZ) for the Dornach field management, R. Schulin (ETHZ) for the scientific support and for allowing the work in Dornach, and V. Barbezat and W. Rosselli (Snow, Forest and Landscape Institute) for the management of Les Abattes field experiment. Maria Greger, University of Stockholm, Sweden, provided the Salix viminalis cuttings (clone 78980).

REFERENCES

1. Adriano, D.C. (2001) Trace elements in terrestrial environments, 2nd edition, Springer-Verlag New York Inc., NY, 867p.
2. Angle, J.S., Baker, A.J.M., Whiting, S.N., Chaney, R.L. (2003) Soil moisture effects on uptake of metals by Thlaspi, Alyssum, and Berkheya, Plant Soil 256, 325-332.
3. Baker, A.J.M. (1981) Accumulators and excluders - strategies in the response of plants to heavy metals, J. Plant. Nutr. 3, 643-654.
4. Baker, A.J.M., and Walker, P.L. (1990) Ecophysiology of metal uptake by tolerant plants in A.J. Shaw (ed.), Heavy metal tolerance in plants: evolutionary aspects, CRC Press Inc., Boca Raton, Fl. pp. 155-177.
5. Bhadra, R., Spanggord, R.J., Wayment, D.G., Hughes, J.B., and Shanks, J.V. (1999) Characterization of oxidation products of TNT metabolism in aquatic phytoremediation systems of Myriophyllum aquaticum, Environ. Sci. Technol. 33, 3354-3361.
6. Breckle, S-W., and Kahle, H. (1992) Effect of toxic heavy metals (Cd, Pb) on growth and mineral nutrition of beech (Fagus sylvatica L.), Vegetatio 101, 43-53.
7. Burken, J.G., and Schnoor, J.L. (1998) Predictive relationships for uptake of organic contaminants by hybrid poplar trees, Environ. Sci. Technol. 32, 3379-3385.
8. Chappell, J. (1998) Phytoremediation of TCE in Groundwater using Populus. Washington, D.C.: Technology Innovation Office, Office of Solid Waste and Emergency Response.
9. Cooper, E.M., Sims, J.T., Cunningham, S.D., Huang, J.W. and Berti, W.R. (1999) Chelate-assisted phytoextraction of lead from contaminated soils, J. Environ. Qual. 28, 1709-1719.
10. Costa, G., and Morel, J.-L. (1994) Cadmium uptake by Lupinus albus (L.): cadmium excretion, a possible mechanism of cadmium tolerance, J. Plant Nutr. 16, 1921-1929.
11. Cui, Y., Wang, Q., Dong, Y., Li, H., and Christie, P. (2003) Enhanced uptake of soil Pb and Zn by Indian mustard and winter wheat following combined soil application of elemental sulphur and EDTA, Plant Soil, in press.
12. Danfors, B., Ledin, S., and Rosenqvist, H. (1998) Short-rotation willow coppice – growers' manual, Swedish institute of agricultural engineering Uppsala.
13. Dickinson, N.M. (2000.) Phytoremediation of industrially-contaminated sites using trees. NATO Series…
14. Dubois, J.P. (1991) Analysis report concerning the metal concentration in the soil of the landfill "Les Abattes" (City of Le Locle), Swiss Federal Institute of Technology, Lausanne (in French), unpublished.
15. Ebbs, S.D., Kochian, L.V. (1997) Toxicity of zinc and copper to Brassica species: implication for phytoremediation. J. Environ. Qual. 26, 776-781.
16. Elkhatib, E.A., Thabet, A.G., and Mahdy, A.M. (2001) Phytoremediation of cadmium contaminated soils: role of organic complexing agents in cadmium phytoextraction, Land Contamination and Reclamation 9, 359-366.
17. Ericsson, E. (1981) Effect of varied nitrogen stress on growth and nutrition in three Salix clones, Physiol. Plant. 51, 423-429.

18. FAC (Eidgenössische Forschungsanstalt für Agrikulturchemie und Umwelthygiene) (1989) Methoden für die Bodenuntersuchungen. Schriftenreihe der FAC 5, Bern-Liebefeld, Switzerland. 267p.
19. Geiger, G., Federer, P., and Sticher, H. (1993) Reclamation of heavy metal-contaminated soils: field studies and germination experiments, J. Environ. Qual. 22, 201-207.
20. Genske, D.D. (2003) Urban land: Degradation – Investigation – Remediation, Springer-Verlag, Berlin, 331p.
21. Hammer, D. (1997) Schwermetallakkumulation durch Weidenstecklinge auf vier Schweizer Böden. Diploma Thesis ETHZ, unpublished.
22. Hammer, D., and Keller, C. (2002) Changes in the rhizosphere of heavy metal accumulating plants as evidenced by chemical extractants, J. Environ. Qual. 31, 1561-1569.
23. Hammer, D., and Keller, C. (2003) Phytoextraction of Cd and Zn with Thlaspi caerulescens in field trials, Soil Use and Management 19, 144-149.
24. Hammer, D., Kayser, A., and Keller, C. (2003) Phytoextraction of Cd and Zn with Salix viminalis in field trials, Soil Use and Management 19, 187-192.
25. Hamon, R. ref of the NATO book?
26. Huang, J.W., Chen, J., and Cunningham, S.D. (1997) Phytoextraction of lead from contaminated soils. in Kruger, E.L. et al. (eds) Phytoremediation of Soils and Water Contaminants. American Chemical Society, Washington. pp. 283-298.
27. Iyengar, S.S., Martens, D.C., and Miller, W.P. (1981) Distribution and plant availability of soil zinc, Soil Sci. Soc. Am. J. 45, 735-739.
28. Kabata-Pendias, A. (2001) Trace elements in soils and plants, 3rd edition. CRC Press, London, UK. 413p.
29. Kayser, A. (1999) Evaluation and enhancement of phytoextraction of heavy metals from contaminated soils, dissertation submitted to the Swiss federal institute of technology Zürich, Diss ETH No. 13563.
30. Kayser, A., Schröder, T.J., Grünwald, A., and Schulin, R. (2001) Solubilization and plant uptake of zinc and cadmium from soils treated with elemental sulfur, Internat. J. Phytorem. 3, 381-400.
31. Kayser, A., Wenger, K., Keller, A., Attinger, W., Felix, H.R., Gupta, S.K., and Schulin, R. (2000) Enhancement of phytoextraction of Zn, Cd and Cu from calcareous soil: the use of NTA and sulphur amendments, Environ. Sci. Technol. 34, 1778-1783.
32. Keller, C., and Védy, J.-C. (1994) Distribution of Cu and Cd fractions in two forested soils, J. Environ. Qual. 23, 987-999.
33. Keller, C., Hammer, D., Kayser, A., Richner, W., Brodbeck, M., and Sennhauser, M. (2003) Root development and heavy metal phytoextraction efficiency: comparison of different plant species, Plant Soil 249, 67-81.
34. Keller, C., Kayser, A., Keller, A., and Schulin, R. (2000) Heavy metal uptake by agricultural crops from sewage-sludge treated soils of the Upper Swiss Rhine valley and the effect of time, in I.K. Iskandar (ed), Environmental Restoration of metals-contaminated soils, CRC Press LLC, Boca Raton, FL. pp. 273-291.
35. Klang-Westin, E., and Eriksson, J. (2003) Potential of Salix as phytoextractor for Cd on moderately contaminated soils, Plant Soil 249, 127-137.
36. Landberg, T., and Greger, M. (1996) Differences in uptake and tolerance to heavy metals in Salix from polluted and unpolluted areas, Applied Geochemistry 11, 175-180.
37. Li, Y.-M., Chaney, R.L., Brewer, E.P., Angle, J.S., Nelkin, J. (2003b) Phytoextraction of nickel and cobalt by hyperaccumulator Alyssum species grown on nickel-contaminated soils, Environ. Sci. Technol. 37, 1463-1468.
38. Li, Y.-M., Chaney, R.L., Brewer, E.P., Roseberg, R., Angle, J.S., Baker, A.J.M., Reeves, R., and Nelkin, J. (2003b) Development of a technology for commercial phytoextraction of nickel: economic and technical considerations, Plant Soil 249, 107-115.

39. Liphadzi, M.S., Kirkham, M.B., Mankin, K.R., and Paulsen, G.M. (2003) EDTA-assisted heavy-metal uptake by poplar and sunflower grown at a long-term sewage-sludge farm, Plant Soil, in press.
40. Lombi, E., Zhao, F.J., Dunham, S.J., McGrath, S.P. (2001) Phytoremediation of heavy-metal-contaminated soils: natural hyperaccumulation versus chemically enhanced phytoextraction. J. Environ. Qual. 30, 1919-1926.
41. McBride, M.B., and Martinez, C.E. (2000) Copper phytotoxicity in a contaminated soil: remediation tests with adsorptive materials, Environ. Sci. Technol. 34, 4386-4391.
42. McLaughlin, M.J., and Henderson, R. (1999) Effect of zinc and copper on cadmium uptake by Thlaspi caerulescens and Cardaminopsis halleri, in extended abstracts of the 5th International Conference on the Biogeochemistry of Trace Elements (ICOBTE). 12-15 July 1999, Vienna, Austria. pp. 886-887.
43. OIS (1998) Ordinance relating to impacts on the soil. Collection of Swiss Federal Legislation SR 814.12.
44. Pahlsson, A.-M.B., 1989. Toxicity of heavy metals (Zn, Cu, Cd, Pb) to vascular plants. Water Air Soil Pollution 47, 287-319.
45. Perronnet, K., Schwartz, C., and Morel, J.L. (2003) Distribution of cadmium and zinc in the hyperaccumulator Thlaspi caerulescens grown on multicontaminated soil, Plant Soil 249, 19-25.
46. Perronnet, K., Schwartz, C., Gérard, E., and Morel, J.L. (2000) Availability of cadmium and zinc accumulated in the leaves of Thlaspi caerulescens incorporated into soil, Plant Soil 227, 257-263.
47. Polomski, J., and Kuhn, N. (1998) Wurzelsysteme, Eidgenössische Forschungsanstalt für Wald, Schnee und Landschaft, Birmensdorf. Paul Haupt, Bern, Switzerland. 290 p.
48. Reeves, R.D., and Baker, A.J.M. (2000) Metal-accumulating plants, in I. Raskin and B.D. Ensley (eds.), Phytoremediation of toxic metals using plants to clean up the environment, John Wiley & Sons, Inc. New-York, NY. pp. 193-230.
49. Robinson, B.H., Fernandez, J.-E., Madejon, P., Maranon, T., Murillo, J.M., Green, S., and Clothier, B. (2003) Phytoextraction: an assessment of biogeochemical and economic viability, Plant Soil 249, 117-125.
50. Rosselli, W., Keller, C., and Boschi, K. (2003) Phytoextraction capacity of trees growing on a metal-contaminated soil, Plant Soil 256, 265-272.
51. Sander, M.-L., and Ericsson, T. (1998) Vertical distribution of plant nutrients and heavy metals in Salix viminalis stems and their implications for sampling, Biomass Bioenergy 14, 57-66.
52. Sauerbeck, D. (1989) Der Transfer von Schwermetallen in die Pflanze, in Beurteilung von Schwermetallkontaminationen im Boden. DECHEMA. Fachgespräche Umweltschutz. Stuttgart am Mainz, Germany. pp. 281-316.
53. Schnoor, J.L., Licht, L.A., McCutcheon, S.C., Wolfe, N.L., and Carreira, L.H. (1995) Phytoremediation of organic and nutrient contaminants, Environ. Sci. Technol. 29, 318A-323A.
54. Schwartz, C., Echevarria, G., Morel, J.-L. (2003) Phytoextraction of cadmium with Thlaspi caerulescens, Plant Soil 249, 27-35.
55. Schwartz, C., Morel, J.-L., Saumier, S., Whiting, S.N., and Baker, A.J.M. (1999) Root architecture of the Zn-hyperaccumulator plant Thlaspi caerulescens as affected by metal origin, content and localisation in soil, Plant Soil 208, 103-115.
56. Siciliano, S.D., and Germida, J.J. (1998) Degradation of chlorinated benzoic acid mixtures by plant-bacteria associations, Environ. Toxicol. Chem. 17, 728-733.
57. van Noordwijk, M., van Driel, W., Brouwer, G., and Schuurmans, W. (1995) Heavy-metal uptake by crops from polluted river sediments covered by non-polluted topsoil, Plant Soil 175, 105-113.
58. Van Splunder, I., Voesenek, L.A.C.J., Coops, H., De Vries, X.J.A., and Blom, C.W.P.M. (1996) Morphological responses of seedlings of four species of Salicaceae to drought, Can. J. Bot. 74, 1988-1995.

59. Vangronsveld, J., and Cunningham, S.D. (1998) Introduction to concepts in J. Vangronsveld and S.D. Cunningham (eds), Metal-contaminated soils: in situ inactivation and phytorestoration, R.G. Landes Compagny, Georgetown, TX. pp 1-15.
60. Wenger, K., Kayser, A., Gupta, S.K., Furrer, G., and Schulin, R. (2002) Comparison of NTA and elemental sulfur as potential soil amendments in phytoremediation, Soil and Sediment Contamination 11, 655-672.
61. Wenzel, W.W., Unterbrunner, R., Sommer, P., and Sacco, P. (2003) Chelate-assisted phytoextraction using canola (Brassica napus L.) in outdoors pot and lysimeter experiments, Plant Soil 249, 83-96.
62. Zhao, F.J., Lombi, E., and McGrath, S.P. (2003) Assessing the potential for zinc and cadmium phytoremediation with the hyperaccumulator Thlaspi caerulescens, Plant Soil 249, 37-43.

CHAPTER 9

LITAVKA RIVER ALLUVIUM AS A MODEL AREA HEAVILY POLLUTED WITH POTENTIALLY RISK ELEMENTS

Some Methods for Studying Polluted Soils

L. BORUVKA[1] AND R. VACHA[2]
[1]*Department of Soil Science and Geology*
Czech University of Agriculture in Prague, Czech Republic
[2]*Research Institute for Soil and Water Conservation in Prague, Czech Republic*

Abstract

The Litavka River alluvium close to Pribram, Czech Republic, represents an area heavily polluted with risk elements. Three sources of soil pollution can be distinguished: naturally increased metal content due to specific composition of parent rocks, atmospheric deposition from a smelter, and floods of water polluted with metal-processing wastes. As such, the alluvium is an exceptional area in the Czech Republic. Therefore, many studies have been conducted there to reveal risk elements behaviour under different conditions, risk for man and for the environment, and principal factors that may take an effect. The studies have been using many different methods. This contribution presents some of these methods and their results obtained in the area. It focuses both on traditional methods and methods that are not very common or widely used. Attention is paid mainly to Ca, Pb, and Zn. A small-scale remediation experiment with different soil amendments to reduce risk element mobility and bioavailability is presented at the end.

1. INTRODUCTION

Remediation of polluted soils represents a necessary and reasonable way of reducing environmental risk. However, any decision on remediation or other management of polluted sites should be preceded by a thorough analysis of the level of pollution, its sources, kind and forms of pollutants, their behaviour, etc. Potential risk of pollutant leaching or input to food

J.-L. Morel et al. (eds.), Phytoremediation of Metal-Contaminated Soils, 267–298.

chain should be assessed. Only the remediation supported by a detailed analysis of the actual state can be successful.

Generally, there are three key points to be solved. First one is indication of the pollution, i.e. delineation of the polluted areas. Second problem is determination of the sources of pollution. It is difficult especially in the areas where more sources act together and it is necessary to assess the contribution of each of them. Third problem is an appropriate evaluation of the risks presented by the pollution. Element behaviour, mobility, bioavailability, toxic effects to plants and other organisms, and other potential risks should be taken into account. Comparison with some limit values seems to be an easy and common way. Setting up environmentally relevant criteria of pollution is, however, another complicated and rather controversial issue.

There is a number of methods available for studying soils polluted with risk elements and solving these problems. Traditional approach consists in chemical analysis. Beside total element content that does not tell a lot about potential risks, we can use a kind of speciation analysis. In addition, different indirect methods can be used, like biomarkers or microbial tests. Magnetic measurement, that reflects the level and distribution of anthropogenic pollution, represents another alternative method. Statistical and other mathematical methods, so-called "data-mining", can also add to the information on pollution. Only appropriate selection and combination of the methods can provide reasonable and desirable results.

The Litavka River alluvium close to Pribram, Czech Republic, represents an area heavily polluted with risk elements. Three sources of soil pollution can be distinguished: naturally increased metal content due to specific composition of parent rocks, atmospheric deposition from a smelter, and floods of water polluted with metal-processing wastes. As such, the alluvium is an exceptional area in the Czech Republic. Therefore, many studies have been conducted there to reveal risk elements behaviour under different conditions, risk for man and for the environment, and principal factors that may take an effect. The studies have been using many different methods. This contribution presents some of these methods and their results obtained in the area. It focuses both on traditional methods and methods that are not very common or widely used. Attention is paid mainly to Ca, Pb, and Zn. A small-scale remediation experiment with different soil amendments to reduce risk element mobility and bioavailability is presented at the end.

2. DESCRIPTION OF THE LOCALITY

2.1. General Description

The whole area belongs to the Czech Massive. The Litavka River flows close to Pribram, a town in Central Bohemia located approximately 50 km to the south from Prague (Fig. 1). Its tributaries start in the Brdy Mountains. The river then flows through Pribram Hills and a valley between the Brdy and Hrebeny Mountains to the north, where it discharges to the Berounka River. The elevation decreases from 700 m at the tributaries to approximately 300 m. Average annual temperature is 6.5 to 7.5ºC, average annual precipitation ranges from 600 to 800 mm. Bedrocks are formed mainly by schists, sandstones, graywackes, and quartzs. These rocks are mostly covered in the alluvia by non-calcareous alluvial sediments, on other areas by geests. Minerals rich in potentially risk elements include galenite (PbS), sphalerite (ZnS), boulangerite ($Pb_5Sb_4S_{11}$), and antimonite (Sb_2S_3). Higher content of these minerals caused naturally increased levels of risk elements in soils. Major soil unit of the alluvium is Fluvisol, at the positions further from the river Cambisol. Most soils are loams and sandy loams.

2.2. Sources of Pollution

As it was mentioned above, risk element content in soils of the area under study is naturally increased due to a specific composition of parent rocks. The background contents are therefore generally higher than in other areas of the Czech Republic. In addition to this geogenic origin, however, there are two major anthropogenic inputs of risk elements to soil. Both are related to metal mining and processing in Pribram. It has a long history. Mining activities have been performed here, according to archaeological discoveries, since the 10th century (Mesto Pribram, 2002). The oldest written note on a local metallurgical plant and mines is dated to 1311. In addition to traditional silver mining, iron ore was also extracted and treated in the region of Pribram from the 10th to the 19th centuries. Metal processing continues nowadays in the joint stock company Kovohute Pribram. Since 1973, only lead waste, mainly used vehicle accumulators, is processed there for raw lead, lead alloys, and lead products. Kovohute Pribram ensures lead recycling as the only company in the Czech Republic.

One way of soil pollution from the metallurgy is through atmospheric deposition. The emission monitoring works since 1969, when 624 tons of lead escaped into the atmosphere (Mesto Pribram, 2002). At that time, lead ores were processed by calcination and in the stack furnace, only with a simple separation equipment based on the principle of gravitation. In 1972, a wet separation processor made in France was introduced into the operation. This device decreased the emissions escaping into the atmosphere substantially, but never worked according to presumptions. A real shift in decreasing Pb emission was achieved in 1982, when Polish bag filters were installed. The last distinctive shift happened after installation of German f in soils. Due to high amounts of Pb and other risk elements emitted in the past,ilters in 1997. Nevertheless, lead deposition was long enough to accumulate metal pollution from this source is not limited exclusively to soil surface or topsoil, but can be

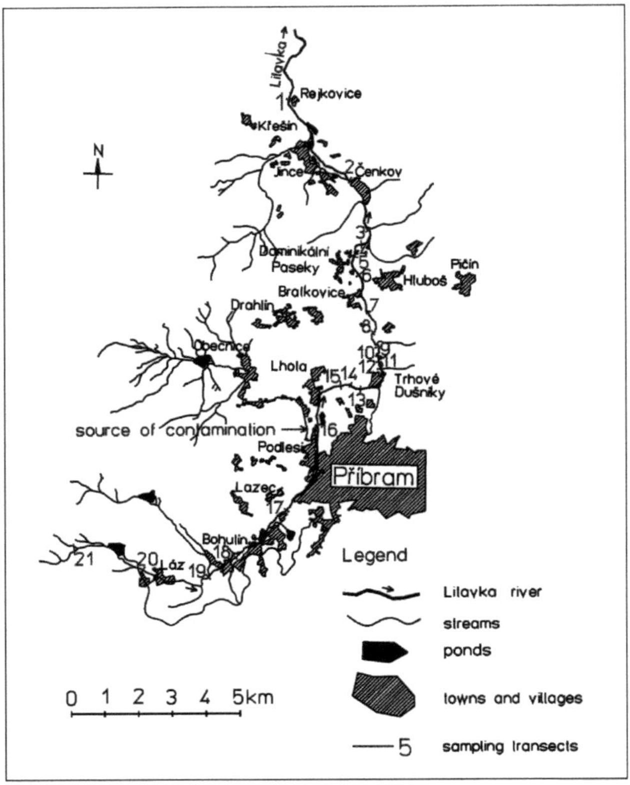

Fig. 1. Map of the Litavka River with the site of the principal source of pollution and sampling transect numbers referred to in several studies presented in this contribution.

found also in deeper layers. Typical for this type of pollution is its widespread spatial distribution, covering relatively large area with slow decrease of the content of polluting elements with the distance from the smelter as the source of pollution.

Another way of anthropogenic enrichment of soil with risk elements from the smelter were floods with contaminated water. In the 1970's and 1980's, walls of waste sedimentation ponds were broken several times. As these ponds were close to the Litavka River, the content of the ponds, containing very high amounts of heavy metals and other risk elements, was released to river water and consequently came with floods to the alluvium. The soils in the alluvium are therefore extremely polluted. In comparison to the pollution caused by atmospheric deposition, this pollution covers relatively narrow strip of land, but the level of pollution is many times higher. Distinction between the different sources of pollution is an important task for the research in the area.

3. BASIC RESEARCH OF SOIL POLLUTION WITH RISK ELEMENTS

3.1. Content and Forms of Risk Elements in Soils

3.1.1. Total and Pseudo-Total Content

Total content of risk elements in soil provides an indicator of the level of pollution, but it does not say anything about the mobility and bioavailability of the risk elements. It cannot serve by itself for the determination of element origin, either. For the determination of total element content in soil, a complete destruction of soil matrix is necessary. Using a combination of strong acids is a common way; hydrofluoric acid should be included to destroy silicate matrix (e.g. Tessier, et al., 1979; Banin, et al., 1981; Brümmer, et al., 1986).

Aqua regia digestion offers a reasonable and relatively simple alternative to the total metal determination (McGrath and Cunliffe, 1985). It provides pseudo-total element content in soil with a limitation in determination of metals strongly bound in silicate structure. The silicate fraction, however, is not very important neither for pollution assessment, because elements of anthropogenic origin are usually rather in more mobile forms, nor for environmental risk evaluation. Tlustos, et al. (1994) found the extraction by aqua regia almost equivalent to the total decomposition in case of Cd, in case of Zn the average amount extracted with aqua regia was, however, only 73.4 % of the total content. Results of those two methods of Zn extraction, however, were not significantly different at significance level 0.01.

Mean, minimum, and maximum content of Cd, P, and Zn determined by aqua regia digestion on a set of 42 soil samples from the Litavka River alluvium is shown in Table 1. From the wide range of values it is apparent that both slightly and heavily polluted sites can be found in the alluvium.

3.1.2. Extractable Content

Extraction methods can provide an insight into the forms in which an element is present in soil. It allows a better assessment of element mobility and actual risk for the environment. Different single extractants are used to release different element forms from soil. The fractions should be, however, defined on an operational basis rather than as exact chemical forms (Brümmer, et al., 1986). The extracted amount depends on the type of extractant, its concentration, ratio soil : extractant, time of interaction, temperature, and other factors (Randall, et al., 1983; Campanella, et al., 1985; Jopony and Young, 1994). Nevertheless, it provides an approximation to real element speciation in soil.

Several extraction schemes are commonly used. We can extract soil with a single extractant, or we can use different extractants on separate soil subsamples (simultaneous extraction) (Young, et al., 1992; Bell, et al., 1991; Ure, et al., 1993). Alternatively, we can apply a sequential extraction. A classical method of sequential extraction was published by Tessier, et al. (1979). Förstner (1993) consideres sequential extraction as the most useful tool for a long-term prediction of negative effects of heavy metals.

Exchangeable or mobile forms of risk elements are usually extracted by solutions of neutral salts (e.g. $Ca(NO_3)_2$, $CaCl_2$, $MgCl_2$, $NaNO_3$, NH_4NO_3) (Tessier, et al., 1979; Brümmer, et al., 1986; Houba and Novozamsky, 1993; Mench, et al., 1994; Gupta and Häni, 1983; Sauerbeck and Styperek, 1985). Organically bound forms can be determined after organic matter destruction by e.g. H_2O_2 or NaOCl (Tessier, et al., 1979; Brümmer, et al., 1986), or they can be extracted by alkalic reagents as NaOH, $Na_4P_2O_7$, $K_4P_2O_7$, NH_4OH

Table 1. Mean, minimum, and maximum metal content released from soil by different extractants and by aqua regia digestion (mg kg^{-1}) (Borůvka and Drábek, 2000).

	0.1M $Ca(NO_3)_2$	0.05M $Na_4P_2O_7$	2M HNO_3	Aqua regia
		Cadmium		
Geom. average	4.46	0.73	21.82	19.09
Minimum	0.14	0.11	0.27	0.30
Maximum	20.33	2.23	130.40	129.00
		Lead		
Geom. average	4.67	30.22	1986	1918
Minimum	1.05	3.53	87	72
Maximum	195.75	174.00	9665	8445
		Zinc		
Geom. average	164.0	49.1	1628	2007
Minimum	0.4	1.2	11	22
Maximum	1554.0	813.8	13060	11739

(Brümmer, et al., 1986; Young, et al., 1992; Krishnamurti, et al., 1995). Another possibility is extraction by complexing reagents as DTPA, EDTA, or Na_2EDTA (Brümmer, et al., 1986; Tlustos, et al., 1994; Mortvedt, et al., 1981; Davies, 1985; Sheppard Thibault, 1992); this fraction is sometimes referred to as potentially mobilizable (Podlesakova, et al., 2001). Metals occluded on Fe/Mn oxides can be extracted by $NH_2OH.HCl$ or oxalic acid (Tessier, et al., 1979; Brümmer, et al., 1986; Young, et al., 1992). Most common methods of risk element extraction and fractionation were recently described in (Quevauviller, 2002).

On a set of more than 300 soil samples from the Litavka River alluvium, simple extraction of Cd, Pb, and Zn with cold 2 M HNO_3 (soil : solution ratio 1 : 10, w/v) was used to determine the level of pollution (Boruvka, et al., 1996). This method should release maximum potentially mobilizable content of metals. In addition, a two-step sequential extraction for the determination of exchangeable (extracted with 0.1 M $Ca(NO_3)_2$) and organically bound (extracted with 0.05 M $Na_4P_2O_7$) forms was also applied (Boruvka, et al., 1997).

Extraction with 2 M HNO_3 is a common method in the Czech Republic and it determines the maximum pool of potentially available metals, so that it can be more important for the evaluation of environmental risk of pollution than total content. The maximum tolerable element contents in soils of the Czech Republic are also set for 2 M HNO_3 extraction (Ministry of the Environment of the Czech Republic, 1994). Following average percentages of metal extractability with nitric acid from the total content were reported (Nemecek, et al., 1995): 85 for Cd, 58 for Mn, 51 for Pb, 42 for Mo, 41 for Cu, 38 for Co, 26 for Zn, 20 for Be, 19 for Ni, 17 for V, 12 for As, and only 9 for Cr. These values, however, are much higher in Fluvisols and in polluted soils. In our study (Borůvka and Drábek, 2000) on samples from the Litavka River alluvium, contents of Cd, Pb, and Zn did not differ significantly between nitric acid extraction and aqua regia digestion. In some cases, the results of 2 M HNO_3 extraction were even slightly higher than the results of aqua regia digestion. These results may be partially explained by soil heterogeneity. Another possible explanation can be seen in a tolerable analytical error (in the range of 10% maximum), in our case mainly due to multiple sample dilution before measurement. The results of 2 M HNO_3 extraction and aqua regia digestion were closely correlated (Fig. 2).

The share of exchangeable form of Cd accounted for 0.3 to 53.5% of the amount determined by aqua regia digestion, corresponding values for Pb ranged from 0.03 to 5.0%, and for Zn from 0.1 to 24.3% (Borůvka and Drábek, 2000) (Table 1). Organically bound fraction accounted for 0.8 to 5.9 % in case of Cd, 0.2 to 4.0% in case of Pb, and 0.6 to 8.9% in case of Zn. The highest share of exchangeable form was found for Cd, share of exchangeable

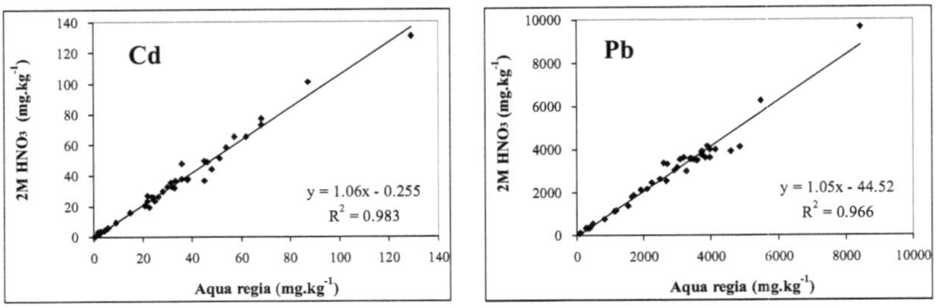

Fig. 2. Linear regression of the relationship between soil metal content extracted by 2 M HNO_3 and amount released by aqua regia digestion.

Pb was very low. Lead was the only metal with the content of organically bound fraction higher than or at least comparable to the content of exchangeable fraction. However, most of Pb was present in other forms. These values confirmed relatively low mobility of Pb in soils in comparison with Cd and Zn (Ramos, et al., 1994).

3.1.3. Determination of the Origin of Risk Elements According to Their Contents

There are generally three ways to distinguish the origin of risk elements in soil based on their content: speciation, profile distribution, and spatial distribution. However, none of them is absolutely reliable. They should rather be combined and, if possible, amended with some additional information like parent rock composition or known anthropogenic loads.

Risk Element Speciation. Exploitation of risk element speciation is based on the following assumptions (Förstner, 1987):

- Pollutant bonding is less stable in polluted materials than in lithogenic ones.
- Interaction between solution and solid phase in polluted systems tends to form more labile associations than in natural systems.
- Remobilisation of pollutants in polluted systems is more effective due to their weaker bonds to soil particles.

The share of mobile or mobilizable risk element forms can thus serve for the distinction between anthropogenic pollution and natural enrichment (Podlesakova, et al., 2001; Vacha, et al., 2002). Unfortunately, the opposite was true in case of the Litavka River alluvium. The share of exchangeable form there generally decreased with increasing level of pollution (Fig. 3). When the sampling sites were separated to those affected by floods of contaminated water and sites influenced only by atmospheric deposition (e.i. less polluted) (Boruvka, et al., 1996), the share of mobile forms was generally lower in the area of floods (Table 2). This could be possibly caused in part by saturation of specific sorption sites and consequently favourable conditions for creation of defined metal compounds in the more polluted areas (Brümmer, et al., 1986). The effect of pH values, that were accidentally higher on the most polluted areas, and organic matter content, that was higher in the area of atmospheric deposition, cannot be omitted, either.

Profile Distribution. Distribution of risk elements in soil profile can indicate their major origin. In the case of anthropogenic origin, especially atmospheric deposition, the risk elements are supposed to accumulate mainly

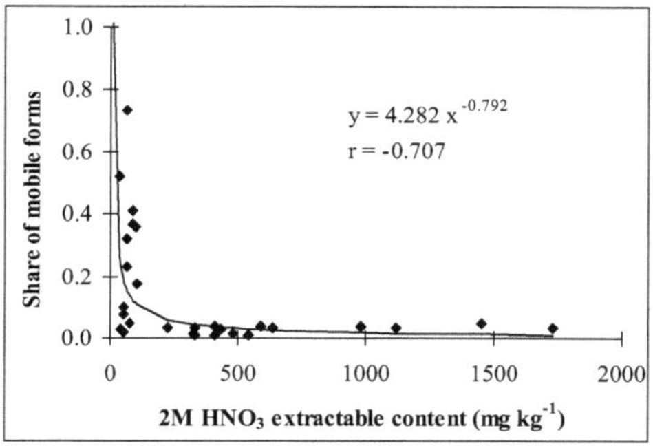

Fig. 3. Relationship between share of the sum of exchangeable and organic Pb fractions and 2M HNO_3 extractable metal content in soil.

Table 2. Average share of exchangeable and organically bound metal fraction on 2M HNO_3 extractable content in soil (geometric means; %).

Depth	Fraction	Area of atm. deposition			Area of floods		
(cm)		Cd	Pb	Zn	Cd	Pb	Zn
0-15	exchangeable	32.07	0.94	12.14	26.44	0.07	11.50
	organically bound	8.39	5.35	5.48	2.82	1.55	3.17
30-40	exchangeable	37.09	1.30	11.83	24.92	0.06	11.17
	organically bound	5.98	2.99	3.46	1.48	0.84	2.91
50-60	exchangeable	33.91	2.43	14.11	25.47	0.10	12.76
	organically bound	5.14	2.33	1.95	1.30	0.82	2.43

in the topsoil. In the case of geogenic origin, on the contrary, the highest content can be expected in the deepest layers (Boruvka, et al., 1996; Vacha, et al., 2002). However, this does not need to be always true. In case of old soil anthropogenic loads, the mobile forms of pollutants could have been leached to the depth. On the other hand, risk elements of lithogenic origin could have been transported upwards by plant roots. Even more complicated the situation is when more different sources are combined, as it is the case of the Litavka River alluvium.

The highest Cd, Pb, and Zn content was generally found in the topsoil, both in the area polluted by flooding water and the area polluted only by atmospheric deposition (Boruvka, et al., 1996). Only one site close to the Litavka River tributaries showed Pb content increasing to the depth, indicating prevailing lithogenic origin. Nevertheless, relatively high Pb content in the depth of 50 to 60 cm, with regard to generally low mobility of this element in the soil profile (Kabata-Pendias and Pendias, 1992), indicated an important influence of parent rock on Pb content in the soil also on other sites. Surface accumulation of metals was measured as the topsoil enhancement ratio, a value computed as the ratio between the content extractable with 2 M HNO_3 in the topsoil (0-15 cm) and in the depth 30 to 40 cm. This ratio was significantly higher in the area influenced by atmospheric deposition where the content in topsoil was in average two to four times higher than in subsoil. In the area affected by floods, the mean ratio was only approximately 1.5 (Boruvka, et al., 1996).

In case of exchangeable fraction of all the three elements, there was no significant difference between the soil layers, which confirmed very high mobility of that form (Boruvka, et al., 1997). Profile distribution of organic fraction showed that there was apparent accumulation of Pb, and to a lower extent also of Cd, in the topsoil organic matter. For Zn, however, there was no significant difference in the organic fraction content between soil layers.

Spatial Distribution. Spatial distribution of risk element content usually corresponds to the pathway of enrichment. In case of lithogenic enrichment, the spatial distribution is related to local geology. Non-point soil pollution is usually widespread. Spatial distribution of point contamination exhibits some structures with maxima close to the point source, and with decreasing level of pollution with the distance from the source (Bradley and Cox, 1987; Podlesakova, et al., 1994; Rieuwerts, 1993). This was expected also in the area under study.

Atmospheric deposition was supposed to influence larger area, with maxima close to the smelter. This will be discussed in more detail in chapter 0 Pollution with contaminated water should be related to the river itself and to the probability of flooding on each site. According to the position and the level of pollution, sites influenced mainly by atmospheric deposition and sites polluted mainly by floods of contaminated water thus could be distinguished (Boruvka, et al., 1996). The former included sites upstream from the setting pits and some sites downstream located more than three meters above the river level where the floods were not probable. Distribution of Cd content extractable with 2 M HNO_3 in the topsoil of the area possibly influenced by polluting floods is shown in Fig. 4. In general, level of pollution was lower on sites more distant from the river. After attaining maxima, the content decreased downstream along the river with the distance from the source. However, the contents were still very high, in comparison with limit values, even on the most distant locations. Similar results were obtained also for Pb and Zn.

3.2. Plant Uptake

3.2.1. Element Content in Plants

Risk element uptake by plants from soil depends on the level of pollution, forms of the risk element in soil, its mobility in the system soil – plant, and also on the plant species. Chemical speciation of risk elements in soils can provide a good indication or approximation of the element uptake by plants. Many authors reported some correlations between the content of risk element extracted from soil by an extractant and its content in plants (e.g. Houba and Novozamsky, 1993; Mench, et al., 1994; Gupta and Häni, 1983). However, no method of soil analysis was found absolutely reliable and

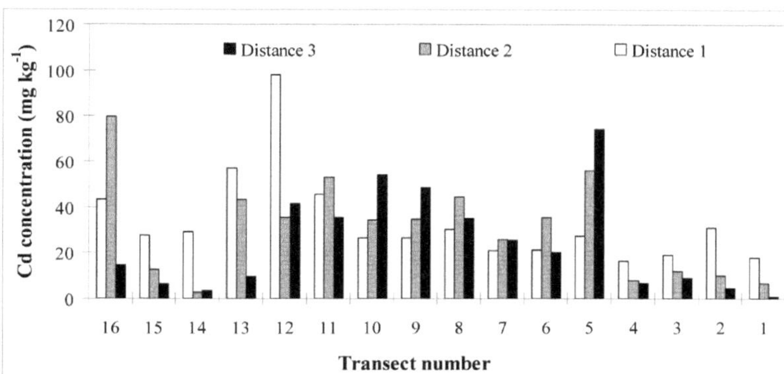

Fig. 4. Distribution of Cd content in the topsoil of the area possibly influenced by polluting floods. Transects 16 to 1 indicate the order of transects down the river from the principal source of pollution (see Fig. 1). Distance 1 is close to the river (up to 5 m), distance 3 is the furthest, having thus lower probability of floods. Averages of both banks were used in the graph.

universal in predicting element uptake by plants. Real uptake can be therefore determined only by plant analysis.

Samples of roots and shoots of spontaneously grown grasses *Poa sp.* and *Festuca sp.* were collected in the alluvium of the Litavka River. In the area of atmospheric deposition as the main source of pollution, mean content of Cd, Pb and Zn in plant shoots was 0.312, 4.42, and 43.4 mg kg^{-1} dry weight, respectively; in roots this content was 2.38, 58.4, and 104.4 mg kg^{-1} dry weight, respectively. In the area influenced by floods of contaminated water, mean content of Cd, Pb and Zn in plant shoots was 1.87 19.4, and 326.5 mg kg^{-1} dry weight, respectively, and in roots this content was 28.9, 465.8, and 1359 mg kg^{-1} dry weight, respectively (Boruvka, et al., 1997). The ratio of metal content in shoots and roots as a measure of metal mobility in plants decreased in the order Zn > Cd > Pb. This ratio was lower in the area of floods than in the sites influenced only by atmospheric deposition.

Table 3. Correlation coefficients of the relationship between metal concentration in plant roots and shoots and concentration of metal fractions in the topsoil (0 to 15 cm) according to the prevailing source of pollution; the concentrations were for calculations transformed to their common logarithms in order to obtain approximately normal distribution of values as they were originally strongly skewed.

Metal	Fraction in the topsoil extracted with	Area polluted by atmospheric deposition		Area polluted by floods	
		Roots	Shoots	Roots	Shoots
Cd	0.1 M $Ca(NO_3)_2$	0.664***	0.238	0.556***	0.308**
	0.05 M $Na_4P_2O_7$	0.368*	–0.061	0.495***	0.154
	2 M HNO_3	0.824***	0.401*	0.609***	0.294**
Pb	0.1 M $Ca(NO_3)_2$	–0.042	–0.150	0.283**	0.035
	0.05 M $Na_4P_2O_7$	0.422*	–0.225	0.452***	0.098
	2 M HNO_3	0.682***	0.454*	0.618***	0.175
Zn	0.1 M $Ca(NO_3)_2$	0.568**	0.616***	0.757***	0.693***
	0.05 M $Na_4P_2O_7$	0.426*	0.331	0.758***	0.523***
	2 M HNO_3	0.663***	0.484**	0.793***	0.586***

*,**,*** Significant at the 0.05, 0.01, and 0.001 probability levels, respectively.

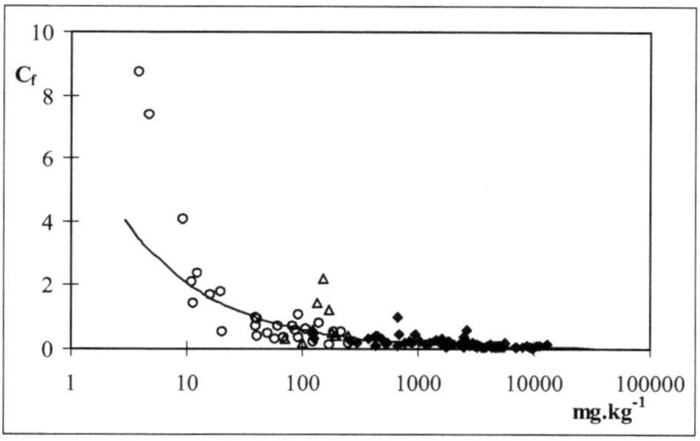

Fig. 5. Relationship between concentration factor (Cf) of Zn and concentration of 2 M HNO3 extractable Zn in topsoil. Circles - sites slightly polluted by atmospheric deposition, rhombs - sites heavily polluted by floods of contaminated water, triangles - other sites (r = -0.867).

The concentration of all the three elements in plant roots was correlated with the content of their determined forms in topsoil, except for exchangeable Pb (Boruvka, et al., 1997) (Table 3). The closest correlation was found for Zn. Better correlation was generally found in the area influenced by floods than in the area affected only by atmospheric deposition.

3.2.2. Concentration Factors

Metal transport from soil to plants was assessed also using concentration factor (C_f) calculated as the ratio of metal concentration in plant shoots and concentration of 2 M HNO_3 extractable metal in the topsoil (depth 0 to 15 cm). Values of C_f for Zn were quite high, geometric mean was 0.198 and in several cases concentration in plant shoot was even higher than concentration in soil. Concentration factor of Pb was, on the other hand, very low, with geometric mean 4.55×10^{-3}. Geometric mean of C_f for Cd was 0.063. Generally, the values of the concentration factor decreased exponentially with increasing level of contamination (Fig. 5). It was probably due to a stronger effect of physiological barriers against metal transport to the aerial parts of plants at high levels of contamination.

3.3. Influence of Soil Properties on Risk Element Behaviour

3.3.1. Effect of Soil Properties on Risk Element Speciation in Soil

Risk element behaviour and speciation in soil depends on soil properties. We have to take this fact into account when assessing the risk of pollution.

The most important properties that control risk element behaviour include especially soil pH, humus content and quality, and clay content (e.g. Brümmer, et al., 1986; Kabata-Pendias and Pendias, 1992). However, the effect of these properties may be different in different soils or at different level and source of pollution. The effect can be analysed by simple correlation and regression. However, since the properties act in mutual interaction, some multivariate methods may be more useful. This will be discussed more in chapter 0 of this contribution.

Most of studied soils in the Litavka River alluvium were acid with relatively high organic matter content. Strong negative correlation was found between soil pH and the share of exchangeable metal fraction on 2 M HNO_3 extractable metal content in soil (Fig. 6). Soil organic matter content was correlated with the share of organically bound fraction of lead and zinc (r = 0.653 for Pb and 0.595 for Zn). Soil texture did not exhibit strong influence on metal speciation in soil.

3.3.2. Effect of Organic Matter Composition

The effect of organic matter represents rather complicated issue. Humic substances of low mobility decrease mobility of some metals, while other substances with smaller molecules can increase metal mobility (Angehrn-Bettinazzi, et al., 1989). Increased mobility due to organic matter com plexation of heavy metals becomes more important especially at higher pH (König, et al., 1986; Waller and Pickering, 1993).

On a set of 20 topsoil samples from the Litavka River alluvium, fractionation of humic substances into weakly and strongly bound fulvic acids (FA) and weakly and strongly bound humic acids (HA) was done, with

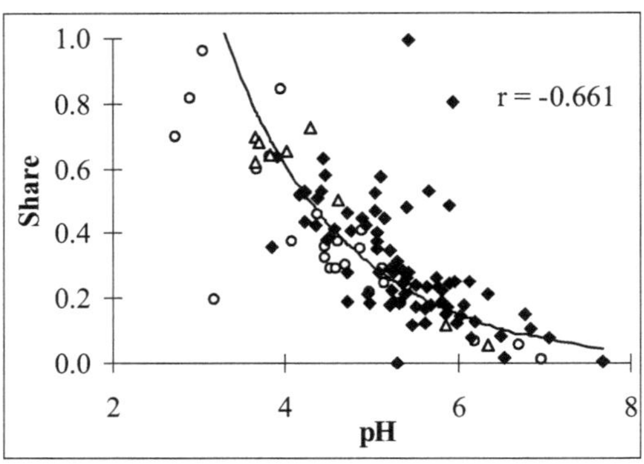

Fig. 6. Relationship between the share of exchangeable fraction of Cd and soil pH. Circles - sites slightly polluted by atmospheric deposition, rhombs - sites heavily polluted by floods of contaminated water, triangles - other sites.

subsequent determination of Cd, Pb, and Zn in each fraction (Boruvka, et al., 2002). From obtained results it is obvious that all three metals are bound more on FA than on HA, probably due to higher amount of functional groups and higher cation exchange capacity of FA (Fig. 7). Relatively higher share of the HA fraction was found for Pb. The very low effect of HA fractions on Zn and Cd binding does not correspond to the share of these substances in soil. Which concerns the separation of weakly and strongly bound FA, there are also differences between the metals under study. More Zn was found in the fraction of strongly bound FA than on the weakly bound fulvic acids. For Pb, in the contrary, the fraction of weakly bound FA was more important. Cadmium was present in the two FA fractions in almost the same amount. The fact that FA are more mobile and easily soluble in a wide pH range suggests that metal binding on these substances may increase metal mobility and bioavailability, especially in the case of Cd and Zn. Increased bioavailability of metals bound on FA was confirmed by fairly close correlation between metal content in plant roots and FA–bound fraction of the metals (data not shown). Relatively higher share of Pb bound on HA compared to Cd and Zn supports literary data on immobilization of this metal through binding on soil organic matter.

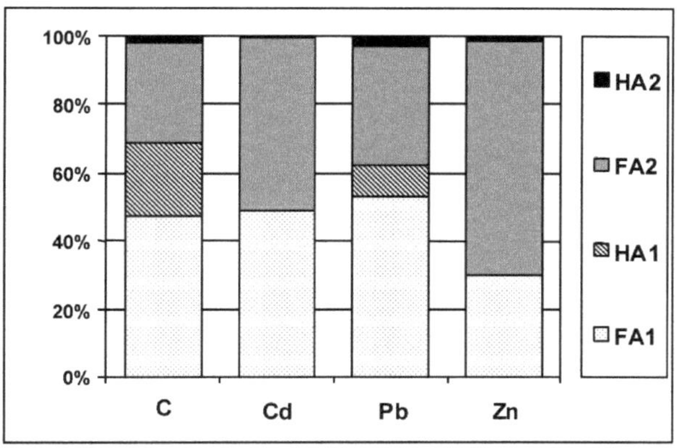

***Fig.** 7.* Distribution of organic C in weakly and strongly bound fulvic acids (FA1 and FA2) and weakly and strongly bound humic acids (HA1 and HA2), and relative distribution of Cd, Pb, and Zn between these fractions of soil humic substances.

4. OTHER METHODS FOR STUDYING SOIL POLLUTION WITH RISK ELEMENTS

4.1. Microbiological and Biochemical Tests

Microbiological and biochemical tests are very sensitive to soil quality degradation. Filip (2002) found that N_2-fixation, total microbial biomass, soil respiration, and dehydrogenase activity could serve as indicators of soil quality. However, the sensitivity of mocroorganisms is not specific to a risk element and is influenced by a number of other factors, such as basic soil properties or seasonal variability. Moreover, the effect of pollution on microbial or biochemical activity does not necessarily have to correspond to plant uptake of risk elements, so that a direct exploitation of microbiological and biochemical tests for the prediction of actual environmental risk is limited (Nemecek, et al., 1998). Nevertheless, these tests can be used as a good indicator of soil degradation. If a decrease of microbial activity is found, however, then detailed research using chemical and other methods should follow.

Different microbiological and biochemical tests were used on a set of soil samples from 6 sites from the Litavka River alluvium (Smejkalova, et al., 2001). The sites included heavily polluted areas (number 8, 12, and 16 in the map on Fig. 1), as well as slightly polluted areas (number 1 and 4). Site number 20 was chosen as a local background. Total numbers of bacteria were highest on the sites 1 and 20 and these numbers decreased towards the source of pollution (Table 4). Numbers of spore-forming bacteria and oligotrophic bacteria showed the same trend. Generally, bacteria are more sensitive to heavy metal pollution of soil than fungi and actinomycetes (Babich and Stotzky, 1983). Presence of pollutants in soil may thus change the composition of microbial community. Promising results were obtained with the parameters of endomycorrhizal colonization (Mikanova, et al., 2001). Also the ratio between C in the microbial biomass and total organic C reflected well the level of pollution (Smejkalova, et al., 2001). It is based on the fact that microorganisms in less polluted soils build in their biomass more C than on polluted areas (Insam and Domsch, 1988).

Among soil enzymatic activities, activity of dehydrogenase and invertase corresponded the best to the level of pollution (Smejkalova, et al., 2001; Mikanova, et al., 2001) (Fig. 8). Less important effect of pollution was found on the activity of amylase and phosphatase. The fact that on transect 1 the activity of some enzymes exceeded the background can be explained by higher pH

Table 4. Total content of heavy metals in soil and numbers of microorganisms per 1 g of dry soil.

Site	Total content (mg kg^{-1})			Total number of bacteria	Spore forming bacteria	Micromycetes	Oligotrophic bacteria
	Cd	Pb	Zn				
1	2.4	113.5	249.8	3.6 x10^6	0.8 x10^6	1.2 x10^5	2.6 x10^7
4	5.4	530.5	407.0	1.4 x10^6	1.6 x10^5	1.2 x10^5	0.9 x10^7
8	113.8	6335.9	12557.4	1.0 x10^6	$< 0^4$	3.4 x10^4	2.9 x10^6
12	59.0	3450.7	6230.8	2.4 x10^6	1.3 x10^5	3.3 x10^4	4.6 x10^6
16	61.3	7040.3	7497.9	7.4 x10^5	$< 10^4$	2.2 x10^4	4.4 x10^6
20	1.9	106.0	202.5	3.7 x10^6	1.0 x10^6	3.4 x10^4	3.3 x10^7

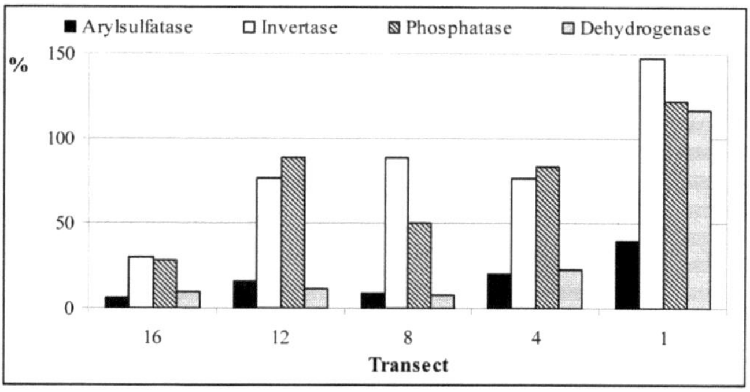

Fig. 8. Relative enzymatic activity in soil from the sites influenced by floods with polluted waters. Activities on transect 20, that is considered a local background, were set to 100%.

values compared to the background transect (number 20). Generally, it is supposed that heavy metals inhibit enzymatic reactions by forming complexes with substrate, blocking reactive functional groups of enzymes, or reacting with the enzyme-substrate complex (e.g. Serra-Wittling, et al., 1995).

4.2. Magnetic Measurements

Measurement of magnetic properties offers a proxy method for the indication of soil pollution. It is based on measurements of the concentration of ferrimagnetic minerals of anthropogenic origin (Petrovsky, et al., 2001). In case of an identical source of heavy metals and magnetic minerals, simple measurements of magnetic susceptibility can well discriminate between polluted and clean areas. Magnetic susceptibility is the ease with which a material can be magnetized at different frequencies of the applied alternating magnetic field. Measurements of magnetic susceptibility of soils have been used in mapping of areas polluted by industrial emissions (Hay, et al., 1997; Heller, et al., 1998; Kapicka, et al., 1999), or by road traffic (Hoffmann, et al., 1999). Methods of measuring magnetic parameters can be found e.g. in Collinson (1983), an overview of the applications in soil pollution research was given in (Petrovsky and Ellwood, 1999; Petrovsky, et al., 2000). These methods can be used both in the laboratory and in the field.

The aim of the study conducted in the Litavka River alluvium was to examine the degree of correlation between magnetic parameters and concentration of heavy metals and to outline polluted area from the unpolluted one (Petrovsky, et al., 2001). It was shown that unpolluted samples contained bigger amount of larger, magnetically softer particles dominated by lithogenic origin, and/or superparamagnetic grains, and/or hematite/goethite grains. However, the increased magnetization ratio in the polluted samples could be not only due to smaller sizes, but also to different mineralogy or mineral chemistry originating from the pollution source. Polluted samples contained significant contributions of smaller and magnetically harder grains, likely originating from industrial processes during the lead ore treatment. Along with heavy metals, these particles were transported from the ash basin deposit into the soil. In addition, the concentration pattern of magnetic minerals agreed with the concentrations of heavy metals studied in deeper soil layers that were not affected by post-depositional changes due to climate or remediation efforts.

4.3. Pedometrics, "data mining"

4.3.1. Multivariate Statistics

Multivariate statistics can describe the behaviour of risk elements in soil and influencing factors more exactly than univariate statistics. It takes into account more factors together, which corresponds better to real situation in soil or in the soil-plant system. Factor analysis for the analysis of correlation structures, and multiple regression for the assessment of the effect of different independent variables on a dependent variable are used most often.

Factor analysis. Factor analysis aims in replacing a large number of correlated variables with a limited number of hypothetical "factors" that are hidden behind the measured parameters and can explain a major part of variability in the data set. Some problems related to factor analysis applications in geochemistry were recently discussed by Reimann et al. (2002). Careful selection of initial variables that will be included into calculation presents a key point of the analysis. Interpretation of the results is another difficult task. Cooperation of soil (or plant) scientist and a statistician is usually useful.

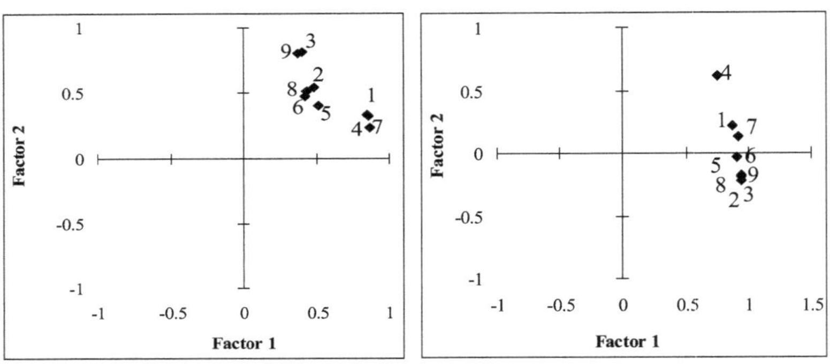

Fig. 9. Projection of metal contents in soil into plane of the first two factors for the area of atmospheric deposition (left) and for the area of flood contamination (after Varimax rotation; right). Points labelled 1, 2, and 3 represent contents of Cd in the layers 0-15 cm, 30-40 cm, and 50-60 cm, respectively, points 4, 5, 6 and 7, 8, 9 are contents of Pb and Zn, respectively, in those layers).

On the soils from the Litavka River alluvium, factor analysis together with correspondence analysis was first applied by Pavel and Kozak (1989), Kozak, et al. (1995). It distinguished between polluting elements (Pb, Cu, Zn, Cd, Sn) and non-polluting ones (Ni, Cr, Co, V). At the same time, the analysis distinguished between polluted and unpolluted sites.

In a more recent study analysing soil pollution with Cd, Pb, and Zn using factor analysis (Boruvka, et al., 1996), three factors accounting for more than 90% of total variation were selected separately for the area of atmospheric deposition and the area affected by floods with contaminated water (Fig. 9). Based on obtained data, the following interpretation of calculated factors was suggested. In the area influenced by atmospheric deposition, the first factor was considered as the factor of homogeneity of metal distribution in soils. Weights of this factor for topsoil metal contents were slightly lower than for subsoil. The second factor was considered as the factor of atmospheric deposition and surface accumulation of metals. That accumulation was the most important for Pb so that weight of that factor was highest for Pb in topsoil. For Zn as the most mobile metal the weight was lowest. Weights of this factor for metal contents in deeper layers were close to 0 or negative. The third factor was supposed to be the factor of parent rock influencing mainly Pb content and also Cd content in deeper layers. In the area polluted by floods, the first factor was supposed to be the factor of surface contamination. The second factor was supposed to be the factor of metal mobility, which was bigger for Zn and Cd than for Pb. The third factor was interpreted as the factor of contamination in the intrinsic soil layers. This factor was distinguished because of a very long time of metal production in the region during which the metals from both atmospheric and liquid deposition could be accumulated in the soil and transformed by pedogenic processes. Factor analysis was applied also to the set of plant concentrations of Cd, Pb, and Zn (Boruvka, et al., 2002).

Multiple regression. Step-wise multiple regression enables to select the most important variables that influence one dependent variable. Final regression models, if enough reliable, can become after some generalisation and validation a part of pedotransfer rules or simulation models. The models are based on a general equation:

$$\mathbf{y} = a + b_1\,\mathbf{x_1} + b_2\,\mathbf{x_2} + \ldots + b_n\,\mathbf{x_n},$$

where $\mathbf{x_1}$ to $\mathbf{x_n}$ are independent variables, e.g. individual basic soil properties, b_1 to b_n are their respective regression coefficients, and a is a constant.

As an example, multiple regression models of Cd, Pb, and Zn content in plant roots ans shoots and their concentration factors as functions of element content in soil (extractable with 2 M HNO_3) are presented (Table 5). Only variables with regression coefficient different from 0 with probability at least 95% were included into the models. Clay content and simple parameters of humus quality were thus omitted. Element content in soil appeared in all the regression models, having positive regression coefficient in the models of the content in plants and negative coefficient in the models of concentration factors. Soil pH was included in the regression models of element content in roots for all the three metals under study, and also in the models of Zn content in shoots and Zn concentration factor. Soil pH showed in all cases negative regression coefficient. A decrease in metal mobility and bioavailability due to higher soil pH corresponds to literary data. Soil organic carbon content appeared in the models of metal content in plant shoots and C_f for all the three metals and Zn content in the roots, in all cases with negative regression coefficient. The effect of C_{org} was attributed to forming sorption sites for metals in soil. The regression models explained a large part of the variability of Zn content in both plant roots (85.1%) and shoots (77.2%), and Cd and Pb content in roots (71.9 and 67.7%, respectively). The explanation of the variability of Cd and especially Pb content in plant shoots was lower (31.0 and 18.0%, respectively). This supports the supposition of the existence of some barriers against metal transport from roots to the above-ground plant parts. Nevertheless, the models of multiple regression provided much better insight to the effect of soil properties on metal transfer to plants than simple regression. It confirms that soil properties influence metal transfer to plants in mutual combination and are inter-dependent.

Table 5. Regression models of metal contents in plants and concentration factors (Cf) as functions of metal content in soil, soil pH, and organic carbon content (Corg).

Metal content/parameter		Constant	Soil content (log mg kg^{-1})	pH	C_{org} (log %)	R^2 (%)	P
Cd	roots (log mg kg^{-1})	0.620	0.843	–0.088	–	71.91	< 0.001
	shoots (log mg kg^{-1})	–0.433	0.460	–	–0.627	31.02	< 0.001
	C_f (log)	–0.996	–1.244	–	–1.444	35.31	< 0.001
Pb	roots (log mg kg^{-1})	–0.402	1.151	–0.181	–	67.73	< 0.001
	shoots (log mg kg^{-1})	0.004	0.377	–	–0.776	18.01	0.005
	C_f (log)	0.008	–1.434	–	–1.787	25.99	< 0.001
Zn	roots (log mg kg^{-1})	1.301	0.728	–0.116	–0.249	85.09	< 0.001
	shoots (log mg kg^{-1})	1.376	0.586	–0.151	–0.363	77.16	< 0.001
	C_f (log)	3.168	–0.952	–0.348	–0.836	72.81	< 0.001

4.3.2. Geostatistics

Application of geostatistical methods on the spatial distribution of risk elements in soils is quite common (e.g. Wopereis, et al., 1988; Goovaerts and Webster, 1994; Webster, et al., 1994; Einax and Soldt, 1995; Markus and McBratney, 1996). Ordinary kriging enables a general description of the distribution and indicates maxima of pollution, usually corresponding to the source of pollution. The advantage compared to traditional interpolation techniques is the fact that together with the map of spatial distribution of estimated element content, we can obtain also a map of estimated error. Based on this, we can do an additional sampling on the sites with high potential error to confirm or disprove the less reliable data. Disjunctive kriging can help to delineate areas with a high risk of overcoming limit values from the areas less polluted or non-polluted. The geostatistical methods can be useful even in case of non-point contamination. Factorial kriging in Goovaerts and Webster (1994) enabled to distinguish between the effect of farm management, acting in a local scale, and the effect of parent rocks, acting at a regional scale. A problem related to the application of geostatistics is a large required number of sampling points (at least 100) for the results to be reliable.

The attempts to apply geostatistics in the Litavka River alluvium were not very successful so far. One reason is that it is a narrow strip of land, so that a one-dimensional geostatistics could be more appropriate. The other problem is a strong anisotropy: in the direction of the river flow, the variability is lower and has a longer range than in the traverse direction. Nevertheless, the geostatistical methods were successfully applied on the data on risk element contents in the whole region (Adamec, 1990). It is clear from the results that the contents increase extremely towards the smelter (Fig. 10). The pollution, in this case mostly from atmospheric deposition, influenced also the subsoil due to a long-term influence of the pollution (Fig. 11). Similar spatial distribution was found for all studied elements.

5. REMEDIATION EXPERIMENT

The aim of a small-plot field experiment in a heavily polluted part of the Litavka River alluvium (near the transect no. 11 on Fig. 1) was to investigate the possibility of use of inorganic and organic sorbents for the immobilisation of some risk elements in soil (Vacha, et al., 2002). The objectives were:

– to choose suitable inorganic and organic materials for the immobilisation of risk elements,

– to evaluate the efficiency of the application of inorganic and organic materials into the soil in decreasing the transfer of risk elements (As, Cd, Pb, Zn) from soil to plants,

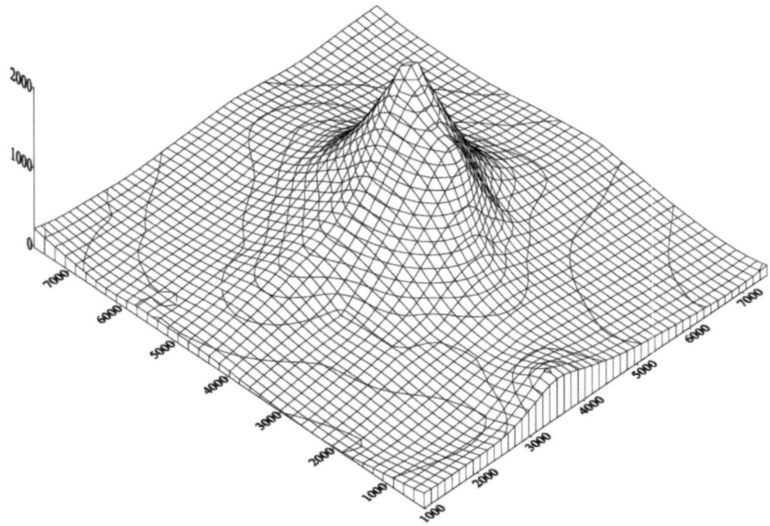

Fig. 10. Lead content in the topsoil of the Pribram region in 3D view (mg kg^{-1}).

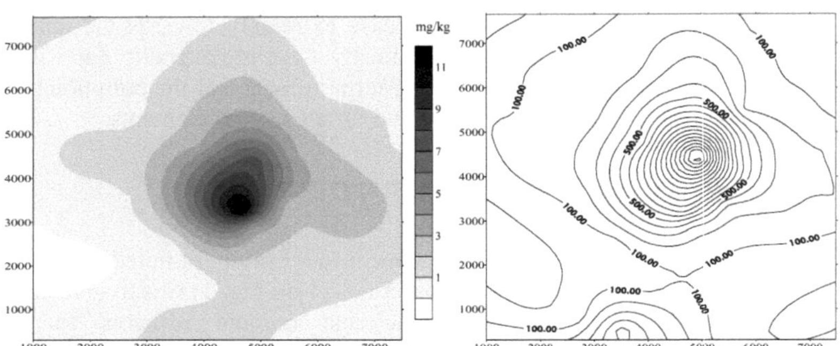

Fig. 11. Lead (left, contour map) and Cd (right, colour scale map) content in the subsoil of the Pribram region (mg kg^{-1}; scales on axes are given in meters).

– to assess the influence of the amendments on speciation of the elements in soil,

– to assess the effect of the change of pH values on the extractability of risk elements from the soils amended with sorbents,

– to reveal other characterics that could significantly influence the process of risk element immobilisation in soil.

5.1. Soil Amendments

Selected organic and inorganic sorbents represent available and economically realistic materials. The inorganic amendments included synthetic zeolite, loamy shale, and dolomitic limestone. Muck, peat, and sludges were used as organic amendments.

5.1.1. Inorganic Sorbents

Synthetic Zeolite. This zeolite was prepared by the use of hydrothermal alteration of the ashes in alkaline conditions by the presence of Na ions (Kovanda, et al., 1996). The zeolite is of P_c type with cubic structure. Chemical formula is $Na_xAl_xSi_{16-x}O_{32} . yH_2O$. The content of selected elements is presented in Table 6.

Loamy Shale. This material was obtained from the Institute of Mineral Raw Materials in Kutna Hora. It has high buffering capacity for Pb, Hg, Cd, Cr, Cu, Ni, Zn, and petroleum substances. The sorption of anions is low, because of the alkaline reaction of the shale. Mean yield of sorption is for Pb 95%, Hg 70%, Cd 70-80%, Cr 80%, Zn 60-80%, and Ni 60%.

Dolomitic Limestone. The content of $CaCO_3 + MgCO_3$ was 95%, the content of $CaCO_3$ is 50% of this amount. The limestone is finely ground. Following content of selected risk elements in the dolomitic limestone was determined: As < 2 mg kg^{-1}, Cd < 0.5 mg kg^{-1}, Hg < 0.045 mg kg^{-1}, Pb < 7 mg kg^{-1}, and Cr < 4.7 mg kg^{-1} . This material is produced and distributed by Krkonosske Vapenky in Kuncice.

5.1.2. Organic sorbents

Muck The muck is material of sapric Histosols. Used muck was collected from a locality near Lysa nad Labem. The content of organic carbon was 7.2%, pH value 7.45. The contents of micro and macroelements are presented in Table 6.

Peat The peat was collected in Hora Sv. Sebestiana in Northern Bohemia. The content of organic carbon was 30.1%, pH value 3.8.

Sludges The sludge originated from sewage treatment plant in Prague - Podbaba. The content of dry matter was 28%. Contents of micro and

Table 6. Total content of elements in the sorbents (mg kg^{-1}).

Element	Synthetic zeolite	Loamy shales	Muck	Sludge
P	500	860	3.4	654[a]
Na	98000	1100	ND	ND
K	1700	12050	40	644[a]
Ca	8100	9620	8350	13028[a]
Mg	50	1069	130	1012[a]
As	11.2	20.8	10	3.3[b]
Be	9.46	10.84	2.32	ND
Cd	0.38	1.52	0.2	2.85[b]
Co	28.4	53.5	3.3	5.6[b]
Cr	119.2	82.6	22	52[b]
Cu	266	71.5	13.6	292[b]
Hg	0.192	0.245	ND	ND
Mn	52.02	59.85	126	222[b]
Mo	ND	ND	0.74	ND
Ni	166	861	9.9	39.1[b]
Pb	21.6	27.5	24.2	130[b]
V	128.3	107.1	44	29[b]
Zn	100.8	109.1	58.6	1113[b]

[a] available;
[b] extractable with 2M HNO_3; ND – not determined

macroelements are presented in Table 6. The application rate was 18 t of fresh sludge per ha.

5.1.3. Experiment Arrangement and Amendment Application

The field experiment was started in 1998. The soil was classified as Fluvisol, with pH 5.3 and organic C content of 4.2%. Following initial element content in soil was found (in mg kg^{-1}): Cd – 11.8, Pb – 820.0, Zn – 795.0, As – 14.8.

All the amendments were incorporated into the depth of humic horizon (20 cm). The area of each individual plot was 1 m x 1 m. Each treatment had four replications. The plots were planted with mustard (in 1998), rye (in 1999), and mustard and oat (in 2000). The leaves (at the stage of the 5th leaf) and the grain of oat were analysed for the content of risk elements. The yield of plants was determined.

5.2. Results of the Experiment and Discussion

The influence of the incorporation of amendments on the behaviour of risk elements and quantity and quality of crops was evaluated. The behaviour of risk elements was assessed using sequential analysis (Zeien and Brümmer, 1989). For Cd, potentially mobilizable fraction was the most important, followed by mobile fraction and fraction bound on Mn oxides (Fig. 12a). Muck and dolomitic limestone were the most efficient in decreasing water-soluble and exchangeable (i.e. mobile) fractions of Cd in soil; synthetic zeolite and loamy shale were less efficient. Peat amendment rather increased the share of mobile fraction of Cd, probably due to its acidity. In case of Pb, fraction bound on Mn oxides and organically bound fraction were the most important. Muck amendment increased the share of organically bound fraction, decreasing the share of potentially mobilizable fraction and fraction bound on Mn oxides. However, the effect of the amendments on Pb speciation was low. In case of Zn, residual fraction was the most important (Fig. 12b). Similarly to Cd, muck and dolomitic limestone were the most efficient in decreasing the share of mobile Zn fraction, followed by synthetic zeolite and loamy shale. In case of As, which was bound mainly on non-crystalline Fe oxides, the effect of used amendments was negligible.

The effect of amendments on risk element speciation was caused to a large extent by changes in soil pH, especially in case of Zn and Cd, as it was confirmed by the use of correlation analysis (Table 7). Soil pH influenced also the element transfer to plants. Nevertheless, the transfer of risk elements was strongly influenced by the crop. The content of Cd and Zn, and to a lower extent of Pb, in the shoots of mustard was reduced after incorporation of the muck, dolomitic limestone, synthetic zeolite, and loamy shale (data not shown). In the contrary, acid peat increased the transfer of Cd, Zn, and Pb into the mustard. The differences in element content between treatments were lower for oat shoots. The differences in the content of As, Cd, Zn and Pb in grains of rye and oat between the treatments were negligible. A positive effect of the amendments (except for acid peat) on the yield of mustard, rye, and oat was found. Generally, remediation of soils polluted with risk elements using different sorbents is possible. It can decrease the share of mobile fraction of the risk elements and thus reduce their bioavailability. The method is environment friendly and relatively cheap. However, it does not solve the problem forever, it only reduces the negative effects of pollution.

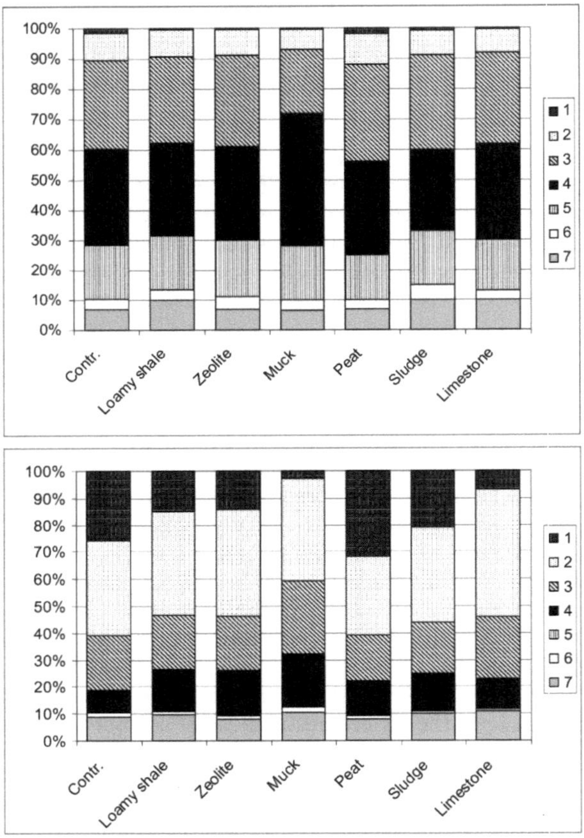

Fig. 12a. Fractionation (by the method described in Zeien and Brümmer, 1989) of risk elements in soil with different amendments. Top – Cd, bottom – Pb (1- mobile fraction, 2 – potentially mobilizable fraction, 3 – fraction bound on Mn oxides, 4 – organically bound fraction, 5 – fraction bound on non-crystalline Fe oxides, 6 – fraction bound on crystalline Fe, 7 – residual fraction).

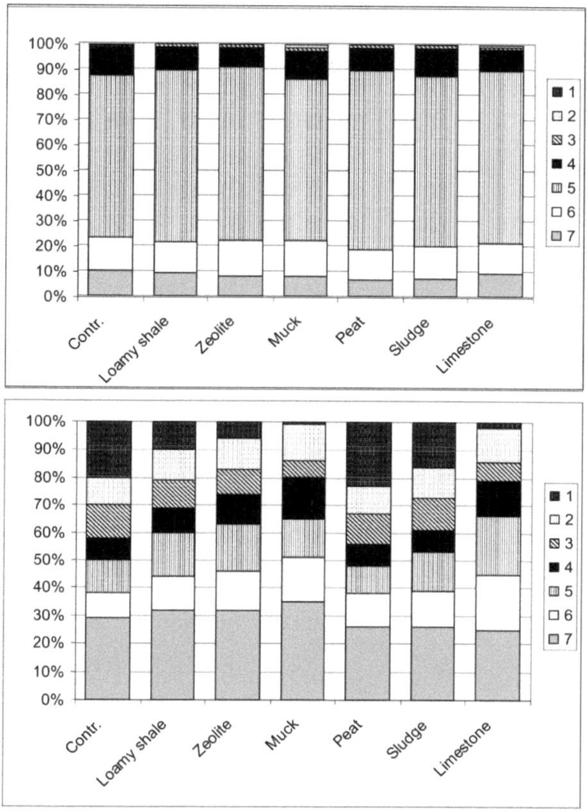

Fig. 12b. Fractionation of risk elements in soil with different amendments. Top – Zn, bottom – As.

Table 7. Correlation coefficients of the relationship between mobile forms of risk elements in soil and their concentrations in plants, and pH values and content of organic carbon (Corg) in soils (remediation trial).

Content in:	Element	pH	C_{org}	Element	pH	C_{org}
Soil – extraction with NH_4NO_3	As	–0.42	0.16	Pb	–0.87	0.30
	Cd	–0.54	–0.13	Zn	–0.97	0.30
Soil – extraction with $CaCl_2$	As	–0.58	0.29	Pb	–0.83	0.26
	Cd	–0.86	0.22	Zn	–0.93	0.32
Soil – extraction with EDTA	As	–0.14	–0.37	Pb	–0.23	–0.35
	Cd	–0.25	–0.04	Zn	–0.57	0.04
Rye-grain	As	–0.51	0.03	Pb	–0.09	0.01
	Cd	0.71	–0.39	Zn	–0.63	–0.03
Oat-grain	As	–0.38	0.08	Pb	0.26	–0.25
	Cd	–0.50	–0.06	Zn	–0.41	0.30
Green oat	As	–0.21	–0.13	Pb	–0.14	0.05
	Cd	–0.65	0.24	Zn	–0.81	0.33
Mustard (1998)	As	–0.55	0.29	Pb	–0.51	0.10
	Cd	–0.90	0.21	Zn	–0.89	0.24
Mustard (2000)	As	–0.50	0.21	Pb	–0.69	0.26
	Cd	–0.79	0.24	Zn	–0.91	0.28

6. CONCLUSIONS

This contribution presented some mehods that can be used in the studies dealing with soils polluted with risk elements. The alluvium of the Litavka River was used as a model area to present exploitation of the methods in practice. Certainly, different methods are useful in different conditions and for different purposes. None of them is universal and absolutely sufficient. Also, the list of presented methods is not complete. There are many other methods suitable for some specific conditions. The methods applicable in an individual situation must be carefully selected and assessed, and if possible combined to fulfil desired objectives. Finally, it must be always kept in mind that results of each analysis of the situation must be correctly interpreted and should be used to find a solution that would decrease the environmental risk of pollution. The decision whether it will be phytoremediation, chemical remediation, or other action is thus the next step.

7. ACKNOWLEDGEMENT

Participation of the following colleagues on the studies and results presented in this contribution is highly acknowledged: E. Petrovsky and A. Kapicka from the Geophysical Institute of the Academy of Science of the Czech Republic, O. Mikanova from the Research Institute of Crop Production in Prague, E. Podlesakova from the Research Institute for Soil and Water Conservation in Prague, S. Kristoufkova, O. Drabek, J. Kozak, and J. Nemecek from the Czech University of Agriculture in Prague, and students M. Smejkalova, M. Vejr, S. Svitakova, N. Roca Pascual, and V. Manez Martinez. The contribution was partially supported by the grant No. 526/02/1516 of the Grant Agency of the Czech Republic.

REFERENCES

1. Adamec, V. (1990): Stanoveni toxickych hladin olova v pude a zpusobu eliminace jejich nezadoucich ucinku (Determination of toxic levels of lead in soil and methods of the elimination of their undesirable effects). *Dissertation.* Research Institute for Soil and Water Conservation, Prague (in Czech).
2. Angehrn-Bettinazzi, C., Thoni, L., and Hertz, J. (1989) An attempt to evaluate some factors affecting the heavy metal accumulation in a forest stand, *Intern. J. Environ. Anal. Chem.* 35, 69-79.
3. Babich, H. and Stotzky, G. (1983) Developing standards for environmental toxicants: The need to consider abiotic environmental factors and microbe-mediated ecological processes, *Environ. Health Prosp.* 49, 247-260.
4. Banin, A., Navrot, J., Noi, Y., and Yoles, D. (1981) Accumulation of heavy metals in arid zone soils irrigated with treated sewage effluents and their uptake by Rhode grass, *J. Environ. Qual.* 10, 536-540.
5. Bell, P.F., James, B.R., and Chaney, R.L. (1991) Heavy metal extractability in long-term sewage sludge and metal salt amended soils, *J. Environ. Qual.* 17, 481-486.
6. Borůvka, L. and Drábek, O. (2000) Extractability of some heavy metal forms from polluted soil, in *Mengen- und Spurenelemente, 20. Arbeitstagung*, Schubert-Verlag, Leipzig, 774-778.
7. Boruvka, L., Drabek, O., and Martinez, V.M. (2002) Distribution of heavy metals between fractions of humic substances in soil, in Anke, M., Muller, R., Schafer, U., and Stoeppler, M. (eds.), *Macro and Trace Elements, Mengen- und Spurenelemente, 21. Workshop*, Schubert-Verlag, Leipzig, 200-203.
8. Boruvka, L., Huan-Wei, Ch., Kozak, J., and Kristoufkova, S. (1996) Heavy contamination of soil with cadmium, lead and zinc in the alluvium of the Litavka river, *Rostl. Vyr.* 42, 543-550.
9. Boruvka, L., Kozak, J., and Kristoufkova, S. (1997) Distribution of Cd, Pb and Zn in plants grown on heavily polluted soils, *Rostl. Vyr.* 43, 249-256.

10. Boruvka, L., Kristoufkova, S., and Kozak, J. (1997) Heavy metal accumulation in plants grown in heavily polluted soils, *Folia Microbiol.* 42, 524-526.
11. Boruvka, L., Kristoufkova, S., Kozak, J., and Huan-Wei, Ch. (1997) Speciation of cadmium, lead and zinc in heavily polluted soils, *Rostl. Vyr.* 43, 187-192.
12. Bradley S.B. and Cox, J.J. (1987) Heavy metals in the Hamps and Manifold Valleys, North Staffordshire, U.K.: Partitioning of metals in flood plain soils, *Sci. Total Environ.* 65, 135-153.
13. Brümmer, G.W., Gerth, J., Herms, U. (1986) Heavy metal species, mobility and availability in soils, *Z. Pflanzenernähr. Bodenk.* 149, 382-398.
14. Campanella, L., Cardarelli, E., Ferri, T., Petronio, B.M., and Pupella, A. (1985) Evaluation of distribution of heavy metals chemical forms in urban sludges, in *Proc. of the Internat. Conf. on "Heavy Metals in the Environment" in Athens*, C.E.P. Consultants Ltd., Edinburgh, 336-338.
15. Collinson, D.W. (1983) *Methods in Rock Magnetism and Palaeomagnetism*, Arrowsmith, Bristol.
16. Davies, B.E. (1985) The plant-availability of Cd, Pb and Zn in soils with reference to soil extractants, pH and redox potential, *J. Sci Food Agric.* 36, 535-536.
17. Einax, J. and Soldt, U. (1995) Geostatistical investigations of polluted soils, *Fresenius J. Anal. Chem.* 351, 48-53.
18. Filip, Z. (2002) International approach to assessing soil quality by ecologically-related biological parameters, *Agriculture, Ecosystems and Environment* 88, 169-174.
19. Förstner, U. (1987) Metal speciation in solid wastes - factors affecting mobility, in Landner, L. (ed.), *Speciation of Metals in Water, Sediment and Soil Systems*, Springer, Berlin, 13-41.
20. Förstner, U. (1993) Metal speciation - general concepts and applications, *Intern. J. Environ. Anal. Chem.* 51, 5-23.
21. Goovaerts, P. and Webster, R. (1994) Scale-dependent correlation between topsoil copper and cobalt concentrations in Scotland, *Eur. J. Soil Sci.* 45, 79-95.
22. Gupta, S. and Häni, H. (1983) Importance of total and biorelevant heavy metals in deciding limits or guidelines on the maximum tolerable metal load in soils, in *Proc. of the Internat. Conf. on "Heavy Metals in the Environment" in Heidelberg*, C.E.P. Consultants Ltd., Edinburgh, 1149-1157.
23. Hay, K.L., Dearing, J.A., Baban, S.M.J., and Loveland, P.J. (1997) A preliminary attempt to identify atmospherically derived pollution particles in English topsoils from magnetic susceptibility measurements, *Phys. Chem. Earth* 22, 207-210.
24. Heller, F., Strzyszcz, Z., and Magiera, T. (1998) Magnetic record of industrial pollution in forest soils of Upper Silesia, Poland, *J. Geophys. Res.* 103, 17767-17774.
25. Hoffmann, V., Knab, M., and Appel, E. (1999) Magnetic susceptibility mapping of roadside pollution, *J. Geochem. Explor.* 99, 313- 326.
26. Houba, V.J.G. and Novozamsky, I. (1993) State of the art and future developments of soil analysis for bioavailability purposes, Discussion Paper for ISO/TC 190/SC 3.
27. Insam, H. and Domsch, K.H. (1988) Relationship between soil organic carbon and microbial biomass on chronosequence of reclamation sites, *Microbial Ecol.* 15, 177-188.
28. Jopony, M. and Young, S.D. (1994) The solid $\leftrightarrow$ solution equilibria of lead and cadmium in polluted soils. *Europ. J. Soil Sci.* 45, 59-70.
29. Kabata-Pendias, A. and Pendias, H. (1992) *Trace Elements in Soils and Plants*, 2nd edition, CRC Press, Boca Raton Ann Arbor, London.
30. Kapicka, A., Petrovsky, E., Ustjak, S., and Machackova, K. (1999) Proxy mapping of fly-ash pollution of soils around a coalburning power plant: a case study in the Czech Republic, *J. Geochem. Explor.* 66, 291-297.
31. König, N., Baccini, P., and Ulrich, B. (1986) Der Einfluß der natürlichen organischen Substanzen auf die Metallverteilung zwischen Boden und Bodenlösung in einem sauren Waldboden, *Z. Pflanzenernähr. Bodenk.* 149, 68-82.
32. Kovanda, F., Kusa, H., and Kolousek, D. (1996) Ion exchange between NH_4^+ saturated zeolites and essential cations of the soil solution (K^+, Na^+, Ca^{2+} and Mg^{2+}), *Ceramics – Silikaty* 41, 61-66.

33. Kozak, J., Janku, J., and Jehlicka, J. (1995) The problems of heavily polluted soils in the Czech Republic: A case study, in Salomons, W., Förstner, U., and Mader, P. (eds.), *Heavy Metals. Problems and Solutions*, Chapter 17, Springer-Verlag, Berlin.
34. Krishnamurti, G.S.R., Huang, P.M., Van Rees, K.C.J., Kozak, L.M. and Rostad, H.P.W. (1995) Speciation of particulate-bound cadmium of soils and its bioavailability, *Analyst* 120, 659-665.
35. Markus, J.A. and McBratney, A.B. (1996) An urban soil study: heavy metals in Glebe, Australia, *Aust. J. Soil Res.* 34, 453-465.
36. McGrath, S.P. and Cunliffe, C.H. (1985) A simplified method for the extraction of the metals Fe, Zn, Cu, Ni, Cd, Pb, Cr, Co and Mn from soils and sewage sludges, *J. Sci. Food Agric.* 36, 794-798.
37. Mench, J.M., Didier, V.L., Löffler, M., Gomez, A., and Masson, P. (1994) A mimicked in-situ remediation study of metal-contaminated soils with emphasis on cadmium and lead, *J. Environ. Qual.* 23, 58-63.
38. Mesto Pribram, Official internet page of the municipality (2002) <http://www.pribram-city.cz/>
39. Mikanova, O., Kubat, J., Mikhailovskaya, N., Vörös, I., and Biro, B. (2001) Influence of pollution with heavy metals on quality of soil in the alluvium of the Litavka river. *Rostl. Vyr.* 47, 117-122.
40. Ministry of the Environment of the Czech Republic (1994) *Order setting up some details of the protection of agricultural land*, No. 13/1994 Coll. (in Czech).
41. Mortvedt, J.J., Mays, D.A., and Osborn, G. (1981) Uptake by wheat of cadmium and other heavy metal contaminants in phosphate fertilizers, *J. Environ. Qual.* 10, 193-197.
42. Nemecek J., Podlesakova, E., and Halova, G. (1995) Background contents of potentially hazardous elements in soils of the Czech Republic (contents extracted with 2M HNO_3), *Rostl. Vyr.* 41, 25-29.
43. Nemecek, J., Podlesakova, E., and Macurova, H. (1998) Indication of soil pollution by microbiological and biochemical methods, *Rostl. Vyr.* 44, 409-417.
44. Pavel, L. and Kozak, J. (1989) Zhodnoceni kontaminace pudy tezkymi kovy metodami vicerozmerne statisticke analyzy. (Evaluation of soil pollution by heavy metals by means of multivariate statistical analysis), *Rostl. Vyr.* 35, 897-904 (in Czech).
45. Petrovsky, E. and Ellwood, B. (1999) Magnetic monitoring of water, air and land pollution, in Maher, B. and Thompson, R. (eds.), *Quaternary Climate, Environment and Magnetism*, Cambridge Univ. Press, UK, 279-322.
46. Petrovsky, E., Kapicka, A., Jordanova, N., and Boruvka, L. (2001) Magnetic properties of alluvial soils contaminated with lead, zinc and cadmium, *J. Appl. Geophys.* 48, 127-136.
47. Petrovsky, E., Kapicka, A., Jordanova, N., Knab, M., and Hoffmann, V. (2000) Low-field magnetic susceptibility: a proxy method of estimating increased pollution of different environmental systems, *Environ. Geol.* 39, 312-318.
48. Podlesakova, E., Nemecek, J., and Halova, G. (1994) Loads of alluvial soils of Labe by hazardous substances, *Rostl. Vyr.* 40, 69-80.
49. Podlesakova, E., Nemecek, J., and Vacha, R. (2001) Mobility and bioavailability of trace elements in soils, in Iskandar, I. K., Kirkham, M. B. (eds.), *Trace Elements in Soil: Bioavailability, Flux, and Transfer*, CRC Press, Boca Raton, 21- 42.
50. Quevauviller, P. (ed.) (2002) Methodologies in Soil and Sediment Fractionation Studies. Single and Sequential Extraction Procedures, Royal Society of Chemistry, Cambridge.
51. Ramos, L., Hernandez, L.M., and Gonzalez, M.J. (1994) Sequential fractionation of copper, lead, cadmium and zinc in soils from or near Donana National Park, *J. Environ. Qual.* 23, 50-57.
52. Randall, C.W. - Donaldson, C.D., Wigington, P.J.Jr., Grizzard, T.J., and Knocke, W.R. (1983) Evaluation of optimal techniques for the partial extraction of heavy metals from soils, in *Proc. of the Internat. Conf. on "Heavy Metals in the Environment" in Heidelberg*, C.E.P. Consultants Ltd., Edinburgh, 257-261.

53. Reimann, C., Filzmoser, P., and Garrett, R.G. (2002) Factor analysis applied to regional geochemical data: problems and possibilities, *Applied Geochem.* 17, 185-206.
54. Rieuwerts, J.S. (1993) Heavy metal contamination of soils in the vicinity of a secondary lead smelter in the Czech Republic, A report submitted in partial fulfilment of the requirements for the MSc and/or the DIC, Imperial College, London, 1993.
55. Sauerbeck, D.R. and Styperek, P. (1985) Evaluation of chemical methods for assessing the Cd and Zn availability from different soils and sources, in Leschber, R., Davis, R.D., and L'Hermite (eds.), *Chemical Methods for Assessing Bio-Available Metals in Sludges and Soils*, Elsevier Applied Science Publ., London - New York, 49-67.
56. Serra-Wittling C., Houot S., and Barriuso E. (1995) Soil enzymatic response to addition of municipal solid-waste compost, *Biol. Fertil. of Soils* 20, 226-236.
57. Sheppard M.I. and Thibault, D.H. (1992) Desorption and extraction of selected heavy metals from soils, *Soil Sci. Soc. Am. J.* 56, 415-423.
58. Smejkalova, M., Mikanova, O., and Boruvka, L. (2001) Vliv kontaminace pudy v nive reky Litavky na slozeni a aktivitu pudni mikroflory (Influence of soil pollution in the alluvium of the Litavka River on the composition and activity of soil microflora), in Rejsek, K. and Houska, J. (eds), *Pedologicke dny 2001*, MZLU, Brno, 140-144 (in Czech).
59. Tessier, A., Campbell, P.G.C., and Bisson, M. (1979) Sequential extraction procedure for speciation of particulate trace metals, *Analytical Chemistry* 51, 844-851.
60. Tlustos, P., Dijk, D. van, Szakova, J., and Pavlikova, D. (1994) Extraction of Cd and Zn from soil by selected extractants, *Rostl. Vyr.* 40, 1107-1121.
61. Ure, A., Quevauviller, P., Muntau, H., and Griepink, B. (1993) Speciation of heavy metals in soils and sediments. An account of the improvement and harmonization of extraction techniques undertaken under auspices of the BCR of the Commission of the European Communities. *Intern. J. Environ. Anal. Chem.* 51, 135-151.
62. Vacha, R., Nemecek, J., and Podlesakova, E. (2002) Significant aspects of geochemical and anthropogenic soil loads by potentially risk elements, *Rostl. Vyr.* 48, 336-342.
63. Vacha, R., Podlesakova, E., Nemecek, J., and Polacek, O. (2002) Immobilisation of As, Cd, Pb and Zn in agricultural soils by the use of organic and inorganic additives. *Rostl. Vyr.* 48, 335-342.
64. Vacha, R., Podlesakova, E., Nemecek, J., and Polacek, O. (2002) Immobilisation of As, Cd, Pb and Zn in the soil by the use of inorganic and organic amendments, Paper no. 554, 17th World Congress of Soil Science, Bangkok, Thailand.
65. Waller, P.A. and Pickering, W.F. (1993) The effect of pH on the lability of lead and cadmium sorbed on humic acid particles, *Chemical Speciation and Bioavailability* 5, 11-22.
66. Webster, R., Atteia, O., and Dubois, J.-P. (1994) Coregionalization of trace metals in the soil in the Swiss Jura, *Eur. J. Soil Sci.* 45, 205-218.
67. Wopereis, M.C., Gascuel-Odoux, C., Bourrie, G., and Soignet, G. (1988) Spatial variability of heavy metals in soil on a one-hectare scale, *Soil Sci.* 146, 113-118.
68. Young, L.B., Dutton, M., and Pick, F.R. (1992) Contrasting two methods for determining trace metal partitioning in oxidized lake sediments, *Biogeochemistry* 17, 205-219.
69. Zeien, H. and Brümmer, G.W. (1989) Chemical extraction for the determining of trace elements forms in the soils, *Mitt. Dtsch. Bodenk. Gesell.* 59, 505-510 (in German).

CHAPTER 10

PLANT UPTAKE OF RADIOCAESIUM – POTENTIAL FOR REMEDIATE RADIOPOLLUTED SOILS

N. GONCHAROVA
Laboratory of Environmental Monitoring,
International Sakharov Environmental University,
23 Dolgobrodskaya Str., 220009 MINSK, Belarus

Abstract

In this work it is studied the impact of some environmental and physiological factors, such as density of sowing, watering, CEC (Cation Exchange Capacity) and presence of AM (*Arbuscular mycorrhizal*) fungal hyphae, on the plant ability to uptake and accumulate radiocaesium by roots. Five plant species were used: pea, wheat, soybean, barley and oats. The obtained results show that different density of sowing and alteration of the soil moisture lead to a considerable alteration of CEC and of the level of radiopollution for all used plants. It was found that the density of sowing increases both the CEC of plant tissues (leaves and stem) and the total content of ^{137}Cs in plants. Also, soil moisture increases the plant uptake of radiocaesium. Concerning the hyphae of AM fungus, it was found that it can absorb radiocaesium from the soil and transfer this radiopollutant to the plant with the following order of magnitude: soybean > pea > barley > wheat > oats. Moreover, it was observed that plant species can differentiate the ^{137}Cs uptake in all experiments. This study indicates that careful selection of plant species could enhance the perspectives of plant uses to remediate radiopolluted lands.

1. INTRODUCTION

Wide scale of pollution of agricultural lands with long-lived radionuclides is harmful to human health (Karaoglou *et al.*, 1996). Plants absorb radionuclides discharged into the environment and they may penetrate into the human body though the food chain. Therefore it is important to determine the factors that affect the uptake of radiocaesium by plants in the soil-plant system.

It is well known that there are mainly two groups of factors, which determine the level of radionuclides accumulation in plants. The first group

J.-L. Morel et al. (eds.), Phytoremediation of Metal-Contaminated Soils, 299–306.

is connected with the determination of the level of bioavailability of pollutants in soils; the second group of factors determines the ability of higher plants to absorb the radionuclides and their capacity to be accumulated into the biomass (Grodzinskiy, 1991; Kra-Vets, 1992 and Goncharova, 2001).

The accent in the investigations related to the soil-plant relationship of radionuclides migration and the control of this process, is mostly put on the soil, because of the dependence of pollutants accumulation in plants upon the type of soil and its physicochemical characteristics.

At the same time, in accordance with modern ecophysiological and radioecological data, different organisms in soil and process in the rhizoshere can play an important role in pollutant migration. In fact, metabolites of soil micro-organisms and, especially, root excretion of higher plants and arbuscular mycorrhizal (AM) fungi increase the rate of destruction of solid forms of pollutants and enhance their leaching from the soil matrix (Suzuki *et al.*, 2001).

The transport of elements into plants through the hyphae of AM fungi has been demonstrated for P and Cu (Li *et al.*, 1991), N (Ames *et al.*, 1983), Zn (Kothari *et al.*, 1991; Burkert and Robson 1994), Cd (Joner and Leyval,

Table 1. Effect of the different density of sowing on cation exchange capacity (CEC) of the stem tissues and on their total content of ^{137}Cs.

Variant	CEC of the stem tissues, meq\100g dry biomass	^{137}Cs content, x 10^4 Bq
Pea		
10 plants per pot	97.6 ± 2.5	2.4 ± 0.2
15 plants per pot	102.5 ± 5.2	2.6 ± 0.4
25 plants per pot	125.6 ± 3.4	3.4 ± 0.4
Wheat		
10 plants per pot	70.6 ± 2.8	2.2 ± 0.3
15 plants per pot	84.8 ± 7.2	2.5 ± 0.2
25 plants per pot	144.7 ± 3.8	4.6 ± 0.5
Barley		
10 plants per pot	54.4 ± 1.8	2.4 ± 0.3
15 plants per pot	74.7 ± 5.4	2.3 ± 0.3
25 plants per pot	195.6 ± 13.8	3.8 ± 0.7
Oats		
10 plants per pot	54.8 ± 2.9	2.6 ± 0.3
15 plants per pot	68.6 ± 5.2	3.8 ± 0.4
25 plants per pot	115.5 ± 3.2	4.6 ± 0.7
Soybean		
10 plants per pot	155.6 ± 5.2	4.2 ± 0.5
15 plants per pot	197.6 ± 7.6	4.6 ± 0.6
25 plants per pot	218.8 ± 8.2	6.2 ± 0.8

Table 2. Effect of different level of the soil moisture on the cation exchange capacity of the stem tissues and ^{137}Cs concentration of plants.

Variant	CEC, meq\100g dry mass	^{137}Cs concentration, Bq\kg
Pea		
Control	86.8 ± 3.1	3.2 ± 0.5
Watering	116.2 ± 2.8	4.2 ± 0.6
Wheat		
Control	50.5 ± 2.9	2.2± 0.7
Watering	122.6 ± 3.8	3.4 ± 0.2
Barley		
Control	46.2 ± 2.8	2.4 ± 0.3
Watering	145.7 ± 3.8	3.8 ± 0.7
Oats		
Control	45.6 ± 1.9	2.6 ± 0.4
Watering	114.5 ± 3.4	3.4 ± 0.8
Soybean		
Control	124.5 ± 3.8	4.2 ± 0.4
Watering	198.7 ± 7.5	6.4 ± 0.6

1997) and Fe (Caris *et al.,* 1998). For other elements, such as Sr and Cs, although their uptake was found to increase by AM colonization (Jackson *et al.,* 1973; Rogers and Williams, 1986), the contribution of the hyphae of the AM fungus to the uptake of these elements has not been elucidated.

The objective of the present study was to investigate the impact of some environmental and physiological factors, such as watering, density of sowing, CEC and presence of AM fungal hyphae on the plant ability to uptake and accumulate radiocaesium by roots. In fact, it is known that some of the main factors that are of importance from the agroecological point of view are the level of watering and moisture of soil and the density of the plants sowing. Finally, this experimental study was extended to take into account some agricultural plant species.

2. MATERIAL AND METHODS

2.1. Experimental and Measurement Systems

The experiments were carried out under greenhouse conditions and have been repeated four times. The type of the used soil was sodi-podzolic sandy loam. Three levels of density of sowing were used, namely 10, 15 and 25 plants per pot, the pot surface was approx. $0.05m^2$ and the quantity of soil was 5 kg per pot (dry soil). The used soil was from a long intensive tillage, in which it was studied the AM fungal populations using the mycorrhizal inoculation method of Suzuki (Suzuki *et al.*, 2000). Investigations have been carried out on five plant species: pea, wheat, soybean, barley and oats. The above plants were 14 days old. Pea, soybean, barley and oats were diploid varieties and wheat was hexaploid. These species have differences in their root systems: soybean and pea (*Fabaceae*) have a "core-type" root system and wheat, barley and oats (*Graminaceae*) have a fibrous root system.

Plants were grown under a 120 $\mu Em^{-2} s^{-1}$ illumination (14h) and 25° C of temperature. The total field moisture capacity of this soil type is normally 30%, but in the condition of this experimental work the total moisture was ranged from 50% to 100%.

^{137}Cs have been brought in the soil in form of soluble salt $^{137}CsNO_3$ with a concentration of 10^{-5} Mol/l and a specific activity of the used solution of 20 kBq/ml.

Cation exchange capacity (CEC) of the intact plant tissues has been defined using the method of Crooke (Knight *et al.,* 1961) as it was modified by Blarney (Blarney, 1990).

Radioactivity of soil and plant samples was measured by gamma-spectrometry using a high – purity Ge(Li) detector. Measurement times were extended until the counting error was reduced to less than 10%.

The uptake of radiocaesium by the plants (concentration) and the relative migration were defined as:

Concentration of ^{137}Cs = Bq /kg of dry weight of plant.

Relative migration (% of migration) = (radioactivity that migrated into the Root Compartment/total amount of radioactivity added to the soil) x 100.

Data were expressed as the mean ± SD (standard deviation).

2.2. Mycorrhizal Colonization and Count of Spore Number

After the radioactivity measurement, the plants roots were cleaned and stained with trypan blue, and the percentages of AM colonization and colonized root length were assessed by the grid-line intercept method (Giovannetti and Mosse 1980). To confirm the penetration of hyphae into the "hyphal compartment" (HC), the number of spores of the AM fungus in the sand of the HC was counted. The spores in each sand sample (10 g on a dry weight basis) were collected by the wet sieving method using 56 and 250 μm sieves.

3. RESULTS AND DISCUSSION

The data in Tables 1 show the effect of different densities of sowing on cation exchange capacity (CEC) of the stem tissues and on their total content of ^{137}Cs. In fact different density of sowing leads to a considerable alteration of CEC and of the level of radiopollution for all species of plants that have been used in our experiments.

Pea show increased values of cation exchange capacity and of radionuclides content (1.3 and 1.4 times respectively) as a response to an increasing density of sowing by 2.5 times; wheat showed 2 times more for both cases; barley more than 3.6 and 1.6 times respectively; oats 2.1 and 1.8 times more and soybean - 1.4 and 1.5 times more respectively. As in the above experiment with different density of sowing, the alteration of the soil moisture leads to modification of CEC of the plant biomass and of caesium concentration, too (Table 2).

Pea show increased values of CEC capacity and of radionuclide concentration accumulation by 1.3 times as a response to the increasing level of the soil moisture by a factor of 2, wheat show values 2.4 and 1.5 times more, barley 3.2 and 1.6 times more, oats 2.5 and 1.3 times more and soybean 1.6 and 1.5 times more respectively. These results confirm that the moisture of soil and the type of plant are very important factors that determine the degree of mobility of any pollutant in the "soil-plant" system.

It is noticeable that there is a difference in the degree of response of different plant species to the factors that we studied. As an example, soybean showed a caesium concentration approximately two times higher that this one of wheat, both in control and watering conditions.

Concerning the hyphae of the AM fungus, it was found that caesium was absorbed and transported through the hyphae (Fig.1).

Our results show that the hyphae of the AM fungus can absorb radiocaesium from the soil and transfer this radiopollutant to the plant. The absorption ability of ^{137}Cs by the AM fungal hyphae may be low. Under field conditions, the actual contribution of the AM fungus to the uptake of this element is related to the density and viability of the hyphae. The hyphal viability may vary not only with the plant-fungus combination but also with the chemical, physical and biological conditions of soil. However, to analyse the real contribution of the AM fungus, additional investigation on the relationship among hyphal viability, hyphal density and uptake of caesium should be carried out.

For ^{137}Cs uptake by the AM fungus the following order has been observed:

Soybean > pea > barley > wheat > oats.

The results also emphasize the need of using different plant species for isolation cultures, because of unknown host-fungus preferences.

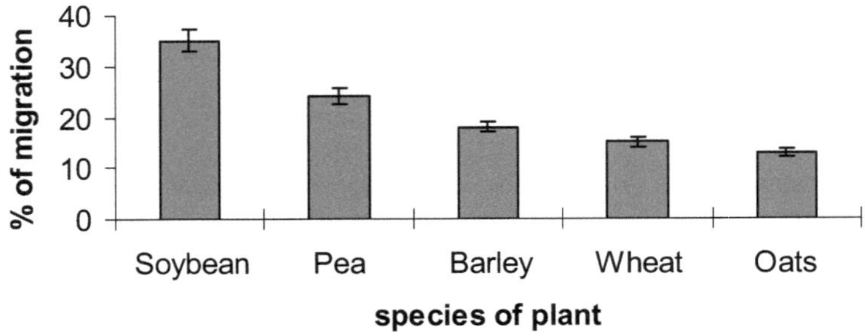

Fig. 1. Relative migration (%) of radiocaesium into the "root compartment" (RC) of various plants (bare shows the standard deviation).

4. CONCLUSIONS

This work was focused on the plant CEC and on the regulation of the uptake of radiocaesium by the plants, and especially on the: effect of different density of sowing, effect of different level of soil moisture, mycorrhiza influence and uptake differences of plant species.We found that the density of sowing increases both the CEC of plant tissues (leaves and stem) and the total content of ^{137}Cs in plants. In a modelling study (Kirk and Staunton, 1989) it was found that the higher the density of roots in the soil the more ^{137}Cs was accumulated by a wide variety of grassland plants. Although the ability to accumulated radionuclides varies among a wide array of plant species occupying different habitats, many plants growing on contaminated soil have been shown to accumulate radionuclides, especially ^{137}Cs and ^{90}Sr (Pinder *et al.*, 1984; Salt *et al.*, 1992,1997).

We also found that the plant uptake of radiocaesium can be dependent from soil moisture. This data agree with the opinion that soil properties, especially soil moisture, play an important role on the process of soil-to-plant transfer of radiocaesium (Arapis *et al.*, 1999).

Moreover, the relative migration of radiocaesium into the "root compartment" is influenced by the AM of plant species. It was found that inoculation with AM increased root biomass which resulted in greater quantities of ^{137}Cs accumulation from the soil (Entry *et al.*, 1998).

Finally, the plant species differentiated the ^{137}Cs uptake in all cases of our experiment. It is of importance to notice that data on variation due to plant species might be integrated into mathematical models predicting the fate of radiocesium in various soil-plant systems and/or assessing the radiological risk. Moreover, a selection of plants species for the purpose of phytoremediation of soil contaminated with high levels of radiocaesium could be foreseen.

REFERENCES

1. Ames (R.N), Reid (C.P.P), Porter (L.K) & Cambardella (C), 1983 – Hyphal uptake and transport of nitrogen from ^{15}N-labelled sources by *Glomus mosseae,* a vesicular-arbuscular mycorrhizal fungus. *New Phytol.,* 95: 381-396.
2. Arapis (G), Chesnakov (A), Ivanova (T), Potapov (V), Sokolik (G), 1999 – Evaluation of dose equivalent rate reduction as a function of vertical migration of ^{137}Cs in contaminated soils. *J. Environ Radioactivity,* 46: 251-263.
3. Blarney (F.P), 1990 - Role of Root Cation Exchange Capacity in Differential Aluminium Tolerance of Litus Species. *J. of Plant Nutrition,* 13, (6): 729-745.
4. Burkert (B) and Robson (A), 1994 - ^{65}Zn uptake in subterranean clover *(Trifolium subterraneum L.)* by three vesicular-arbuscular mycorrhizal fungi in root-free sandy soil. *Soil Biol. Biochem.*, 26: 1117-1124.
5. Caris (C), Hordt (W), Hawkins (H.-J), Romheld (V) & George (E), 1998 – Studies of iron transport by arbuscular mycorrhizal hyphae from soil to penut and sorghum plants. *Micorrhiza,* 8: 35-39.
6. Eutry (J.A.), Watrud (L.C.), Relves (M), 1999 – Accumulation of ^{137} Cs and ^{90}Sr from contaminated soil by three grass species inoculated with mycorrizal fungi. *Environmental pollution,* 104: 449-457.
7. Giovannetti (M) and Mosse (B), 1980 – An evaluation of techniques for measuring vesicular-arbuscular mycorrhizal infection in roots. *New Phytol.,* 84: 480-500.
8. Goncharova (N), 2001 –The transport of ^{137}Cs and ^{90}Sr across root cell membranes.
9. *Revue de Cytologie et Biologie vegetales-Le Botaniste,* 23 (1/2): 15- 20.
10. Grodzinskiy (D.M.), 1991- Anthropogenic radionuclide anomaly and plants. Kiev, Lybid: 158 p. (in Russian).
11. Jackson (N.E.), Miller (R.H.) & Franklin (R.E.), 1973 – The influence of vesicular-arbuscular mycorrhizae on uptake of ^{90}Sr from soil by soybeans. *Soil Biol. Biochem.,* 5: 205-211.
12. Joner (E.J) and Leyval (C), 1997 – Uptake of ^{109}Cd by roots and hyphae of *Glomus mosseae*/ Trifolium subterraneum mycorrhiza from soil amended with high and low concentration of cadmium. *New Phytol.,* 135, 353-360.
13. Karaoglou (A), Desmet (G), Kelly (G.N) & Menzel (H.G.), 1996 – The radiological consequences of the Chernobyl accident. EUR 16544 EN. European Commission and the Belarus, Russian and Ukrainian Ministries of Chernobyl Affairs, Emergency Situations and Health.
14. Knight (A.M.), Crooke (W.M.), Inkson (R.H.E.), 1961 - Cation-exchange capacities of tissues of higher and lower plants and their related uronic acid contents. *Nature,* 192: 142-143.
15. Kirk (G.J.D), Staunton, (S), 1989 – On predicting the fate of radioactive cesium in soil beneath grass land. *Journal of Soil Science,* 40:71-84.
16. Kothari (S.K.), Marshner (H) & Romheld (V), 1991 – Contribution of MA mycorrhizal hyphae in acquisition of phosporus and zinc by maize grown in calcareous soil. *Plant Soil,* 131: 177-185.
17. Kravets (A.P), 1992 - Physiological factors absorption nonbiogenic ions by higher plant. *Physiol. Plant.,* 85, (5): 238-239.
18. Li (X,-L), George (E) & Marshner (H). 1991a - Extension of phosphorus depletion zone in MA – mycorrhizal white clover in a calcareous soil. *Plant Soil,* 136: 41-48.
19. Li (X,-L), George (E) & Marshner (H). 1991b - Acquisition of phosphorus and copper by MA – mycorrhizal hyphae and root–to–shoot transport in white clover. *Plant Soil,* 136: 49-57.
20. Pinder III (J.E.), Mc Leod (K. W), Alberts (J.J)., Adriano (D.C.), Corey (J.C.) 1984 – Uptake of ^{244}Cm and ^{238}Pu and other radionuclides by trees inhabiting a contaminated floodplain. *Health Physics,* 47: 375-384.

21. Rogers (R.D.) and Williams (S.E.) 1986- Vesicular – arbuscular mycorrhiza: Influence on plant uptake of cesium and cobalt. *Soil Biol. Biochem.,* 18: 371-376.
22. Salt (C.A), Kay (J.W), Donaldson (L), Woolsey, (J.M).1997 – The influence of defoliation intensity, season, and leaf age on radiocaesium concentrations in *Agrostis capillaries*. *Journal of Applied Ecology,* 34: 1177-1189.
23. Salt (C.A), Mages (E.S), Elston (D.A), 1992 – effects of season, grazing intensity and diet composition on the radiocaesium in take by sheep on a re-seeded hill pasture. Journal of Applied Ecology, 29: 378-387.
24. Suzuki (H), Kumagai (H), Oohashi (K), Sakamoto (K), Inubushi (K), Enomoto (S) & Ambe (F), 2000 - Uptake of 15 trace elements in arbuscular mycorrhizal marigold measured by the multitracer technique. *Soil Sci. Plant Nutr.,* 46: 283-289.
25. Suzuki (H.), Kumagai (H), Oohashi (K), Sakamoto (K), Inubushi (K) & Enomoto (S). 2001- Transport of Trace Elements through the Hyphae of an Arbuscular Mycorrhizal Fungus into Marigold Determined by the Multitracer Technique. *Soil Sci. Plant Nutr.,* 47(1): 131-137.

CHAPTER 11

STUDY OF CD ACCUMULATION BY THLASPI CAERULESCENS UNDER PRESENCE OF GLUCURONIC ACID CONTAINING EXOPOLYSACCHARIDES

A. LOPAREVA
International Sakharov Environmental University, Dolgobrodskaya str., 23, 220009, Minsk, Belarus

Abstract

The problem of environmental pollution by heavy metals is actual nowadays. Phytoremediation – use of plants to extract or to stabilize environmental pollutants – is an attractive alternative to phisico-chemical methods due to its low cost and high efficiency .The exudates of bacteria in plant rhizosphere play an important role in plant-microbial interactions. The role of plant exudates on the behavior of heavy metals in soil and mechanism of its accumulation by plants needs more deep understanding. Microbially driven processes are essential also for mineral weathering , where a complex role is played by microbial polysaccharides and proteinaceous structures, which can either inhibit mineral dissolution under some conditions by forming biofilms or promote chemical weathering by producing extracellular polymers of acidic nature and increase metal bio- and phytoavailability and thus affecting the fertility of soils.

In this work influence of exopolysaccharides produced by symbiotic bacteria Rhizobiaceae on Cd in two forms in plant rhizosphere and it's accumulation in shoots and roots of metal hyperaccumulator plant *Thlaspi caerulescens* was studied.

1. INTRODUCTION

The problem of soil pollution by heavy metals because of various human and natural activities received much attention recently. Heavy metals are difficult to remove from the environment and unlike many other pollutants cannot be chemically or biologically degraded. Cd environmental pollution mainly from mining and smelting, dispersal of sewage sludge and the use of phosphate fertilizers is an increasing problem. Agricultural soils are mainly

J.-L. Morel et al. (eds.), Phytoremediation of Metal-Contaminated Soils, 307–314.

contaminated with Cd from the excessive use of phosphate fertilizers, dispersal of sewage sludge and atmospheric deposition (Mejáre, Bülow, 2001). Cd is readily taken up by many crops including cereals, potatoes, vegetables and fruits.

Phytoremediation – use of plants to extract or to stabilize environmental pollutants – is an attractive alternative to phisico-chemical methods due to its low cost and high efficiency. One of possible strategies is phytoextraction of heavy metals by the hyperaccumulator plants. Phytoextraction is the use of plants to remove toxic elements of excess of contaminated environments (Baker AJM, et al., 1994). The metal is extracted from soil and translocated to harvestable parts of the plant.

The exudates of bacteria in plant rhizosphere play an important role in plant-microbial interactions. The role of plant exudates on the behavior of heavy metals in soil and mechanism of its accumulation by plants needs more deep understanding (Shumate, et al., 1985). Among 400 species of hyperaccumulator plant known to the moment, Thlaspi caerulescens Presl is widely investigated as it is the single known Cd-hyperaccumulator (McGrath, 1998). Microbially driven processes are essential also for mineral weathering (Banfield, et al., 1999), where a complex role is played by microbial polysaccharides and proteinaceous structures, which can either inhibit mineral dissolution under some conditions by forming biofilms or promote chemical weathering by producing extracellular polymers of acidic nature and increase metal bio- and phytoavailability and thus affecting the fertility of soils. The role of rhizobium polysaccharides in plant-host interactions is quite well studied (Dudman, 1984), but their capacity to absorb heavy metals was investigated only recently (Mamaril, et al., 1989).

The effectiveness of phytoextraction depends on the capacity of hyperaccumulator plants to grow, to develop their root system, to accumulate available metals into vegetable parts of plant (Schwartz, et al., 1999). Increasing metal phytoavailability may be beneficial for phytoextraction and recovery using (hyper) accumulating plants, whereas an essential reduction in toxic metal phytoavailability (phytostabilization) may be promising for planting cultural crops.

2. MATERIALS AND METHODS

In this work influence of exopolysaccharides produced by symbiotic bacteria Rhizobiaceae on Cd in two forms in plant rhizosphere and it's accumulation in shoots and roots of metal hyperaccumulator plant Thlaspi caerulescens was studied. During our previous study (Lopareva, et al., 2003) from Medicago plants were isolated mutant Rhizobiaceae strains producing new EPS with perspective properties. One, named strain 5 (Str5) consists of 50% glucuronic acid and of glucose and galactose. The second strain used in this study is glucuronan (Glu) consists of 100% glucuronic acid (patented and studied at LPMV, Amiens, France). EPS were obtained in fermentor, and precipitated by ethanol and lyophilized. Two polysaccharides (0.1 g/L) were injected in Thlaspi caerulescens J&C Presl (Brassicaceae) rhizosphere in sterile conditions during 5 weeks on 90 ml modified media Hoagland and Arnon (1938) in hydroponics pots. Experiment was conducted in hydroponics in phytotron with controlled standard conditions. Cd was added in two forms: a) Cd metal Cd0 280 mg/L (Prolabo, Cd shots, 99.5%) and b) $Cd(NO_3)_2$ 100 µg/L. Cd pur solubilisation experiment was conducted in 100 ml plastic pots during 24 hours at 28°C with manual shaking 3 times per day. Metal concentrations and nutritive media were the same as in hydroponics experiment with plants. Morphological characterization of plants was performed after scanning plant roots in special equipment (WinRhizo scan system, Regent Instruments Image Analysis system) (Allison, et al., 1991). Total Cd content was analyzed after sulfuric acid mineralization by atomic absorption spectrometry (Spectraa 220, Varian, Zeeman effect oven). ANOVA, Tukey multiple comparisons test and Lilliefors test for normality were realized with statistical software STATISTIKA 5.0.

3. RESULTS AND DISCUSSIONS

Influence of EPS of different content on morphological parameters and plants dry weight were analyzed. Root surface area of plants from Cd salt contaminated media treated with glucuronan was 1.4 times greater than Str5 ($p < 0.05$, ANOVA, Tukey multiple comparisons test). Roots length of plants from Cd0 contaminated media treated with str5 was 1.6 times greater than with glucuronan (Fig. 1).

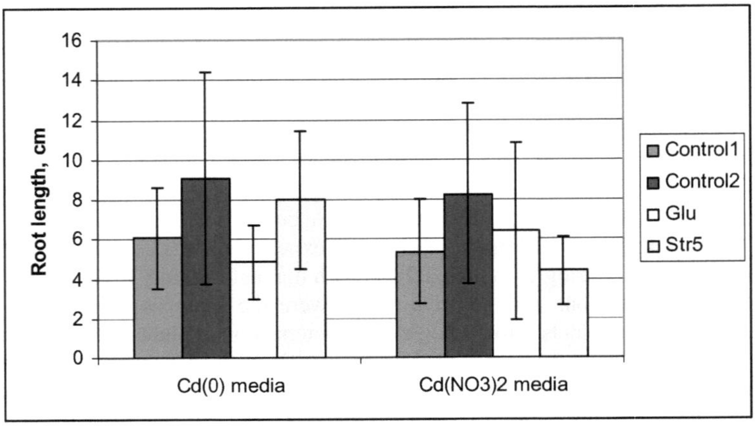

Fig. 1. Root length of *Thlaspi caerulescens* plants, growing on hydroponic media containing Cd in two forms: Cd0 – metal shots and $Cd(NO_3)_2$ in presence of EPS glucuronan (Glu), Str5. Control 1 – media without EPS, Control 2 – media without EPS and without Cd.

Comparison of root morphology in differently contaminated media showed only one significant difference for treatment with EPS from Str5: root's length and surface area for plants treated with Str 5 when growing in nutritive media with Cd shots was 1.7-1.8 times greater than the same ones growing on Cd(NO3)2 ($p < 0.01$ ANOVA, Tukey multiple comparisons test). So influence of strain 5 EPS on plant growth and development depends on metal form and possible metal concentration, as Cd concentration in Cd pure media highly exceeded Cd concentration in Cd salt media. EPS presence in rizhosphere did not influence plant dry weight.

Experiments with influence of EPS on metal dissolution in free form (Cd0) showed that 100% glucuronic acid EPS significantly stimulate metal dissolution compared to control (Fig. 2).

This effect is more significant in distilled water (UHQ) than in nutritive media for plants. In water Cd concentration after 24 hours treatment with glucuronan was 3.8 times greater than with EPS from strain 5, in nutritive media this ratio was only 1.3. Str5 containing 50% of glucuronic acid was not as efficient in Cd dissolution.

Our results are in accordance with a study of Welch and colleagues (Welch, et al., 1999) who demonstrated that, like low molecular weight organic ligands, the high molecular weight acid polysaccharides can also affect mineral weathering reactions. Acid polysaccharides tended to accelerate the dissolution reaction, and increase the total amount of material released to solution compared to inorganic conditions.

Treatment of plants with glucuronan and EPS from strain 5 stimulated Cd dissolution from pure metal form (Cd0) and its concentration in plant

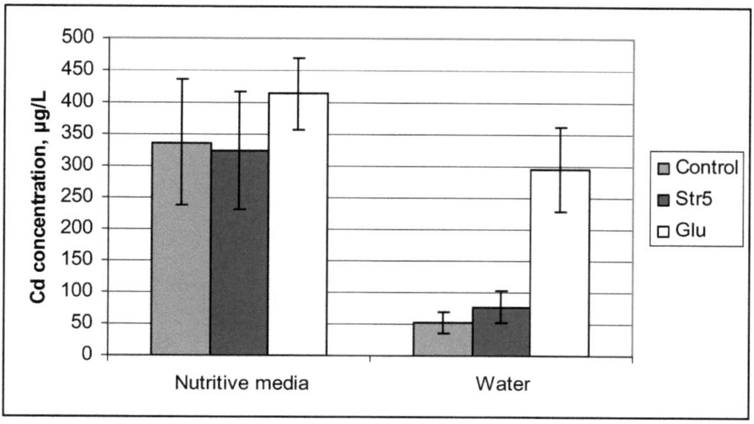

Fig. 2. Cd0 - metal in free form dissolution during 24 hours treatment with EPS in distilled water and nutritive media.

rhizosphere and roots – in final solution after 5 weeks treatment Cd concentration was 6-7 times greater in presence of Str5 and glucuronan than in control pots (Fig. 3a).

Despite the fact that polysaccharide presented in solution stimulated dissolution of pure metal and Cd concentration in solution was high, this

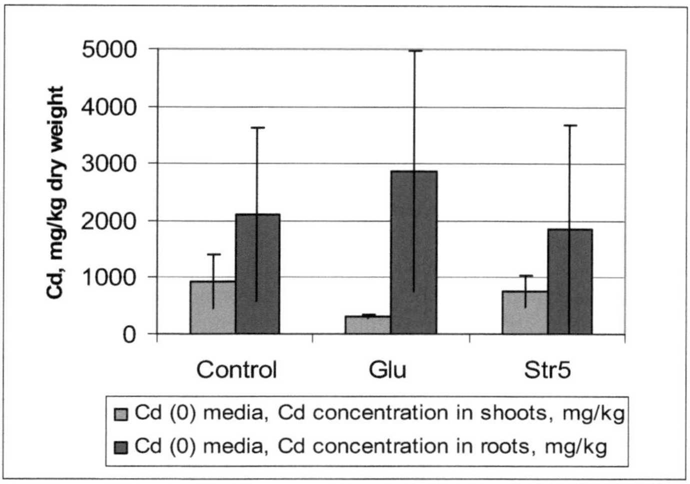

Fig. 3a. *Cd concentration in biomasse (roots and shoots) of hyperaccumulator plant T. Caerulescens grown on hydroponic media containing Cd0 metal in presence of EPS: Glu and Str5; and control without EPS.*

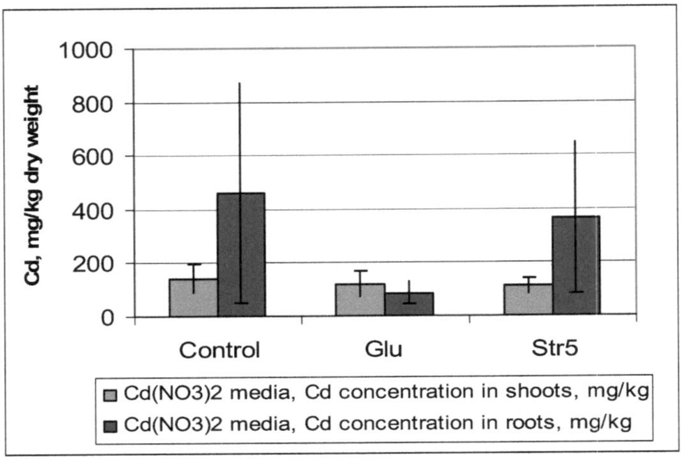

Fig. 3b. Cd concentration in biomasse (roots and shoots) of hyperaccumulator plant *T. Caerulescens* grown on hydroponic media containing $Cd(NO_3)_2$ in presence of EPS: Glu and Str5; and control without EPS.

liberated Cd did not transfer to shoots. Cd concentration in shoots treated with glucuronan was 3 times less than treated with control. But Cd concentration in roots was slightly (1.6 times) greater than in control case.

In experiment with Cd salt containing media presence of glucuronan decreased Cd concentration both in shoots and roots comparing to control and to Str5 effect (Fig. 3b). EPS from strain 5 which have less glucuronic acid resulted in no significant difference from control. It is interesting to remark that glucuronan which enhanced pure metal solubilisation, in experiment with metal salt led to decrease of Cd concentration even in roots. Possibly this effect can be explained by the fact that Cd salt concentration was low and EPS binding sites complexed free Cd ions.

This result can be regarded in two ways. At first, the aim of our experiment was to find how the presence of biopolymers which help to dissolve pure metal will influence Cd accumulation by T.caerulescens – a hyperaccumulator plant. But the fact that glucuronan, the polymer consisting of glucuronic acid chains, can regulate Cd accumulation in plant shoots let us propose further experiments with crop plants in soil in double culture with leguminous plants – hosts of rhizobium bacteria, producing EPS.

So we tested two polysaccharides with different glucuronic acid content for influence on metal in two forms: Cd0 and Cd(NO3)2. The EPS did increased Cd concentration in solution after treatment, glucuronan (consist of 100% glucuronic acid) was more efficient than EPS from Str5 (50% glucuronic acid).

But this relationship is complex and need more investigation as higher Cd ion concentration in nutritive media entailed decrease of Cd concentration in shoots of metal hyperaccumulator T.caerulescens. EPS stimulated Cd concentraton in plant rhizosphere.

Biosorption processes are mainly reactions of pseudo-ionic exchange, when metal ion is exchanged to contre-ion of biomasse. Several functional groups play role in biosorption process depending on pH and metal properties. Efficiency of metal sorption by bacteria link with the fact that they produce extracellular polysaccharides (EPS) containing uronic acids (Cotoras, et al., 1992).

4. ACKNOWLEDGEMENT

Author is grateful to Professors Josiane and Bernard Courtois for EPS (glucuronan, property of LPMV-CNRS FRE 2779, IUT Amiens and Str5) used in this study. The study was supported by NATO basic fellowship "Effects of cultivation in association of heavy metal hyperacumulator plants and leguminous plants in contaminated soils" and realised in LSE, ENSAIA/INPL INRA, Nancy, France.

REFERENCES

1. Allison, J. D., Brown, D. S., Novo-Gradac, K. J. MINTEQA2/ PRODEFA2, A geochemical assessment model for environmental systems: Version 3.0 Users manual. EPA/ 600/3-91/021. US Environmental Protection Agency, Athens, GA. 1991
2. Baker, A.J.M., Reeves R.D., Hajar, A.S.M. Heavy metal accumulation and tolerance in British populations of the metallophyte Thlaspi caerulescens J & C Presl (Brassicaceae) New Phytol 1994; 127: 61-68
3. Banfield, J.F., Barker, W.W., Welch, S.A., and Taunton, A. Biological impact on mineral dissolution: application of the lichen model to understanding mineral weathering in the rhizosphere// Proc. Natl. Acad. Sci. USA 96.1999. p. 3404-3411
4. Cotoras, D., Millar M., Viedma P., Pimentel J. and Mestre A. Biosorption of metal ions by Azobacter vinelandii //World J. Microbiol. Biotechnol. 1992. - 8: 319-323
5. Dudman, W.F. The polysaccharides and oligosaccharides of Rhizobium and the role in the infection process. In: Veeger C & Newton WE (Eds) Advances in Nitrogen Fixation Research. Nijhoff/Dr. W. Junk Publishers, The Hague, Netherlands. 1984.: 397-404
6. Lopareva, A., N. Pawlicki-Julian, N. Goncharova, B.Courtois, J.Courtois.Chracherization of EPS from Rizobium of plants grown on the area contaminated with radionuclides // Proceedings of Working Group 3, Soil-Plant Relationships. XXXIII Annual Meeting of ESNA, Viterbo, Italia, 27- 31 August 2003. p. 42-46
7. Mamaril, J.C., Paner, E.T., Capuno, V.T., Trinidad, L.C. Reduction of heavy metal concentration in liquids by Rhizobium polysaccharides //Asean J Sci. Technol. Dev. 1989. - 6 (2): 75-86
8. McGrath, S.P. Phytoextraction for soil remediation. In: Plants that hyperaccumulate heavey metals, ed. Brooks RR.1998. p. 261-289
9. Mejáre, M. and L. Bülow Metal-binding proteins and peptides in bioremediation and phytoremediation of heavy metals//Trends in Biotechnology 2001.Vol. 19 No. 2 p. 67

10. Schwartz, C., Morel, J.L., Saumier, S., Withing, S.N., Baker, A.J.M. Root structures of the Zn-hyperaccumulator seedling Thlaspi caerulescens have affected by metal origin, content and localization in soil. Seedling Soil 1999; 208: 103-115
11. Shumate, S.E II & Strandberg, G.W. Accumulation of metals by microbial cells. Comprehensive Biotechnology. 1985. - Vol. 4. Pergamon Press, New York
12. Welch, S.A., Barker W.W., and J.F. Banfield .Microbial extracellular polysaccharides and plagioclase dissolution // Geochimica et Cosmochimica Acta,1999 Vol. 63, No. 9, p. 1405-1419

CHAPTER 12

SUNFLOWER GROWTH AND TOLERANCE TO ARSENIC IS INCREASED BY THE RHIZOSPHERIC BACTERIA *PSEUDOMONAS FLUORESCENS*

SHILEV, S.[1,3], A. FERNÁNDEZ1, M. BENLLOCH[2] AND E.D. SANCHO[1]
[1]*Departamento de Microbiología,*
[2]*Departamento de Agronomía, E.T.S.Ingenieros Agrónomos y Montes,*
Universidad de Córdoba, Apto. 3048,
1480-Córdoba, Spain
[3]*Department of Microbiology and Environmental Biotechnologies,*
Agricultural University of Plovdiv, 12 Mendeleev Str.,
4000-Plovdiv, Bulgaria

Abstract

Phytoremediation is not a very well known process. Some of the mechanisms involved in accumulation and detoxification of the metals are still unidentified. Anywhere, the role of soil microbes in the improvement of the phytoremediation technique seems to be very important.

The bacteria *Pseudomonas fluorescens* had a quite positive influence on sunflower plant growth and arsenic accumulation. The plants grown in presence of arsenic showed a reduction in growth with a decrease in leave surface, height, fresh and dry weight. In the treatments with inoculated *Ps. fluorescens* was observed a higher growth rate of the plants, compared with not inoculated treatments.

1. INTRODUCTION

The contamination of soils with heavy metals, have gradually increased over the past decades. These metals are difficult to remove and cannot be chemically or biologically destructed. The phytoremediation is a cost-effective and promising method to overcome this problem (Whiting, S.N. et al., 2001). We had reported (Alcántara et al., 2001 & 2002) that rhizospheric bacteria obtained from the contaminated area of Aznalcóllar (Southern Spain) promote sunflower growth and arsenic accumulation. The isolated rhizobacteria was previously studied with respect to its tolerance to metals (Shilev, S. et al., 2001). This work is a part of a multidisciplinary

J.-L. Morel et al. (eds.), Phytoremediation of Metal-Contaminated Soils, 315–326.

research project on phytoremediation carried out by the EMIR-UCO research group (Agronomy and Forest Science High Technical School, University of Cordoba, Spain; Shilev et al., 2001 & 2005; Ruso et al., 2001).

2. MATERIAL AND METHODS

Sunflower seeds (cv. Sun-Gro 393, Eurosemillas, Córdoba, Spain) were grown under controlled conditions in pots containing a soil: sand mixture (2:1) in the absence (control plants) or presence of arsenic (5-20 mg l^{-1}, as Na_2HAsO_4). The plants were periodically watered with nutrient medium and visually evaluated.

We carried out the experiment in a greenhouse for 35 days. The plants had been inoculated every two days with the rhizobacteria *Pseudomonas fluorescens* (10^9 cells per liter of soil), isolated from the contaminated area of Aznalcóllar and identified on the basis of carbon source utilization (Garland and Mills, 1991) by using MicroLog® System (Biolog Inc.). The control plants were treated with the exhausted bacteria growth medium. Completely randomized design with four replicates per treatment was performed. The number of viable bacteria populations was periodically determined using the method of plate dilution technique. Plant growth (height, leaf area, fresh and dry weight), chlorophyll content (SPAD index), volume of the exudates and toxicity symptoms were measured and visually evaluated. Arsenic content was determined by atomic absorption spectrometry VARIAN SpectrAA-100, in roots, leaves, steams and exudates.

3. RESULTS AND DISCUSSION

The bacteria *Pseudomonas fluorescens* had a quite positive influence on sunflower plant growth and arsenic accumulation. The plants grown in presence of arsenic showed a reduction in growth with a decrease in leave surface, height, fresh and dry weight. In the treatments with inoculated *Ps. fluorescens* was observed a higher growth rate of the plants, compared with not inoculated treatments.

We had also studied the changes in the number of viable bacteria in the soil during the experiment (Table 1).

In the beginning, when arsenic was added (day 10th) the number of bacteria was the same in all treatments. At harvest, in the higher doses of arsenic (20 mg l^{-1}) the viable bacteria were 10 fold more in the treatments with the tolerant bacteria, in relation to treatments without bacteria added.

Pseudomonas fluorescens affect also the water transport in the plant, promoting the translocation of arsenic to the shoots, thus the inoculated plants showed higher volume of exudates per gram of fresh weight (99%

Table 1. Number c.f.u. (colony forming units) per gram of air-dried soil, counted during the experiments.

Treatments	Days after germination		
	10	25	35
Control	$2,4.10^7$	$1,1.10^8$	$1,9.10^8$
Control + B	$2,5.10^7$	$1,7.10^8$	5.10^8
As 5 mg l^{-1}	$4,9.10^7$	$3,8.10^8$	10^8
As 5 mg l^{-1} + B	$1,3.10^7$	9.10^7	10^8
As 20 mg l^{-1}	3.10^7	$9,5.10^6$	3.10^7
As 20 mg l^{-1} + B	$2,4.10^7$	$1,5.10^8$	$5,6.10^8$

Table 2. Accumulation of arsenic in μg.g-1 DW, mean ± standard error (n = 4).

Treatments	Leaves	Roots
Control	0.28±0.2	3.18±0.46
Control + B	0.17±0.1	4.94±1.4
As 5mg l^{-1}	8.82±0.65	4.7±2
As 5mg l^{-1} + B	14.5±0.24	23±0.46
As 20mg l^{-1}	9.9±1.9	26.7±8.9
As 20mg l^{-1} + B	20.4±3.5	245.6±15.2

more than the treatments without *Pseudomonas*). On the other hand, this bacteria increased significantly the accumulation of the arsenic in the plants (up to 5 and 9 folds, in roots; and up to 1,5 folds in leaves) in the treatments with As (5 mg.l^{-1} and 20 mg.l^{-1}, respectively; Table 2).

In general, the phytoremediation is not a very well known process. Some of the mechanisms involved in accumulation and detoxification of the metals are still unidentified. Anywhere, the role of soil microbes in the improvement of the phytoremediation technique seems to be very important.

4. ACKNOWLEDGEMENT

This work was supported by the Consejería de Medio Ambiente de la Junta de Andalucía and FEDER (project 1FD97-2101)

REFERENCES

1. Alcántara, E., R. Barra, M. Benlloch, A. Ginhas, J.V. Jorrín, A. Lora, M.A. Ojeda, M. Puig, A. Pujadas, R. Requejo, J. Romera, E.D. Sancho, S. Shilev and M. Tena. 2002. Fitorremediación de suelos contaminados del área de Aznalcóllar. M. Benlloch, E. Sancho and M. Tena (eds.). Universidad de Cordoba. Spain. ISBN 84-7801-651-1

2. Garland, J.L. and Mills. 1991. Classification and characterization of heterotrophic microbial community-level carbon source utilization. Appl. Environ. Microbiol. 57. 2351-2359.
3. Ruso, J., J. Zapata, M. Hernández, M.A. Ojeda, M. Benlloch, E. Prats, M. Tena, R. López and J. Jorrin. 2001. Toxic metals accumulation and total soluble phenolics in sunflower and tobacco plants. Minerva biotecnologica. 13:1 93-95.
4. Shilev, S., A. Fernández, E. Sancho. 2005. Assessment of the tolerance of sunflower seedlings grown in presence of arsenite and arsenate. J. Environ. Prot. Ecology. (In press)
5. Shilev, S., M. Benlloch, E. Sancho. 2003. Utilization of rhizobacteria *Pseudomonas fluorescens* in phytoremediation strategies, p. 39. *In*: Vanek, T. and J.–P. Schwitzguebel (eds.). Phytoremediation Inventory COST Action 837 view. Published by VOCHB AVČR. Czech Republic.
6. Shilev, S., J. Ruso, M. Puig. M. Benlloch, J. Jorrin and E.D. Sancho. 2001. Rhizospheric bacteria promote sunflower (Helianthus annuus L.) plant growth and tolerance to heavy metals. Minerva biotecnologica. 13:1 37-39.
7. Whiting, S.N., M.P. de Souza, N. Terry. 2001. Rhizosphere bacteria mobilize Zn for hyperaccumulation by *Thlaspi caerulescens*. Environ. Sci. Technol., 35: 3144-3150.

CHAPTER 13

NATURAL ZEOLITES EFFECT ON HEAVY METALS IMMOBILIZATION IN SOIL

C. H. BARBU, L. OPREAN AND C. SAND
"Lucian Blaga" University
Bd. Victoriei 10, Sibiu
2400, ROMANIA

Abstract

The problem of soil decontamination is a very acute one in Romania, due to the local conditions.

Such a region that suffers much because pollution with heavy metals is the Copsa Mica area, where historical and actual contamination has extremely large values. The fact that this is one of the poorest regions in the country prevents the landowners to take any action against the heavy metals entering the food chain. The paper presents our field researches on the introduction into the arable layer of zeolite (clinoptilolitic) volcanic tuffs (ZVT) and the effects in time on the heavy metals uptake and on the yield of some cultivated plants, in a heavily polluted area in Romania.

1. INTRODUCTION

One of the major problems of soils polluted with heavy metals (Pb, Cd, Zn, Cu, Mn) is their use for farming, considering that, no matter of edible plants or forages, these metals enter into a food chain having at its end the humans.

The problem of soil decontamination is a very acute one in Romania, due to the local conditions. Thus, for more than 50 years both factories (including the heavy pollutant smelters) and land was state-owned. After 1990, the land has been given back to the former owners and the factories sold, without stipulating the obligation of the company owners to remediate the polluted soils. Even so, in many cases, pollutants continue to spread over these soils, because of a multitude of reasons.

Such a region that suffers much because pollution with heavy metals is the Copsa Mica area, where historical and actual contamination has extremely large values. The fact that this is one of the poorest regions in the country prevents the landowners to take any action against the heavy metals entering the food chain.

A possibility to prevent the entering of these heavy metals is the soil decontamination by classical methods, but the amount of soil to be decontaminated is immense, the costs reflecting this. Another possibility would be the phytoremediation, especially phytoextraction (Barbu and Grovu, 2001) of these metals, but this requires a very long time and the range of hyperaccumulators existing in Romania is very narrow. The third possibility could be the transformation or the solubilization of these metals (using chellators) and making them to leave the rhizosphere, but without reaching the groundwater, which is very difficult to achieve, also existing the risk of solubilizing the useful metals (Kabata-Pendias and Pendias, 2001). Last, but not least, a choice can be the immobilization of these heavy metals, in order to prevent their uptake by the cultivated plants or their reaching the groundwater. In this respect, there are being done many researches for finding affordable materials, with a low price, able to immobilize great amounts of heavy metals. Among the tested materials there are numerous types of clays, phosphates (apatite), volcanic or fly ash, dolomite, as well as natural or synthetic zeolites (Leppert, 1990; Chlopecka and Adriano, 1997).

The paper presents our field researches on the introduction into the arable layer of zeolite (clinoptilolitic) volcanic tuffs (ZVT) and the effects in time on the heavy metals uptake and on the yield of some cultivated plants, in a heavily polluted area in Romania.

2. MATERIALS AND METHODS

2.1. Cultivated Plants

During the first years there have been used two species of *Brassicaceae* (family known for its great uptake capacity of heavy metals), largely used for human consumption: headed cabbage - *Brassica oleracea* var. *capitata* and turnip cabbage - *Brassica oleracea* var. *gongylodes*. In the second year it was used a vegetable known as non-accumulator of heavy metals, onion - *allium cepa*. There were analyzed both the exterior parts (unwashed and washed with tap water), that usually are peeled and given to the domestic animals (chickens, pigs) and the interior ones, used as food.

2.2. Experimental Place

The experiments took place in a private vegetable garden, situated in Micăsasa, 11 km West from the pollution source (SOMETRA smelter). The geographical and meteorological characteristics are:

- altitude: 254 m;
- Average annual temperature: 9.6°C;
- Annual insolation period: 1900 hours;
- Annual precipitations: 598.6 mm;
- Average annual air humidity: 75%;
- Annual atmospheric calm: 41.4%;
- Annual average wind direction: NE-16.83%; SW-11.95%; E-10.48%; W-9.11%.

The soil sampling and the analyzes were conducted according to the classical methods provided by the Romanian Standards (STAS) in force, described briefly as follows:

- Sampling: STAS 7184/1-84;
- Humidity: drying at 105°C and weighing (STAS 7184/8-79);
- Texture: Kacinsky method (granulometry);
- pH: in 1:2.5 water suspension, OP - 206 pH-meter (STAS 7184/13-88);
- Alkaline-earth carbonates: Scheibler method, with HCl 1:3 (STAS 7184/16-80);
- Humus: Determination of the organic carbon content by the Walkey-Black method and conversion - factor 1.72 (STAS 7184/21-82);
- Phosphorus: Egner-Riehm-Domingo method, with ammonium acetate-lactate (STAS 7184/19-82);
- Potassium: with a PHLAPHO 4 flame-photometer, in ammonium acetate-lactate (STAS 7184/18-80);
- Heavy metals: with a PERKIN-ELMER atomic absorption spectrometer, after mineralization with *aqua regia.*

2.3. Application of the Zeolite and the Fertilizer

The land, with slight slope (2%), was hoed and parceled in autumn, creating three experimental zones, 1.25 m x 2.00 m each: one where natural Romanian zeolites, (cost: USD 0.1/kg) having the diameter of 0 - 0.4 mm, and 81% clinoptilolite, were incorporated, in a 2 kg/sqm dose, the second with 2 kg/sqm zeolites and 2 kg/sqm semifertilized manure, the third remaining as control. The zeolite characteristics are presented in table 1.

Table 1. Chemical composition (% oxides) of the Romanian zeolite volcanic tuff.

SiO_2	Al_2O_3	Fe_2O_3	CaO	MgO	K_2O	Na_2O	TiO_2	Calcination loss
69.98	12.60	1.04	4.80	0.80	2.20	0.40	0.18	9.61

In the second year, there were added neither zeolites, nor fertilizer.

2.4. Plant Cultivation

The seedlings, obtained in the own greenhouse, were planted at the beginning of May at 50 x 50 cm (headed cabbage) - 15 pieces and 35 x 35 cm (turnip cabbage) - 28 pieces on each lot. During the vegetation period there have been done the usual works (watering, hoe). No phase fertilization or chemical pest control were made. The cropping was made in mid-August, weighing the yield of each lot.

In the second year, on all the parcels there were made beds at 20 cm distance and the onion bulbs were planted at 4 - cm distance, in April. The cropping was made in mid-July, in the meanwhile the plants being thinned out twice.

2.5. Sampling and Analysis of the Vegetal Material

For analyze there have been taken the following quantities of plants, originating both from the experimental place and from the control places:

- 1.00 kg turnip cabbages, unwashed, for the whole analyze;
- 1.00 kg turnip cabbages, washed with 3 L tap water, then peeled, for the separate analyze of the peel and the core;
- 1.00 kg headed cabbage. The exterior leaves were taken out and were analyzed, unwashed;
- 1.00 kg headed cabbage. The exterior leaves were taken out, washed with tap water and then analyzed;
- the interior cabbage leaves from these 2.00 kg were analyzed, unwashed, after extracting the stalks. The stalks were also analyzed.
- 1.00 kg onion from each parcel. The exterior foils, the middle foils and the core have been analyzed (being extracted from polluted soil, the results depend on the amount of soil on bulbs, thus the amounts of heavy metals there were determined only on washed vegetables).

After weighing and air drying for 72 hours, all the samples were oven dried at 105°C, for 6 hours and then re-weighed, thus determining the humidity, too.

From each vegetal sample it was taken 1.0000 g of dried material that was mineralized with *aqua regia* then evaporate, until a white residue. The residue was dissolved in HNO_3 4%, then filtered. The determination of the heavy metals were performed on a PERKIN ELMER atomic absorption spectrometer, with cavitary cathodes at $\lambda = 283.3$ nm for Pb and $\lambda = 229.0$ nm for Cd. The analysis of the heavy metals in the plants from the experimental zone was done in triplicate and for the control plants, in duplicate.

3. RESULTS AND DISCUSSIONS

3.1. Soil Characteristics

The soil characteristics in the experimental zone are shown in table 2. It can be seen that, between the initial soil and the control one, there are minor differences, due to the plant cultivation. Adding zeolites have significant effects on soil: increase in humus amount, blocking of mobile potassium and phosphorus, making them less leachable. The effect on texture and soil pH is not major (the decrease with 0.1. units being due to the micorrhyzal activity). Adding zeolites and organic fertilizers results in an accentuated pH decrease, and in an increase of the humus amount and in the mobile potassium and phosphorus. Texture is not modified in all three cases (sandy-clay-loam).

Table 2. Physical and chemical characteristics of the studied soils.

	pH	Carbonates (%)	Humus (%)	P_2O_5 (ppm)	K_2O (ppm)
		Initially - May 2001			
	7.48	6.23	3.03	110	460
		Intermediately - August 2001			
Control soil	7.46	6.24	2.96	102	443
Soil + ZVT	7.36	6.63	3.34	73	480
Soil + ZVT + fertilizer	6.99	6.44	3.91	463	522
		Finally - July 2002			
Control soil	7.40	6.20	2.90	80	210
Soil + ZVT	7.33	6.55	3.28	60	220
Soil + ZVT + fertilizer	7.30	6.40	3.50	228	240

3.2. Vegetable Yields

The yields for headed and turnip cabbage on the three experimental lots are presented in table 3. The weight of the headed cabbage was 1.00-1.80 kg/piece and of the turnip cabbage 0.40-0.60 kg/piece.

Table 3. Yields (Kg/m2) obtained on the three experimental lots.

	Control	ZVT	% Control	ZVT + fertilizer	% Control	% ZVT
Headed cabbage	9.250 ± 0.520	10.650 ± 0.580	15.3	10.900 ± 0.590	17.8	2.3
Turnip cabbage	5.700 ± 0.320	6.550 ± 0.360	14.9	6.650 ± 0.370	17.5	1.5

These yields confirm our expectations, i.e. the use of ZVT brings an increase in yield; in the case of the use of semifermented organic fertilizer together with ZVT, this increase is relatively greater, but considering the increase of the Pb and Cd uptake (see below), the use of such a fertilizer is not recommended on the soils polluted with heavy metals.

The onion on the parcels with ZVT and with ZVT and fertilizer was more vigorous than on the control soil, thus the number of plants to be taken out before maturity was greater. Finally, on all the three lots the yields were about 7 kg/m^2.

3.3. Amount of Heavy Metals in Plants

The amount of heavy metals in the studied plants and in the reference ones, as well as in the respective soils, are presented in the tables 4 and 5, together with the Romanian maximal admitted values (MAV). Considering that the average water content in turnip cabbage is 91% and in headed cabbage, 90%, the amount of heavy metals in the fresh vegetables can be calculated dividing to 10. The average humidity of the onion core was 75%, thus the amount of heavy metals can be calculated by dividing to 4.

Table 4. Amount of Pb (mg/kg D.W) in the studied soils (MAV = 100) and plants (MAV = 5).

	Control	ZVT	% Control	ZVT + fert.	% Control	% ZVT
Turnip cabbage, peels, unwashed	32.5 ± 2.7	30 ± 2.6	–7.7	37.5 ± 2.9	+15.4	+23
Turnip cabbage, peels, washed	27 ± 2.5	20 ± 2.3	–26	29 ± 3.3	+7.4	+45
Turnip c., core	25 ± 2.6	22.5 ± 2.6	–10	26 ± 3.1	+4.4	+16
Headed cabbage, ext. leaves, unwashed	42.5 ± 4.3	38.5 ± 4.0	–9.4	45 ± 4.5	+5.9	+16.9
Headed cabbage, ext. leaves, washed	19 ± 2.0	18 ± 2.0	–5.3	21 ± 2.2	+10.5	+16.7
Headed cabbage, interior leaves	17.5 ± 1.9	15 ± 1.7	–14.3	19 ± 2.1	+8.6	+26.7
Headed cabbage, stalk	10 ± 1.3	8.5 ± 1.3	–15	11 ± 1.2	+11	+30.6
Onion, exterior foils	3.75 ± 0.8	2.8 ± 0.5	–25.3	2.9 ± 0.4	–22.7	+3.6
Onion, middle foils	1.05 ± 0.4	0.85 ± 0.4	–19	0.9 ± 0.3	–14.3	+5.9
Onion, core	0.3 ± 0.15	0.25 ± 0.1	–16.7	0.25 ± 0.1	–16.7	0

Table 5. Amount of Cd (mg/kg D.W.) in the studied soils (MAV = 6) and plants (MAV = 2).

	Control	ZVT	% Control	ZVT + fert.	% Control	% ZVT
Turnip cabbage, peels, unwashed	2.5 ± 0.3	2.2 ± 0.3	–12	3.0 ± 0.4	+20	+36
Turnip cabbage, peels, washed	2 ± 0.25	1.7 ± 0.2	–15	2.1 ± 0.24	+5	+23.5
Turnip c., core	1.75 ± 0.20	1.5 ± 0.2	–14.3	1.8 ± 0.20	+2,9	+20
Headed cabbage, ext. leaves, unwashed	2.25 ± 0.3	1.75 ± 0.25	–22.2	2.4 ± 0.35	+6.7	+37
Headed cabbage, ext. leaves, washed	0.8 ± 0.2	0.65 ± 0.15	–18.8	0.9 ± 0.2	+12.5	+38.5
Headed cabbage, interior leaves	0.75 ± 0.15	0.6 ± 0.1	–26.7	0.8 ± 0.15	+6.7	+33.3
Headed cabbage, stalk	0.6 ± 0.15	0.5 ± 0.15	–16.7	0.65 ± 0.2	+8.3	+30
Onion, exterior foils	1.25 ± 0.25	1.05 ± 0.2	–16	1.15 ± 0.3	–8	+9.5
Onion, middle foils	0.7 ± 0.17	0.6 ± 0.15	–14.3	0.65 ± 0.2	–7	+8.3
Onion, core	0.1 ± 0.05	0.9 ± 0.25	–10	0.9 ± 0.24	–10	0

NOTES:
The edible parts of the plants were underlined;
% Control = percent variation towards the control;
% ZVT = percent variation towards the soil with ZVT only.

It can be seen that in the case of the use of ZVT only, the Pb and Cd uptake is reduced compared to the control, both by adsorption and cationic exchange. This reduction capacity is maintained for the second year, too. The addition of natural semifermented fertilizer has conducted to the decrease of the soil pH in the first year, thus to the increase of the Pb and Cd solubility and bioavailability.

In the second year, the acidification effect of the fertilizer disappears and the amount of up-taken metals is almost the same for the vegetables cultivated on soil with ZVT only and on soil with ZVT and fertilizer.

4. CONCLUSIONS

1. The use of zeolite volcanic tuffs on the soils contaminated with heavy metals results in decreasing their uptake in headed cabbage - *Brassica*

oleracea var. *capitata*, turnip cabbage - *Brassica oleracea* var. *gongylodes* and onion - *Allium cepa*, as well as increasing their yields.

2. The immobilization effect last during the second year of vegetables cultivation, too.

3. The use of zeolite volcanic tuffs combined with semifermented organic fertilizer leads to the pH decrease and to the increase of the Pb and Cd availability, manifested in the increase of these elements concentration in plants. This effect disappears in the second year after application.

4. The yield increase in the case of the use of zeolite volcanic tuffs combined with semifermented organic fertilizer does not justify their use on soils contaminated with heavy metals, thus being preferred the exclusive use of zeolite volcanic tuffs, accessible and cheap materials.

5. The significant reduction of the heavy metals amount in the cultivated plants enable us to continue the experiments, in order to determine the durability in time of the zeolites addition, the effect upon other species and to increase the immobilization degree of heavy metals in soil. The further experiments will also use fertilizers (natural or chemical), that will not acidify the soil.

REFERENCES

1. Barbu, C.H., Grovu, R. Phytoremediation of the soils sontaminated with heavy metals in Copsa Mica area - Romania, in Kostecki, P.K., Calabrese, E.J. and Dragun, J. (Editors), Contaminated Soils, Volume 6, 69-78, CRC Press, Amherst, MA, 2001.
2. Kabata-Pendias, A., Pendias, H. Trace elements in soils and plants, 3rd ed., CRC Press, Boca Raton, FL, U.S.A., 2001.
3. Leppert, D. Heavy metal sorption with clinoptilolite zeolite: alternatives for treaing soil and water, Trans. Soc. Min., Metall., Explor. 288: 604-608, 1990.
4. Chlopecka, A., Adriano, D.C. Inactivation of metals in polluted soils using natural zeolite and apatite, in Iskandar, I. and Pierzynski, G. (editors) Proceedings of the 4th Int. Conf. Biogeochem. Trace Elements, Berkley, CA, 1997.

CHAPTER 14

THE ARSENIC-PHYTOREMEDIATION POTENTIAL OF GENETICALLY MODIFIED *PSEUDOMONAS* SPP.

O. I. SIZOVA, V. V. KOCHETKOV AND A. M. BORONIN
Institute of Biochemistry and Physiology of Microorganisms,
Russian Academy of Sciences (RAS),
Pushchino, Moscow oblast, 142290 Russia

Abstract

Gene constructions rendering recombinant bacteria resistant to arsenic and increasing their ability to dissolve soil phosphates and/or arsenates were created by cloning the *ars* operon and the gene of citrate synthase from a chromosome of the strain *Pseudomonas aeruginosa* PA01. Genetically modified variants of the strain *Pseudomonas aureofaciens* BS1393 have been constructed that are resistant to high concentrations of arsenic and dissolve poorly soluble phosphates and/or arsenates. The recombinant strains *P. aureofaciens* BS1393(pUCP22:: *ars*RBC) and *P. aureofaciens* BS1393(pUC P22:: *glt*A) exerted positive effects on the survival of sorghum (*Sorghum saccharatum* L.) and its ability to accumulate arsenic.

1. INTRODUCTION

Biological accumulation of arsenic by plants is a promising way of its removal from polluted soils. In recent years, accumulation and redistribution of arsenic in plants have been investigated, as well as the mechanisms of detoxification of its compounds (Pickering, et al., 2000; Schmoger, et al., 2000;. Hartley-Whitaker, et al., 2001; Lyubun, et al., 2001). However, virtually no attention has been given to the involvement of rhizosphere microorganisms in the phytoremediation of arsenic-polluted soils.

The bioavailability of arsenic salts is of great importance for the process of phytoremediation of arsenic-polluted soils. The bulk of arsenic in soil is absorbed by soil colloids, hampering its availability to plants. Some bacterial strains belonging to the genus *Pseudomonas* can solubilize phosphates (Kumar and Singh, 2001). Since some properties of arsenic are close to those of phosphorus (Gamayurova, 1993), it is conceivable that rhizosphere bacteria solubilizing phosphoric compounds can increase the bioavailability of poorly soluble arsenic compounds in soil. Furthermore, phosphates can

J.-L. Morel et al. (eds.), Phytoremediation of Metal-Contaminated Soils, 327–334.

increase the bioavailability of soil arsenates, thus stimulating the supply of arsenic to plants (Peryea, 1989; Fitz and Wenzel, 2002; Chen, et al., 2002).

The goal of the present study was to develop gene constructions providing rhizosphere bacteria with (1) tolerance to high concentrations of arsenite/arsenate and (2) the ability to transform bound arsenic into a form available to plants; we also focused on the influence of genetically modified *Pseudomonas* strains on the growth and arsenic accumulation ability of sorghum plants growing on arsenic-polluted soil.

2. MATERIALS AND METHODS

The bacterium *Pseudomonas aureofaciens* BS1393 (VKM V-2188 D) was chosen for the study because of its high antagonistic activity against a broad spectrum of phytopathogenic fungi and bacteria. Our previous studies demonstrated that *P. aureofaciens* BS1393 exceeded other bacteria in its ability to favor the survival of sorghum plants (*Sorghum saccharatum* L.) on soil polluted with arsenic (Sizova, et al., 2002).

Bacteria were grown on L-broth, L-agar, M9 synthetic medium (Maniatis, 1982) (supplemented with glycerol as a source of carbon and energy), and synthetic medium with hydroxyapatite (HAP) (Goldstain, 1986). The media and supplements were autoclaved at 0.5 atm for 30 min.

Bacteria were grown at 30°C in petri dishes or Erlenmeyer flasks (at 150 rpm) containing 100 ml liquid medium each.

Nucleotide sequences of the citrate synthase gene (*glt*A) and arsenite operon (*ars*RBC) of *P. aeruginosa* PA01 (Ac. no. AE 004091, GeneBank) were used to develop specific primers for polymerase chain reaction (PCR). The primers for the genes *glt*A and *ars*RBC were developed at the Institute of Biochemistry and Physiology of Microorganisms of the RAS. PCR was performed using RCR Express thermal cycler manufactured by Hybaid (UK). The chromosomal DNA of *P. aeruginosa* PA01 was used as a template for PCR.

Cloning of PCR products into a shuttle vector, isolation of plasmid and chromosomal DNA, and transformation of the recipient strain with the plasmid DNA were performed according to standard protocols (Maniatis, et al., 1982). The transformants containing the vector with the cloned citrate synthase (EC 4.1.3.7) gene were selected in LB medium supplemented with X-gal, IPTG, and gentamicin (concentration, 20 μg/ml medium). The transformants containing the vector with the cloned arsenite operon were selected by their tolerance to arsenic (500 μg/ml). Expression of citrate synthase was evaluated on a mineral medium supplemented with HAP. The bacteria were grown on a liquid medium for 12 h. Then, a drop (5 μl) of the culture was carefully placed onto agar medium. The bacteria were grown for 3 days, until a translucent zone appeared around the colonies, indicating citrate synthesis.

Minimum inhibitory concentrations (MICs) and maximum tolerated concentrations (MTCs) for sodium arsenite and arsenate were determined in L-broth as described earlier by Podol'skaya et al. (2002).

Inoculation of sorghum seed (cultivars Pishchevoe-69) with the bacteria and preparation of soil were performed as described previously (Sizova, et al., 2002). The plants were grown on gray forest soil. Arsenic was added into the soil as an aqueous solution of sodium arsenite up to a final concentration of 100 μg/kg. Each experimental variant was repeated ten times. Noninoculated plants grown on soil polluted with arsenic were used as controls. Laboratory tests with plants were carried out in a climate light chamber under the following conditions: light period, 12 h at 25°C; dark period, 12 h at 20°C. The duration of the experiment was 35 days.

Plant and soil arsenic was determined on an Elemental ICP-MS PQ2 + VG mass spectrometer (UK). Samples were prepared for analysis according to the standard protocol recommended for this device.

3. RESULTS AND DISCUSSION

Bacteria are known to be actively involved in global cycles of arsenic circulation in the biosphere, and arsenic-resistant microorganisms are found among representatives of various taxons. Two mechanisms of bacterial resistance to arsenic are known. The first (and the best-studied) mechanism is associated with the bacterial enzyme arsenate reductase, which reduces pentavalent arsenic to the trivalent form in the cytoplasm. Reduced arsenic is excreted from the cell via a special membrane protein (porin). This mechanism of arsenic resistance, determined by chromosomal and plasmid genes, has been described in many bacteria, such as *Escherichia coli*, *Staphylococcus xylosus*, *Thiobacillus ferrooxidans*, and *Pseudomonas* sp. (Cervantes, et al., 1994; Silver and Phung Le, 1996; Mukhopadhyay, et al., 2002).

The second mechanism is associated with bacterial arsenite oxidase. This enzyme oxidizes trivalent arsenic to the pentavalent form, which is also excreted from the cell via a membrane protein. The enzyme arsenite oxidase has been studied and described in *Alcaligenes faecalis* (Anderson, et al., 1992; Santini, et al., 2000).

To produce genetically modified strains resistant to arsenic and capable of dissolving phosphates, we used the genes responsible for the tolerance to arsenic and the gene of citrate synthase, which is located on a chromosome of *P. aeruginosa* PA01. Based on the nucleotide sequences of the arsenite operon – which contains three genes, (1) the transcription regulator *ars*R, (2) *ars*B, encoding a membrane protein, and (3) *ars*C, encoding arsenate reductase, and the gene of citrate synthase (*glt*A) of this strain, we developed specific primers with artificially inserted endonuclease restriction sites for these genes. The following primers for the arsenite operon were developed: *ars* forward, 5'-GGGGAAGCTTGCAATACATACGGAT-3', containing the *Hind*III restriction endonuclease site, and *ars* reverse, 5'-GGTAGTTGTTAACACATGCTTCCTCCG-3', containing the *Hpa*I restriction endonuclease site. The primers for the citrate synthase gene were *glt* forward, 5'-ACGACCCGGGCTAGTTGTCTGGAAG-3', containing the *Sma*I restriction endonuclease site, and *glt* reverse, 5'-GCTGAATTCGCCTTACCTGCAAAA-3', containing the *Eco*RI restriction endonuclease site.

DNA fragments produced by PCR using the primers described above were cloned into the shuttle vector pUCP22 (*E. coli – P.seudomonas* sp., Ap^r, Gm^r, 4729 b.p.) at the endonuclease restriction sites artificially inserted into the primer sequences. The fragment of about 1700 b.p. amplified by PCR from the primers for the gene of citrate synthase was cloned into the vector at the *Sma*I and *Eco*RI endonuclease restriction sites (Fig. 1a). The fragment of about 2200 b.p. amplified by PCR from the primers for the arsenite operon was cloned into the vector at the *Hind*III and *Sma*I endonuclease restriction sites (Fig. 1b). Cloning resulted in the constructions pUCP22::*glt*A (6420 b.p., phosphate solubilization) and pUCP22::*ars*RBC (6920 b.p., tolerance to arsenite/arsenate).

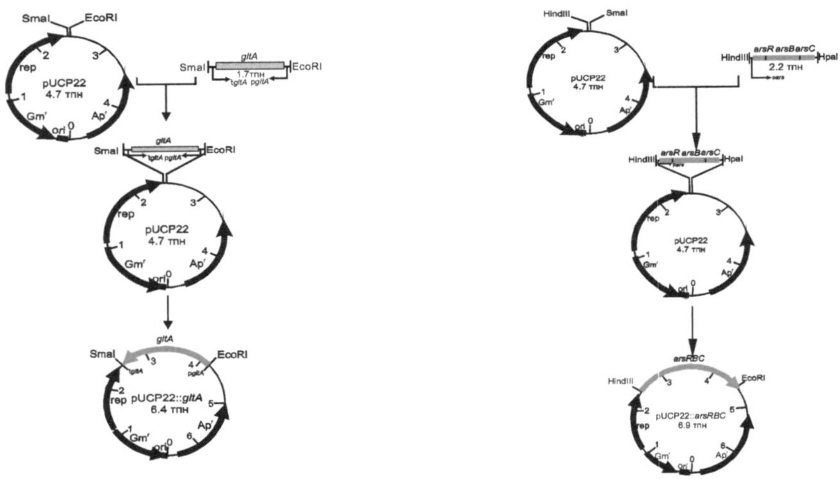

a) b)

Fig. 1. Development of the gene constructions pUCP22::*glt*A (a) and pUCP22::arsRBC (b).

Strain *P. aureofaciens* BS1393 was transformed with the genetic constructions produced by the cloning experiments. The transformation resulted in the following strain variants: strain *P. aureofaciens* BS1393(pUCP22::*ars*RBC), resistant to arsenite/arsenate, and strain *P. aureofaciens* BS1393(pUCP22::*glt*A), capable of dissolving phosphates.

MICs and MTCs were measured for sodium arsenite and arsenate in the original and recombinant strains (Table 1).

Table 1. Minimum inhibitory concentrations (MICs) and maximum tolerated concentrations (MTCs) of sodium arsenite and arsenate for the original and recombinant *Pseudomonas aureofaciens* strains.

Strain	Sodium arsenite		Sodium arsenate	
	MTC, mM	MIC, mM	MTC, mM	MIC, mM
*P. aureofaciens*BS1393	0.7	1	1	40
*P. aureofaciens*BS1393 (pUCP22::*glt*A)	0.8	1	1	42
P. aureofaciens BS1393 (pUCP22::*ars*RBC)	0.6	8	150	250

We hypothesized that treatment of seeds with the recombinant strains produced would defend plants against phytopathogenic microorganisms under the conditions of soil pollution and enhance the biological accumulation of arsenic by plants by increasing the solubility of soil arsenates. To test this hypothesis, the original and plasmid-containing variants of *P. aureofaciens* BS1393 were used for inoculating sorghum seeds.

The experiment demonstrated that all of the bacterial strains used for the seed inoculation positively influenced the seed germination and the height of sorghum plants grown on arsenic-polluted soil (data not shown).

The plants grown from the seeds treated with the recombinant strains (inoculated plants) were characterized by better survivability (Fig. 2) in soil containing arsenic (in comparison with control, noninoculated plants grown from untreated seed). The positive effect of the arsenic-resistant strain on the plant survivability is likely accounted for by the selective advantage of the strain (acquired via tolerance to arsenic), which consists in its ability to exert a prolonged stimulating effect on the plant growth and development. Phosphorus is a nutritional element for plants, necessary for their normal growth and development. The enhanced survivability of plants treated with the phosphate-solubilizing strain is attributable to the ability of this strain to provide the plants with phosphorus in a form available for absorption. All of the plants inoculated with the original strain *P. aureofaciens* BS1393 died during the experiment, likely due to the sensitivity of this strain to arsenic.

Arsenic content in soil and sorgo plants was analyzed on day 35 of the experiment (Table 2). The data demonstrate that the accumulation of arsenic by inoculated plants was maximum (almost 30% higher than in noninoculated plants) when the seed were treated with the recombinant phosphate-dissolving strain. This fact may be accounted for by the production of citric acid, which favors dissolving of bound soil phosphorus, thereby increasing biological availability of arsenic and stimulating its absorption by plants

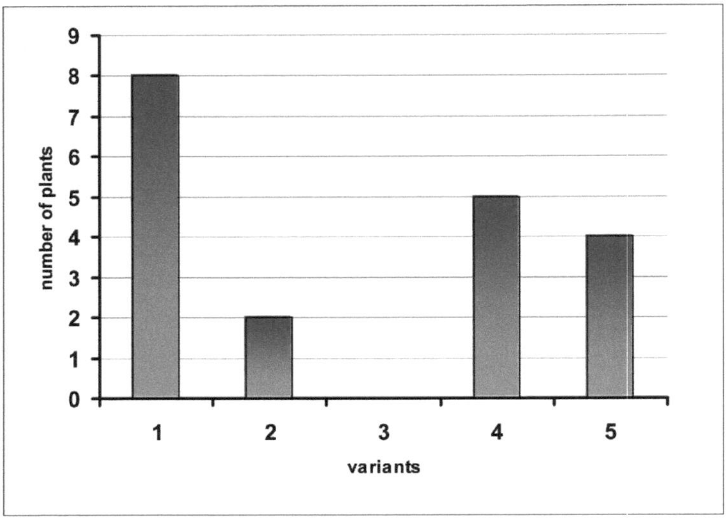

1 - Control (noninoculated plants);
2 - Seeds inoculated with *P. aureofaciens* BS1393;
3 - Seeds inoculated with *P. aureofaciens* BS1393(pUCP22::*glt*A);
4 - Seeds inoculated with *P. aureofaciens* BS1393(pUCP22::*ars*RBC).

Fig. 2. Effect of seed treatment with bacteria on sorghum plant survivability (%) at an arsenic concentration of 100 mg/kg soil.

(Peryea, 1989; Chen, et al., 2002). The amount of arsenic accumulated by the plants treated with the arsenic-tolerant recombinant was 20% higher than in the control. The strain increased the content of the available arsenic in soil (due to the activity of arsenate reductase). Under the conditions of phosphorus deficiency, however, the plants grew more slowly and, most likely, their accumulation of arsenic was weak. Control plants grown on the polluted soil accumulated 81.27 mg/kg arsenic, but in this variant of the experiment most of them died during the first 10 days.

Table 2 shows that the soil in the control experiment contained the greatest amount of arsenic (35.78 mg/kg). After growth of the plants that were inoculated with the strain *P. aureofaciens* BS1393(pUCP22::*glt*A), the level of arsenic in the soil was 17% lower than in the control. After growth of the plants that were inoculated with the strain *P. aureofaciens* BS1393(pUCP22::*ars*RBC), the soil contained amounts of arsenic about 8% lower than in the control. Since all the plants whose seeds were treated with the original strain *P. aureofaciens* BS1393 died during the experiment, we could not evaluate the effect of this strain on the accumulation of arsenic by the plants.

Table 2. Arsenic content in soil and sorghum plants.

Varian	Arsenic, mg/kg dry mass soil	plant
Control (noninoculated plants)	35.78	81.27
P. aureofaciens BS1393 (pUCP22::*gltA*)	26.17	114.16
P. aureofaciens BS1393 (pUCP22::*arsRBC*)	33.06	104.57

Thus, the gene constructions developed in the present study render rhizosphere bacteria tolerant to high arsenite/arsenate concentrations and capable of transforming bound arsenic into a form available to plants. Genetically modified *Pseudomonas aureofaciens* strains have been produced that positively affect the viability, growth, and development of plants on soils polluted with arsenic and favor accumulation of arsenic by plants.

This study was supported by the grant of Russian Federal Science and Technology Program "Research and Development in the Priority Fields of Science and Civil Engineering," subprogram "Novel Methods in Bioengineering" (state contract no. 43.073.1.1.2502).

REFERENCES

1. Anderson, G.R., Williams, J., and Hille, R., *J. Biol. Chem.*, 1992, vol. 267, no. 33, pp. 23674–23682.
2. Cervantes, C., Guangyong, J.I., Ramirez, J.L., and Silver, S., *FEMS Microbiol. Rev.*, 1994, vol. 15, no. 2, pp. 355–367.
3. Chen, T.B., Fan, Z.L., Lei, M., et al., *Chinese Sci. Bull.*, 2002, vol. 47, no. 22, pp. 1876–1879.
4. Fitz, W.J. and Wenzel, W.W., *J. Biotechnol.*, 2002, vol. 99, no. 2, pp. 259–278.
5. Gamayurova, V.S., *Mysh'yak v ekologii i biologii* (Arsenic in Ecology and Biology), Moscow: Nauka, 1993.
6. Goldstain, A.N., *Am. J. Altern. Agriculture*, 1986, vol. 1, no. 2, pp. 51–57.
7. Hartley-Whitaker, J., Ainsworth, G., Wooijs, R., et al., *Plant Physiol.,* 2001, vol. 126, no. 2, pp. 299–306.
8. Kumar, V. and Singh, K.P., *Bioresour. Technol.*, 2001, vol. 76, no. 1, pp. 173–175.
9. Lyubun, Y.V., Kosterin, P.V., Zakharova, E.A., et al., *J. Soil Sediments*, 2001, vol. 2, no. 3, pp. 143–147.
10. Maniatis, T., Fritsch, E.F., and Sambrook, J., *Molecular Cloning: A Laboratory Manual*, New York: Cold Spring Harbor, 1982.
11. Mukhopadhyay, R., Rosen, B.P., Phung Le, T., and Silver, S., *FEMS Microbiol. Rev.*, 2002, vol. 26, no. 2, pp. 311–325.
12. Peryea, F.J., *Hortscience*, 1989, vol. 33, no. 3, pp. 826–829.
13. Pickering, I.J., Prince, R.C., George, M.J., et al., *Plant Physiol.*, 2000, vol. 122, no. 4, pp. 1171–1178.
14. Podol'skaya, V.I., Gruzina, T.G., Ul'berg, Z.R., et al., *Prikl. Biokhim. Mikrobiol.*, 2002, vol. 38, no. 1, pp. 57–62.

15. Santini, J., Sly, L., Schnagl, R., and Macy, J., *Appl. Env. Mikrobiol.*, 2000, vol. 66, no. 1, pp. 92–97.
16. Schmoger, M.E., Oven, M., and Grill, E., *Plant Physiol.*, 2000, vol. 122, no. 3, pp. 793–801.
17. Silver, S. and Phung Le, T., *Annu. Rev. Microbiol.*, 1996, vol. 50, pp. 753–789.
18. Sizova, O.I., Kochetkov, V.V., Validov, Sh.Z., et al., *J. Soil Sediments*, 2002, vol. 2, no. 1, pp. 19–23.

CHAPTER 15

ROUND TABLE DISCUSSION: SUSTAINABLE MANAGEMENT OF NATURAL PLANT RESOURCES FOR PHYTOREMEDIATION

ROSALIND F. KING[1] AND ANNA ROYLE[2]
[1]School of Biological Sciences, Jones Building, Liverpool University, L69 3GS, United Kingdom
[2]Biological and Earth Sciences, John Moores University, Byrom Street, Liverpool, L3 3AF, United Kingdom

Abstract

A round table discussion was held at the Advanced Study Institute and is summarised in this chapter. The panel members Dr R. Reeves, Dr. C. Schwartz, Dr. D Belluck, Dr D. Parker and Dr. M. McLaughlin gave five short presentations. These were designed to cover a range of phytoremediation issues including the protection of useful species and their habitats, the legal aspects of this and the introduction of exotic plant species into the environment. The advantages and disadvantages of phytoremediation were then discussed with relation to alternative remediation strategies. The discussion was then open to the floor, to allow for participation amongst all conference members. The focal point of the discussion was the practicalities of managing rare hyperaccumulator plants populations in the wild and on polluted sites. Other areas of concern were raised, including the potential for genetic modification of hyperaccumulators, the extent of agronomic knowledge of hyperaccumulators and the feasibility of using hyperaccumulators to remediate polluted field sites. The discussion was concluded by Dr Jean-Louis Morel, who was optimistic about the future of phytoremediation but acknowledged that there was a need for extended research before it would be practical to apply this new technology on a large scale.

PROTECTING RARE AND POTENTIALLY USEFUL SPECIES - *Roger Reeves*

This presentation emphasised the rarity of hyperaccumulator species. Out of the 290 000 or so species of vascular plants currently described, only four hundred and forty species have been identified as being hyperaccumulators. Though globally rare, hyperaccumulators have the highest abundance on

J.-L. Morel et al. (eds.), Phytoremediation of Metal-Contaminated Soils, 335–342.

metal-rich sites, although they still only constitute one to two percent of the plant species present there.

The limited range of some hyperaccumulator species was also pointed out. Some are found in only one region or on one site. Examples of this rarity include *Phyllanthus insulae-japen* (a nickel hyperaccumulator) which has only been found once, on the island of Japen, off the coast of New Guinea, *Centaurea ptosimopappoides,* which has only been found in a few $100km^2$ and *Peltaria (Leptoplax) emarginata,* which is found in only around $8km^2$ in Greece. This limited range leaves the plants vulnerable to extinction, which can easily be caused by events such as fires or urbanisation.

Some populations of hyperaccumulators are so small that there is a question as to whether they can be considered a population at all. An example of this occurs on the Isle of Rhum, where only seven *Thlaspi caerulescens* individuals could be found. Another problem is that different populations of plants of a species have been found to have different properties so it is not necessarily the case that two populations of the same species, situated two miles apart will have the same metal uptake characteristics. The loss of one population may therefore mean the loss of a particular trait which cannot be found again in another population of the same species.

The threats to hyperaccumulator existence were outlined and these include:

- Mining and reworking of old mine wastes (and hence clearing of vegetation of the site).
- Pollution. For example, waste discharges from industrial facilities built close to hyperaccumulators populations.
- Fires, as the range of the hyperaccumulator species may be very small.
- Urbanisation – the building of houses or hotels over the habitats of the plants as in the case of an *Alyssum corsicum* population in Greece.
- Agriculture, horticulture and forestry. For example in Canada, vineyards are often planted on ultramafic rock where hyperaccumulators are naturally found.
- Recreational uses such as off-road vehicles driving in the Red Hills of California.
- Botanists and collectors who can sample a population to extinction or down to unsustainable numbers.
- Soil remediation as the established vegetation is cleared and the soil conditions are changed.

With all these threats to hyperaccumulators there is a definite need to attempt to preserve sites on which they are found. Dr. C. Swartz addressed this issue in his presentation.

PROTECTING CONTAMINATED SITES - *Christophe Schwartz*

The issues of protecting polluted sites are certainly controversial, yet contaminated sites can be important habitats due to the rare populations they can contain. It is therefore important that some of these sites be preserved. The reclamation of contaminated sites usually involves the excavation of soil and hence populations growing on the sites are decimated. An example of this was the rehabilitation of a lead mining site upon which *Thlaspi* species were found. Reclamation of the site started in 1994 and revegetation began in 1996. The reclamation involved the addition of non-polluted soils to the site and, consequently, by 1998 there were no *Thlaspi* left on the site.

Some industrial sites are contaminated, but as they are private property, the industries may not allow access to the site, even for collection of plant samples or observation.

Disturbed and contaminated soils occur naturally as well as being manmade and, given time, the natural ecosystems adapt to these conditions. However, to the human eye these sites may appear derelict and can become dumping grounds for all sorts of waste, reducing the quality of the site even further.

In France some of the naturally occurring metalliferous areas are kept aside as nature reserves because of the rare species that inhabit them. These may include hyperaccumulator species as in the national reserve on a polluted site where *Arabidopsis halleri* and *Aremeria maritima* are protected. These areas are relatively safe from destruction but there are many sites housing hyperaccumulator species which are not, and vital populations may be lost. Some species may even become extinct.

Thus we have seen that hyperaccumulators are rare and may be in need of protection. If they become a marketable commodity there may be competition for them and a further threat may arise due to the collection of plants and their seeds. David Belluck went on to explain about the potential legal protection that can be given to these plants.

LEGAL ASPECTS OF PLANT RESOURCE COLLECTION - *David Belluck*

Bioprospecting is the term given to the process of searching for plants with particular properties – such as hyperaccumulators. If hyperaccumulators were to be used for commercial phytomining, or sold as remediation tools, there would be a commercial interest in the area. Any business thinking of entering the market would look up on the plants as it would any other product. So could plants such as *Thlaspi* be patented?

As far as the literature is concerned plants cannot be taken from a field and patented. However, if the bioprospectors breed a new variety of the plant then this variety can be patented. To apply a patent and receive maximum patent protection the bioprospector must prove that the plant they have produced is a "novel" plant or plant culture. Plant cell culture could also be used to find better cell lines which accentuate desired features.

But if we get this patent process underway and begin to use new plants to phytoremediate polluted lands we may be introducing species to areas they would not naturally have been able to colonise. Dr. Parker will address this issue along with other environmental concerns.

INTRODUCING NEW PLANT SPECIES INTO THE ENVIRONMENT - *David Parker*

If hyperaccumulators are to be used then it is likely that the species introduced to the site are not native to the area of the country. Some species, such as the slow growing, *Thlaspi*, may not cause a problem when introduced to a new area. Other species may be more problematic, such as *Brassica juncea* introductions, which have led to the plant becoming an invasive weed species in some countries.

The growth of hyperaccumulators also poses ecotoxicological concerns. Some species may mobilise metals in the soils which were previously immobile. Concentration of metals into the plant tissues may lead to concerns of transfer to the food chain. If the hyperaccumulators are not harvested then leaf drop can cause levels of metals to be hugely elevated in surface soils as leaves with very high metal concentrations decompose.

The final issue is that of genetically modified plants. The need for easily grown, high biomass hyperaccumulators has been addressed and this may result in transgenic plants being introduced along with the risks associated with them. In some countries genetically modified plants are thought of as a good thing, in some countries there is strong resistance to the technology. Genetic modification of traditional agricultural species or other plants may not have a future in these resistant countries.

Thus there is the potential for using hyperaccumulators as a phytoremediation tool, but there may be risks involved. The final speaker, Dr. McLaughlin, summed up the advantages and disadvantages of phytoremediation and compared these to other remediation strategies.

TO PHYTOREMEDIATE OR NOT TO PHYTORE MEDIATE - *Mike Mclaughlin*

The initial advantages to phytoremediation are clear, it is low cost, with little soil disturbance, and eventually helps to stabilise and improve the soil. The public are also amenable to this method which appears to them less destructive than other methods.

However, this method is unproven in the field. Although many lab studies have been done, they are very controlled and hence the results obtained may not be applicable to field-scale work. Phytoremediation is also a slow process and may require continual harvesting with little thought having gone into what might happen to the end product. Another consideration is that the plants are part of the ecosystem and hence the system is not "closed" and the pollution can leave the site. The public may not be accepting of phytoremediation should the need for transgenic or exotic plants arise.

Immobilisation is an alternative strategy. The advantages of this include it being a low cost solution, and being proven to work in the field. It leads to an improvement in soil quality, stabilises soils and no management is needed as nothing is being removed from the site or accumulated in the biomass. However the disadvantages include the fact that the metal pollutants remain in the soil. The long-term stability of the process is not resolved and it therefore may be unsustainable. Nutrient deficiency in the ecosystem may occur as some soil amendments result in the leaching of nutrients such as phosphate, while metal leaching can lead to contamination of surrounding water sources.

Therefore, although phytoremediation is an interesting prospect, there has been insufficient work done for it to be used safely in the field yet. *In situ* remediation by soil amendments may be a safer option at present as it is still low cost whilst stabilising and improving the soil. It has also been proven in the field. The side effects may include phosphate deficiency as with red mud amendments which immobilise phosphate, but this is far less worrying than the increased metal availability that is associated with hyperaccumulators.

The audience were then allowed to pose questions to the panel in the open discussion session that followed the presentations.

OPEN DISCUSSION

The first issue to be raised was whether to protect polluted land as a habitat refuge or to allow it to be remediated. The panel answered by saying that, where small areas of land were involved, there would be a strong argument for preserving the land, especially in areas which have been 'polluted' for 500 years or more. Examples of such areas which are already preserved include a serpentine rich area in the Czech Republic and some disused quarries in the UK. The floor suggested that the plants occurring on these polluted sites could be recultivated, and the panel pointed out that it may be difficult to recreate local conditions for cultivation and the plants may therefore lose their hyperaccumulator properties if not kept in similar conditions to that which is was accustomed to.

The issue of using crop plants as phytoremediators was then raised. It was pointed out that in some countries the locals did not like to see the crop plants unused and often raided phytoremediation projects sites in order to collect the crops. This can be a serious problem as the biomass may be contaminated, putting locals at more risk than when the contamination was just in the soil. The panel said that new regulations, to ensure contaminated crops were not consumed, needed to be introduced with regards to current scientific knowledge, and acknowledged that many people are still wary of the practicalities of phytoremediation. Phytoremediation may be a better tool if used in conjunction with other methods of soil remediation.

The point was made that it is easier and faster to simply excavate and remove polluted soil and the panel noted that soil was a limited resource and that urban areas, with higher land value, could afford this method of remediation. However, with garden soils and rural areas it may be a better option to phytoremediate. The idea of creating a package to sell to market gardeners to help them with slightly contaminated soil was mentioned, but this was thought to be unfeasible. The concern with this was that the public would want the contamination to be removed fast and as such would not accept the slower process of phytoremediation. There would also be issues with producing sufficient seeds from hyperaccumulators.

It was then asked if any field-scale phytoremediation examples existed. The panel gave the example of remediation at a nickel smelter in Canada that was currently underway. The aim was to reduce nickel concentrations by ten fold in five years. An *Alyssum* species was imported from Europe with the hope that the one-year lab trials performed on it would extrapolate on to the five-year field trials.

The floor queried as to how fast the available pool of metals in the soil became replenished and the panel commented that it has yet to be proven on a field scale. One example was quoted from a smelter in the south of France where hyperaccumulators were unknowingly planted on the site. It was later discovered that they were accumulating high concentrations of heavy metals in their biomass, which was being concentrated in the litter layer during leaf loss in autumn. Thus there is an increase in bioavailable metal concentrations in the top soil and the site now needs to be managed by either removing the plants, harvesting the biomass each year or fencing off the site. Experiments are underway to attempt to remove the phytoavailable metal pool and to observe how this pool is replenished over time.

It was also pointed out that the concept of phytoremediation was fairly recent and that there is great potential for metal extraction by plants. Domestication of agricultural plants had taken thousands of years so it may also take many years to optimise this new technique unless genetic modification is used as opposed to traditional plant breeding practices.

This raised the issue of genetically modified phytoremediation crops. The question was raised as to why people in the USA appeared to not have a concern with genetic modification of plants. The panel answered by saying that the USA had adopted a good public relations strategy and the public therefore did not see GM crops as being problematic. However, the point was made that there were inevitable problems for GM crops, such as the need for more pesticide use and this may be a problem if hyperaccumulators were ever to be genetically modified. There was the general consensus that genetic modification of hyperaccumulators may not happen within the near future, as we know very little about the genetics of hyperaccumulators and there may be little funding available for such research. Therefore traditional crop breeding may be a better option at the present time.

The floor then asked if any good agronomic practices existed for seed production of hyperaccumulators. The panel answered that it was still under development. It was also pointed out that once plenty of seeds could be produced we would no longer need to protect polluted sites. However, as yet it has proved difficult to produce seeds and so collection from wild populations is currently the only viable method of obtaining seeds..There is a gap in the knowledge of what happens to wild grown seeds when transferred into healthy soils. Questions arise as to whether the resultant crops will be hyperaccumulators, if they will seed and whether the seeds will be viable or not. Current machinery for seeding and harvesting is unsuitable for hyperaccumulators so there is a need for new technological development. There is also little knowledge on water supply needs and the problems which may be encountered with weeds and insects. There is still a long way to go before hyperaccumulators can be produced as an agricultural crop.

The floor enquired as to the number of hyperaccumulators identified so far and was told that four hundred and forty species had so far been identified. Approximately three hundred were nickel accumulators and over two hundred were tropical plants about which little else was known. Some species had very high biomass, but it was unknown how long it had taken for them to grow that large. There has been no attempt to crop any of the species identified by Dr. Reeves. Many of the potential candidates are small and there are very few with a multi-element capacity. Most are element specific, e.g. *Alyssum* copes well only with high nickel concentrations and not with elevated zinc or cobalt concentrations. It was also reminded that the legal issues addressed by Dr. Belluck would come into play here as the only seeds available are from the wild.

The issues of ethical research were raised. It was pointed out that the objectives in the Rio de Janeiro Convention on biological diversity must be adhered to, to allow the 'host' country to benefit in any commercial profits arising from hyperaccumulators found within their borders. The issues that need to be addressed in this area are those of land ownership and whether entrepreneurs would be allowed exclusivity to any techniques or resources they developed. It was pointed out that in France a species can be protected if found to be rare or of scientific interest. This helps limit seed collection.

The final point raised was whether phytoremediation was useful or not. Those in favour of immobilisation techniques agreed that in certain situations phytoremediation could be a valuable tool. However, there is a need for research into the agronomy of phytoremediation and for hyperaccumulators and other phytoremediator plants to become commercially available. There is the problem that they appear to have only a narrow scope in that they can only be used in mildly contaminated soil. There is also the need to separate research from the commercial aspect. The research challenge is to get a good size plant for agronomic use and then apply it to the commercial setting. The consensus was that more time and research was required and caution was needed when applying phytoremediation to field situations.

There are few remediation techniques for metals. Phytoremediation is likely to be a viable technique but we must be cautious, as there are many risks associated with it.

The idea of phytoremediation needs good publicity. This worked successfully with N-fixers and the same principle could to be applied to phytoremediation. Regulations need to be enforced to prevent the public from using contaminated biomass. Commercial scale phytoextraction will require farming expertise and the technological barrier will need to be overcome. The issues of what to do with the metal rich biomass must be addressed. Although some could be used for fertilizer in metal deficient soil, one must be careful not to pollute and there is still likely to be an excess of harvested biomass to dispose of.

Thus there are still a lot of issues to clear up before this technology can be applied safely to the field. Dr. Morel summed up the discussion by emphasising that phytoremediation has a future, but that people, the agricultural industry and land owners need to be convinced of this. Habitats need to be protected until we have ways of producing seed agriculturally, and new technologies need to be developed. A good metal removal strategy needs to be developed and there is optimism that is will be found one day. In the mean time though, there is the definite need to legally protect natural sources of hyperaccumulators until we have the technology to manage them effectively.

CHAPTER 16

WORKING GROUP DISCUSSION:THE EFFICIENCY AND VIABILITY OF PHYTOREMEDIATION

A. ROYLE
Biological and Earth Sciences, John Moores University, Byrom Street, Liverpool, L3 3AF, United Kingdom

Phytoremediation is a long-term, complex biological process. There is not one single measure of the efficiency of a process such as phytoremediation and therefore many aspects must be considered separately, and in conjunction, to reach a conclusion.

Efficiency is an arbitrary term unless related to an end point. If the end point is use of the land for gardens then the clean-up must be more complete than if a car park was to be constructed on the site. Unless the end use is specified then the efficiency of the process cannot be calculated. The main measurements of efficiency, for example, may be the rate of degradation of organics, or the stabilisation, extraction, or volatilisation of metals, the duration of the remediation process or the cost of the process. In the case of contamination with organic chemicals the degradation products and metabolites must be considered as they may be more toxic than the original chemical. There was a consensus that defined and internationally adopted guidelines for parameters such as monitoring and analytical tools and methods etc. are needed. These are necessary for international comparability of research and remediation and also efficiency measures. Accepted and tested models would also be needed both for the terrestrial and aquatic environments.

It is accepted that phytoremediation is not a remediation technology which can be applied to all contaminated sites, or even the majority of them. The process is slow and can be metal specific. Phytoremediation is suggested as a viable technique when the following parameters are satisfied: the site is of low economic value, time constraints do not apply, a low cost solution is required, the main metal pollution is with only one or two metals and confined to the surface layers, the labile pool is the most toxic form of the metal, metals and other contaminants are not found at phytotoxic levels; and when there is an infrastructure present to safely treat and dispose of the contaminated biomass which may be produced. These parameters apply to phytoremediation as a stand-alone technology. It was not overlooked that coupling phytoremediation with other techniques such as physical and/or chemical treatment can also be viable in many cases. Amendments which increase availability of metals may also be used with phytoremediation technology, but with care, as large increases in availability can lead to

J.-L. Morel et al. (eds.), Phytoremediation of Metal-Contaminated Soils, 343–345.

leaching of metals. If phytostabilisation is the goal then amendments can be added to further decrease the availability of metals in the soils. These amendments can include waste products from other processes.

The main focus of discussion concerned heavy metals, rather than organics, as this was the focus of the NATO ASI. Hyperaccumulators were considered to have considerable potential for phytoremediation, though perhaps more in the future than at the present time. Where contamination with one or two metal species occurs – as in mine tailing or from smelting fallout – hyperaccumulators could be used to remove the metals from the soils. This would only be a viable technology if the plants were harvested before leaf drop so as to avoid severe contamination of the surface layer of the soil. A facility must also be available in close proximity to the plants to treat and dispose of the harvested material, otherwise costs of transport will become too high and the process will cease to be viable. Phytomining was discussed in this context and thought to be a very promising technology if the necessary measures were in place to support it. However, at present this only appears to have been proven to be viable for nickel mining.

Agricultural land, especially in western countries, is experiencing more and more problems with heavy metal accumulation in soils due to use of sludges and phosphate fertilisers. Phytoremediation, by rotation of crops with accumulators or hyperaccumulators, is a very real option with there really only being two other options. The first option is chemical stabilisation but this may itself add more metals to the system. The other option is for the growth of food crops which are excluders on these soils. Sunflowers are one such example.

Phytoremediation of farmland has met resistance in several countries where it was tried due to farmers disliking growing a crop simply to throw it away, or preferring the use of chemical stabilisers. Many farmers demand to be paid to grow accumulator crops on their land.

The wide use of accumulating and hyperaccumulating crops also faces another pitfall and that is the issue of the introduction of exotic species to an area or country. Indian mustard is an accumulator of metals but is also an invasive weed in many countries and several countries have rejected it as a possible species for phytoremediation because of this. *Thlaspi* species are unlikely to act in this invasive manner because they are slow growing and adapted to a very specific environment, but these characteristics also count against them for real use in phytoremediation at present. As this subject area is so new and comparatively unexplored there is a huge opportunity for expansion. At present biomass species have relatively low metal uptake and hyperaccumulating species have low biomass yields. There is potential for plant breeding and genetic modification to create species with high accumulating capabilities and also high biomass. There is also the possibility of engineering plants that are easy to grow but do not run wild in new habitats.

It was concluded that phytoremediation can be viable and efficient when metal levels need to be lowered by only a small amount. When the metal contamination is by one or two elements only it may also be efficient – especially with hyperaccumulators. However it was the opinion of the group that a lot more work needs to be done both in terms of research and informing industry and the potential market as a whole about the technology. It was the opinion of the group that with further research, and perhaps using genetic modification, both the efficiency and the viability of phytoremediation as a competitive remediation technology will be much improved.

The group concluded that phytostabilisation was more viable at present than phytoextraction in soils of very high metal due to the very large time scales involved.

THEMATIC INDEX

MIX
Papier aus verantwortungsvollen Quellen
Paper from responsible sources
FSC® C105338

If you have any concerns about our products,
you can contact us on
ProductSafety@springernature.com

In case Publisher is established outside the EU,
the EU authorized representative is:
Springer Nature Customer Service Center GmbH
Europaplatz 3, 69115 Heidelberg, Germany

Printed by Libri Plureos GmbH
in Hamburg, Germany